APEX CALCULUS I

Version 3.0

Gregory Hartman, Ph.D.

Department of Applied Mathematics

Virginia Military Institute

Contributing Authors

Troy Siemers, Ph.D.

Department of Applied Mathematics

Virginia Military Institute

Brian Heinold, Ph.D.

Department of Mathematics and Computer Science

Mount Saint Mary's University

Dimplekumar Chalishajar, Ph.D.

Department of Applied Mathematics

Virginia Military Institute

Editor

Jennifer Bowen, Ph.D.

Department of Mathematics and Computer Science

The College of Wooster

Contents

Table of Contents   **iii**

Preface   **v**

1 Limits   **1**
   1.1 An Introduction To Limits   1
   1.2 Epsilon-Delta Definition of a Limit   9
   1.3 Finding Limits Analytically   18
   1.4 One Sided Limits   30
   1.5 Continuity   37
   1.6 Limits Involving Infinity   46

2 Derivatives   **57**
   2.1 Instantaneous Rates of Change: The Derivative   57
   2.2 Interpretations of the Derivative   71
   2.3 Basic Differentiation Rules   78
   2.4 The Product and Quotient Rules   85
   2.5 The Chain Rule   96
   2.6 Implicit Differentiation   106
   2.7 Derivatives of Inverse Functions   117

3 The Graphical Behavior of Functions   **123**
   3.1 Extreme Values   123
   3.2 The Mean Value Theorem   131
   3.3 Increasing and Decreasing Functions   136
   3.4 Concavity and the Second Derivative   144
   3.5 Curve Sketching   152

4 Applications of the Derivative   **159**
   4.1 Newton's Method   159
   4.2 Related Rates   166

    4.3   Optimization 173

    4.4   Differentials 180

5 Integration   **189**

    5.1   Antiderivatives and Indefinite Integration 189

    5.2   The Definite Integral 199

    5.3   Riemann Sums 210

    5.4   The Fundamental Theorem of Calculus 228

    5.5   Numerical Integration 240

6 Techniques of Antidifferentiation   **255**

    6.1   Substitution 255

A Solutions To Selected Problems   **A.1**

Index   **A.11**

PREFACE
A Note on Using this Text

Thank you for reading this short preface. Allow us to share a few key points about the text so that you may better understand what you will find beyond this page.

This text is Part I of a three–text series on Calculus. The first part covers material taught in many "Calc 1" courses: limits, derivatives, and the basics of integration, found in Chapters 1 through 6.1. The second text covers material often taught in "Calc 2:" integration and its applications, along with an introduction to sequences, series and Taylor Polynomials, found in Chapters 5 through 8. The third text covers topics common in "Calc 3" or "multivariable calc:" parametric equations, polar coordinates, vector–valued functions, and functions of more than one variable, found in Chapters 9 through 13. All three are available separately for free at www.apexcalculus.com. These three texts are intended to work together and make one cohesive text, *APEX Calculus*, which can also be downloaded from the website.

Printing the entire text as one volume makes for a large, heavy, cumbersome book. One can certainly only print the pages they currently need, but some prefer to have a nice, bound copy of the text. Therefore this text has been split into these three manageable parts, each of which can be purchased for under $15 at Amazon.com.

A result of this splitting is that sometimes a concept is said to be explored in a "later section," though that section does not actually appear in this particular text. Also, the index makes reference to topics and page numbers that do not appear in this text. This is done intentionally to show the reader what topics are available for study. Downloading the .pdf of *APEX Calculus* will ensure that you have all the content.

For Students: How to Read this Text

Mathematics textbooks have a reputation for being hard to read. High–level mathematical writing often seeks to say much with few words, and this style often seeps into texts of lower–level topics. This book was written with the goal of being easier to read than many other calculus textbooks, without becoming too verbose.

Each chapter and section starts with an introduction of the coming material, hopefully setting the stage for "why you should care," and ends with a look ahead to see how the just–learned material helps address future problems.

Please read the text; it is written to explain the concepts of Calculus. There are numerous examples to demonstrate the meaning of definitions, the truth of theorems, and the application of mathematical techniques. When you encounter a sentence you don't understand, read it again. If it still doesn't make sense, read on anyway, as sometimes confusing sentences are explained by later sentences.

You don't have to read every equation. The examples generally show "all" the steps needed to solve a problem. Sometimes reading through each step is helpful; sometimes it is confusing. When the steps are illustrating a new technique, one probably should follow each step closely to learn the new technique. When the steps are showing the mathematics needed to find a number to be used later, one can usually skip ahead and see how that number is being used, instead of getting bogged down in reading how the number was found.

Most proofs have been omitted. In mathematics, *proving* something is always true is extremely important, and entails much more than testing to see if it works twice. However, students often are confused by the details of a proof, or become concerned that they should have been able to construct this proof on their own. To alleviate this potential problem, we do not include the proofs to most theorems in the text. The interested reader is highly encouraged to find proofs online or from their instructor. In most cases, one is very capable of understanding what a theorem *means* and *how to apply it* without knowing fully *why* it is true.

Interactive, 3D Graphics

New to Version 3.0 is the addition of interactive, 3D graphics in the .pdf version. Nearly all graphs of objects in space can be rotated, shifted, and zoomed in/out so the reader can better understand the object illustrated.

As of this writing, the only pdf viewers that support these 3D graphics are Adobe Reader & Acrobat (and only the versions for PC/Mac/Unix/Linux computers, not tablets or smartphones). To activate the interactive mode, click on the image. Once activated, one can click/drag to rotate the object and use the scroll wheel on a mouse to zoom in/out. (A great way to investigate an image is to first zoom in on the page of the pdf viewer so the graphic itself takes up much of the screen, then zoom inside the graphic itself.) A CTRL-click/drag pans the object left/right or up/down. By right-clicking on the graph one can access a menu of other options, such as changing the lighting scheme or perspective. One can also revert the graph back to its default view. If you wish to deactive the interactivity, one can right-click and choose the "Disable Content" option.

Thanks

There are many people who deserve recognition for the important role they have played in the development of this text. First, I thank Michelle for her support and encouragement, even as this "project from work" occupied my time and attention at home. Many thanks to Troy Siemers, whose most important contributions extend far beyond the sections he wrote or the 227 figures he coded in Asymptote for 3D interaction. He provided incredible support, advice and encouragement for which I am very grateful. My thanks to Brian Heinold and Dimplekumar Chalishajar for their contributions and to Jennifer Bowen for reading through so much material and providing great feedback early on. Thanks to Troy, Lee Dewald, Dan Joseph, Meagan Herald, Bill Lowe, John David, Vonda Walsh, Geoff Cox, Jessica Libertini and other faculty of VMI who have given me numerous suggestions and corrections based on their experience with teaching from the text. (Special thanks to Troy, Lee & Dan for their patience in teaching Calc III while I was still writing the Calc III material.) Thanks to Randy Cone for encouraging his tutors of VMI's Open Math Lab to read through the text and check the solutions, and thanks to the tutors for spending their time doing so. A very special thanks to Kristi Brown and Paul Janiczek who took this opportunity far above & beyond what I expected, meticulously checking every solution and carefully reading every example. Their comments have been extraordinarily helpful. I am also thankful for the support provided by Wane Schneiter, who as my Dean provided me with extra time to work on this project. I am blessed to have so many people give of their time to make this book better.

A_PE_X – Affordable Print and Electronic teXts

A$_E^P$X is a consortium of authors who collaborate to produce high–quality, low–cost textbooks. The current textbook–writing paradigm is facing a potential revolution as desktop publishing and electronic formats increase in popularity. However, writing a good textbook is no easy task, as the time requirements alone are substantial. It takes countless hours of work to produce text, write examples and exercises, edit and publish. Through collaboration, however, the cost to any individual can be lessened, allowing us to create texts that we freely distribute electronically and sell in printed form for an incredibly low cost. Having said that, nothing is entirely free; someone always bears some cost. This text "cost" the authors of this book their time, and that was not enough. *APEX Calculus* would not exist had not the Virginia Military Institute, through a generous Jackson–Hope grant, given the lead author significant time away from teaching so he could focus on this text.

Each text is available as a free .pdf, protected by a Creative Commons Attribution - Noncommercial 4.0 copyright. That means you can give the .pdf to anyone you like, print it in any form you like, and even edit the original content and redistribute it. If you do the latter, you must clearly reference this work and you cannot sell your edited work for money.

We encourage others to adapt this work to fit their own needs. One might add sections that are "missing" or remove sections that your students won't need. The source files can be found at github.com/APEXCalculus.

You can learn more at www.vmi.edu/APEX.

1: Limits

Calculus means "a method of calculation or reasoning." When one computes the sales tax on a purchase, one employs a simple calculus. When one finds the area of a polygonal shape by breaking it up into a set of triangles, one is using another calculus. Proving a theorem in geometry employs yet another calculus.

Despite the wonderful advances in mathematics that had taken place into the first half of the 17th century, mathematicians and scientists were keenly aware of what they *could not do.* (This is true even today.) In particular, two important concepts eluded mastery by the great thinkers of that time: area and rates of change.

Area seems innocuous enough; areas of circles, rectangles, parallelograms, etc., are standard topics of study for students today just as they were then. However, the areas of *arbitrary* shapes could not be computed, even if the boundary of the shape could be described exactly.

Rates of change were also important. When an object moves at a constant rate of change, then "distance = rate $\times$ time." But what if the rate is not constant – can distance still be computed? Or, if distance is known, can we discover the rate of change?

It turns out that these two concepts were related. Two mathematicians, Sir Isaac Newton and Gottfried Leibniz, are credited with independently formulating a system of computing that solved the above problems and showed how they were connected. Their system of reasoning was "a" calculus. However, as the power and importance of their discovery took hold, it became known to many as "the" calculus. Today, we generally shorten this to discuss "calculus."

The foundation of "the calculus" is the *limit.* It is a tool to describe a particular behavior of a function. This chapter begins our study of the limit by approximating its value graphically and numerically. After a formal definition of the limit, properties are established that make "finding limits" tractable. Once the limit is understood, then the problems of area and rates of change can be approached.

1.1 An Introduction To Limits

We begin our study of *limits* by considering examples that demonstrate key concepts that will be explained as we progress.

Consider the function $y = \frac{\sin x}{x}$. When x is near the value 1, what value (if any) is y near?

While our question is not precisely formed (what constitutes "near the value

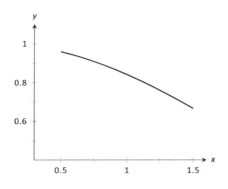

Figure 1.1: $\sin(x)/x$ near $x = 1$.

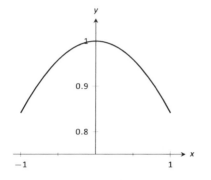

Figure 1.2: $\sin(x)/x$ near $x = 0$.

x	$\sin(x)/x$
0.9	0.870363
0.99	0.844471
0.999	0.841772
1	**0.841471**
1.001	0.84117
1.01	0.838447
1.1	0.810189

Figure 1.3: Values of $\sin(x)/x$ with x near 1.

1"?), the answer does not seem difficult to find. One might think first to look at a graph of this function to approximate the appropriate y values. Consider Figure 1.1, where $y = \frac{\sin x}{x}$ is graphed. For values of x near 1, it seems that y takes on values near 0.85. In fact, when $x = 1$, then $y = \frac{\sin 1}{1} \approx 0.84$, so it makes sense that when x is "near" 1, y will be "near" 0.84.

Consider this again at a different value for x. When x is near 0, what value (if any) is y near? By considering Figure 1.2, one can see that it seems that y takes on values near 1. But what happens when $x = 0$? We have

$$y \to \frac{\sin 0}{0} \to \text{``} \frac{0}{0} \text{''}.$$

The expression "0/0" has no value; it is *indeterminate*. Such an expression gives no information about what is going on with the function nearby. We cannot find out how y behaves near $x = 0$ for this function simply by letting $x = 0$.

Finding a limit entails understanding how a function behaves near a particular value of x. Before continuing, it will be useful to establish some notation. Let $y = f(x)$; that is, let y be a function of x for some function f. The expression "the limit of y as x approaches 1" describes a number, often referred to as L, that y nears as x nears 1. We write all this as

$$\lim_{x \to 1} y = \lim_{x \to 1} f(x) = L.$$

This is not a complete definition (that will come in the next section); this is a pseudo-definition that will allow us to explore the idea of a limit.

Above, where $f(x) = \sin(x)/x$, we approximated

$$\lim_{x \to 1} \frac{\sin x}{x} \approx 0.84 \quad \text{and} \quad \lim_{x \to 0} \frac{\sin x}{x} \approx 1.$$

(We *approximated* these limits, hence used the "$\approx$" symbol, since we are working with the pseudo-definition of a limit, not the actual definition.)

Once we have the true definition of a limit, we will find limits *analytically*; that is, exactly using a variety of mathematical tools. For now, we will *approximate* limits both graphically and numerically. Graphing a function can provide a good approximation, though often not very precise. Numerical methods can provide a more accurate approximation. We have already approximated limits graphically, so we now turn our attention to numerical approximations.

Consider again $\lim_{x \to 1} \sin(x)/x$. To approximate this limit numerically, we can create a table of x and $f(x)$ values where x is "near" 1. This is done in Figure 1.3.

Notice that for values of x near 1, we have $\sin(x)/x$ near 0.841. The $x = 1$ row is in bold to highlight the fact that when considering limits, we are *not* concerned

Notes:

with the value of the function at that particular x value; we are only concerned with the values of the function when x is *near* 1.

Now approximate $\lim_{x\to0} \sin(x)/x$ numerically. We already approximated the value of this limit as 1 graphically in Figure 1.2. The table in Figure 1.4 shows the value of $\sin(x)/x$ for values of x near 0. Ten places after the decimal point are shown to highlight how close to 1 the value of $\sin(x)/x$ gets as x takes on values very near 0. We include the $x = 0$ row in bold again to stress that we are not concerned with the value of our function at $x = 0$, only on the behavior of the function *near* 0.

This numerical method gives confidence to say that 1 is a good approximation of $\lim_{x\to0} \sin(x)/x$; that is,

$$\lim_{x\to0} \sin(x)/x \approx 1.$$

Later we will be able to prove that the limit is *exactly* 1.

We now consider several examples that allow us explore different aspects of the limit concept.

Example 1   **Approximating the value of a limit**
Use graphical and numerical methods to approximate

$$\lim_{x\to3} \frac{x^2 - x - 6}{6x^2 - 19x + 3}.$$

Solution   To graphically approximate the limit, graph

$$y = (x^2 - x - 6)/(6x^2 - 19x + 3)$$

on a small interval that contains 3. To numerically approximate the limit, create a table of values where the x values are near 3. This is done in Figures 1.5 and 1.6, respectively.

The graph shows that when x is near 3, the value of y is very near 0.3. By considering values of x near 3, we see that $y = 0.294$ is a better approximation. The graph and the table imply that

$$\lim_{x\to3} \frac{x^2 - x - 6}{6x^2 - 19x + 3} \approx 0.294.$$

This example may bring up a few questions about approximating limits (and the nature of limits themselves).

1. If a graph does not produce as good an approximation as a table, why bother with it?

2. How many values of x in a table are "enough?" In the previous example, could we have just used $x = 3.001$ and found a fine approximation?

x	$\sin(x)/x$
-0.1	0.9983341665
-0.01	0.9999833334
-0.001	0.9999998333
0	**not defined**
0.001	0.9999998333
0.01	0.9999833334
0.1	0.9983341665

Figure 1.4: Values of $\sin(x)/x$ with x near 1.

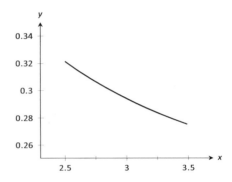

Figure 1.5: Graphically approximating a limit in Example 1.

x	$\frac{x^2-x-6}{6x^2-19x+3}$
2.9	0.29878
2.99	0.294569
2.999	0.294163
3	**not defined**
3.001	0.294073
3.01	0.293669
3.1	0.289773

Figure 1.6: Numerically approximating a limit in Example 1.

Notes:

Graphs are useful since they give a visual understanding concerning the behavior of a function. Sometimes a function may act "erratically" near certain x values which is hard to discern numerically but very plain graphically. Since graphing utilities are very accessible, it makes sense to make proper use of them.

Since tables and graphs are used only to *approximate* the value of a limit, there is not a firm answer to how many data points are "enough." Include enough so that a trend is clear, and use values (when possible) both less than and greater than the value in question. In Example 1, we used both values less than and greater than 3. Had we used just $x = 3.001$, we might have been tempted to conclude that the limit had a value of 0.3. While this is not far off, we could do better. Using values "on both sides of 3" helps us identify trends.

Example 2   Approximating the value of a limit
Graphically and numerically approximate the limit of $f(x)$ as x approaches 0, where

$$f(x) = \begin{cases} x+1 & x < 0 \\ -x^2+1 & x > 0 \end{cases}.$$

SOLUTION   Again we graph $f(x)$ and create a table of its values near $x = 0$ to approximate the limit. Note that this is a piecewise defined function, so it behaves differently on either side of 0. Figure 1.7 shows a graph of $f(x)$, and on either side of 0 it seems the y values approach 1. Note that $f(0)$ is not actually defined, as indicated in the graph with the open circle.

The table shown in Figure 1.8 shows values of $f(x)$ for values of x near 0. It is clear that as x takes on values very near 0, $f(x)$ takes on values very near 1. It turns out that if we let $x = 0$ for either "piece" of $f(x)$, 1 is returned; this is significant and we'll return to this idea later.

The graph and table allow us to say that $\lim_{x \to 0} f(x) \approx 1$; in fact, we are probably very sure it *equals* 1.

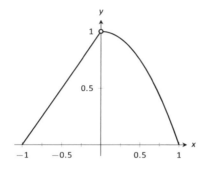

Figure 1.7: Graphically approximating a limit in Example 2.

x	$f(x)$
-0.1	0.9
-0.01	0.99
-0.001	0.999
0.001	0.999999
0.01	0.9999
0.1	0.99

Figure 1.8: Numerically approximating a limit in Example 2.

Identifying When Limits Do Not Exist

A function may not have a limit for all values of x. That is, we cannot say $\lim_{x \to c} f(x) = L$ for some numbers L for all values of c, for there may not be a number that $f(x)$ is approaching. There are three ways in which a limit may fail to exist.

1. The function $f(x)$ may approach different values on either side of c.

2. The function may grow without upper or lower bound as x approaches c.

3. The function may oscillate as x approaches c.

Notes:

We'll explore each of these in turn.

Example 3   **Different Values Approached From Left and Right**

Explore why $\lim\limits_{x \to 1} f(x)$ does not exist, where

$$f(x) = \begin{cases} x^2 - 2x + 3 & x \le 1 \\ x & x > 1 \end{cases}.$$

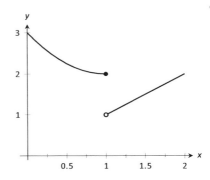

Figure 1.9: Observing no limit as $x \to 1$ in Example 3.

SOLUTION   A graph of $f(x)$ around $x = 1$ and a table are given Figures 1.9 and 1.10, respectively. It is clear that as x approaches 1, $f(x)$ does not seem to approach a single number. Instead, it seems as though $f(x)$ approaches two different numbers. When considering values of x less than 1 (approaching 1 from the left), it seems that $f(x)$ is approaching 2; when considering values of x greater than 1 (approaching 1 from the right), it seems that $f(x)$ is approaching 1. Recognizing this behavior is important; we'll study this in greater depth later. Right now, it suffices to say that the limit does not exist since $f(x)$ is not approaching one value as x approaches 1.

Example 4   **The Function Grows Without Bound**

Explore why $\lim\limits_{x \to 1} 1/(x-1)^2$ does not exist.

SOLUTION   A graph and table of $f(x) = 1/(x-1)^2$ are given in Figures 1.11 and 1.12, respectively. Both show that as x approaches 1, $f(x)$ grows larger and larger.

We can deduce this on our own, without the aid of the graph and table. If x is near 1, then $(x-1)^2$ is very small, and:

$$\frac{1}{\text{very small number}} = \text{very large number}.$$

Since $f(x)$ is not approaching a single number, we conclude that

$$\lim\limits_{x \to 1} \frac{1}{(x-1)^2}$$

does not exist.

Example 5   **The Function Oscillates**

Explore why $\lim\limits_{x \to 0} \sin(1/x)$ does not exist.

SOLUTION   Two graphs of $f(x) = \sin(1/x)$ are given in Figures 1.13. Figure 1.13(a) shows $f(x)$ on the interval $[-1, 1]$; notice how $f(x)$ seems to oscillate

x	$f(x)$
0.9	2.01
0.99	2.0001
0.999	2.000001
1.001	1.001
1.01	1.01
1.1	1.1

Figure 1.10: Values of $f(x)$ near $x = 1$ in Example 3.

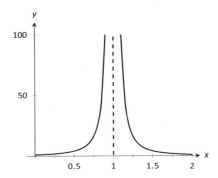

Figure 1.11: Observing no limit as $x \to 1$ in Example 4.

x	$f(x)$
0.9	100.
0.99	10000.
0.999	$1. \times 10^6$
1.001	$1. \times 10^6$
1.01	10000.
1.1	100.

Figure 1.12: Values of $f(x)$ near $x = 1$ in Example 4.

Notes:

near $x = 0$. One might think that despite the oscillation, as x approaches 0, $f(x)$ approaches 0. However, Figure 1.13(b) zooms in on $\sin(1/x)$, on the interval $[-0.1, 0.1]$. Here the oscillation is even more pronounced. Finally, in the table in Figure 1.13(c), we see $\sin(x)/x$ evaluated for values of x near 0. As x approaches 0, $f(x)$ does not appear to approach any value.

It can be shown that in reality, as x approaches 0, $\sin(1/x)$ takes on all values between -1 and 1 infinite times! Because of this oscillation,

$$\lim_{x \to 0} \sin(1/x) \text{ does not exist.}$$

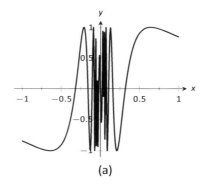

(a)

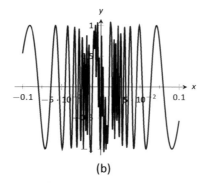

(b)

x	$\sin(1/x)$
0.1	-0.544021
0.01	-0.506366
0.001	0.82688
0.0001	-0.305614
$1. \times 10^{-5}$	0.0357488
$1. \times 10^{-6}$	-0.349994
$1. \times 10^{-7}$	0.420548

(c)

Figure 1.13: Observing that $f(x) = \sin(1/x)$ has no limit as $x \to 0$ in Example 5.

Limits of Difference Quotients

We have approximated limits of functions as x approached a particular number. We will consider another important kind of limit after explaining a few key ideas.

Let $f(x)$ represent the position function, in feet, of some particle that is moving in a straight line, where x is measured in seconds. Let's say that when $x = 1$, the particle is at position 10 ft., and when $x = 5$, the particle is at 20 ft. Another way of expressing this is to say

$$f(1) = 10 \quad \text{and} \quad f(5) = 20.$$

Since the particle traveled 10 feet in 4 seconds, we can say the particle's *average velocity* was 2.5 ft/s. We write this calculation using a "quotient of differences," or, a *difference quotient*:

$$\frac{f(5) - f(1)}{5 - 1} = \frac{10}{4} = 2.5\text{ft/s}.$$

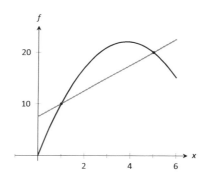

Figure 1.14:   Interpreting a difference quotient as the slope of a secant line.

Notes:

This difference quotient can be thought of as the familiar "rise over run" used to compute the slopes of lines. In fact, that is essentially what we are doing: given two points on the graph of f, we are finding the slope of the *secant line* through those two points. See Figure 1.14.

Now consider finding the average speed on another time interval. We again start at $x = 1$, but consider the position of the particle h seconds later. That is, consider the positions of the particle when $x = 1$ and when $x = 1 + h$. The difference quotient is now

$$\frac{f(1+h) - f(1)}{(1+h) - 1} = \frac{f(1+h) - f(1)}{h}.$$

Let $f(x) = -1.5x^2 + 11.5x$; note that $f(1) = 10$ and $f(5) = 20$, as in our discussion. We can compute this difference quotient for all values of h (even negative values!) except $h = 0$, for then we get "0/0," the indeterminate form introduced earlier. For all values $h \neq 0$, the difference quotient computes the average velocity of the particle over an interval of time of length h starting at $x = 1$.

For small values of h, i.e., values of h close to 0, we get average velocities over very short time periods and compute secant lines over small intervals. See Figure 1.15. This leads us to wonder what the limit of the difference quotient is as h approaches 0. That is,

$$\lim_{h \to 0} \frac{f(1+h) - f(1)}{h} = ?$$

As we do not yet have a true definition of a limit nor an exact method for computing it, we settle for approximating the value. While we could graph the difference quotient (where the x-axis would represent h values and the y-axis would represent values of the difference quotient) we settle for making a table. See Figure 1.16. The table gives us reason to assume the value of the limit is about 8.5.

Proper understanding of limits is key to understanding calculus. With limits, we can accomplish seemingly impossible mathematical things, like adding up an infinite number of numbers (and not get infinity) and finding the slope of a line between two points, where the "two points" are actually the same point. These are not just mathematical curiosities; they allow us to link position, velocity and acceleration together, connect cross-sectional areas to volume, find the work done by a variable force, and much more.

In the next section we give the formal definition of the limit and begin our study of finding limits analytically. In the following exercises, we continue our introduction and approximate the value of limits.

Notes:

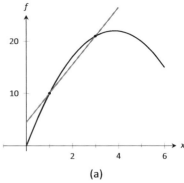
(a)

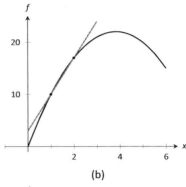
(b)

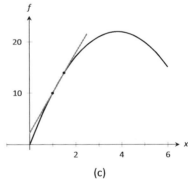
(c)

Figure 1.15: Secant lines of $f(x)$ at $x = 1$ and $x = 1 + h$, for shrinking values of h (i.e., $h \to 0$).

h	$\frac{f(1+h)-f(1)}{h}$
-0.5	9.25
-0.1	8.65
-0.01	8.515
0.01	8.485
0.1	8.35
0.5	7.75

Figure 1.16: The difference quotient evaluated at values of h near 0.

Exercises 1.1

Terms and Concepts

1. In your own words, what does it mean to "find the limit of $f(x)$ as x approaches 3"?

2. An expression of the form $\frac{0}{0}$ is called ____.

3. T/F: The limit of $f(x)$ as x approaches 5 is $f(5)$.

4. Describe three situations where $\lim_{x \to c} f(x)$ does not exist.

5. In your own words, what is a difference quotient?

Problems

In Exercises 6 – 16, approximate the given limits both numerically and graphically.

6. $\lim_{x \to 1} x^2 + 3x - 5$

7. $\lim_{x \to 0} x^3 - 3x^2 + x - 5$

8. $\lim_{x \to 0} \dfrac{x + 1}{x^2 + 3x}$

9. $\lim_{x \to 3} \dfrac{x^2 - 2x - 3}{x^2 - 4x + 3}$

10. $\lim_{x \to -1} \dfrac{x^2 + 8x + 7}{x^2 + 6x + 5}$

11. $\lim_{x \to 2} \dfrac{x^2 + 7x + 10}{x^2 - 4x + 4}$

12. $\lim_{x \to 2} f(x)$, where

$$f(x) = \begin{cases} x + 2 & x \le 2 \\ 3x - 5 & x > 2 \end{cases}.$$

13. $\lim_{x \to 3} f(x)$, where

$$f(x) = \begin{cases} x^2 - x + 1 & x \le 3 \\ 2x + 1 & x > 3 \end{cases}.$$

14. $\lim_{x \to 0} f(x)$, where

$$f(x) = \begin{cases} \cos x & x \le 0 \\ x^2 + 3x + 1 & x > 0 \end{cases}.$$

15. $\lim_{x \to \pi/2} f(x)$, where

$$f(x) = \begin{cases} \sin x & x \le \pi/2 \\ \cos x & x > \pi/2 \end{cases}.$$

In Exercises 16 – 24, a function f and a value a are given. Approximate the limit of the difference quotient, $\lim_{h \to 0} \dfrac{f(a + h) - f(a)}{h}$, using $h = \pm 0.1, \pm 0.01$.

16. $f(x) = -7x + 2, \quad a = 3$

17. $f(x) = 9x + 0.06, \quad a = -1$

18. $f(x) = x^2 + 3x - 7, \quad a = 1$

19. $f(x) = \dfrac{1}{x + 1}, \quad a = 2$

20. $f(x) = -4x^2 + 5x - 1, \quad a = -3$

21. $f(x) = \ln x, \quad a = 5$

22. $f(x) = \sin x, \quad a = \pi$

23. $f(x) = \cos x, \quad a = \pi$

1.2   Epsilon-Delta Definition of a Limit

This section introduces the formal definition of a limit. Many refer to this as "the epsilon–delta," definition, referring to the letters ε and δ of the Greek alphabet.

Before we give the actual definition, let's consider a few informal ways of describing a limit. Given a function $y = f(x)$ and an x-value, c, we say that "the limit of the function f, as x approaches c, is a value L":

1. if "y tends to L" as "x tends to c."

2. if "y approaches L" as "x approaches c."

3. if "y is near L" whenever "x is near c."

The problem with these definitions is that the words "tends," "approach," and especially "near" are not exact. In what way does the variable x tend to, or approach, c? How near do x and y have to be to c and L, respectively?

The definition we describe in this section comes from formalizing **3**. A quick restatement gets us closer to what we want:

3′. If x is within a certain *tolerance level* of c, then the corresponding value $y = f(x)$ is within a certain *tolerance level* of L.

The traditional notation for the x-tolerance is the lowercase Greek letter delta, or δ, and the y-tolerance is denoted by lowercase epsilon, or ε. One more rephrasing of **3′** nearly gets us to the actual definition:

3″. If x is within δ units of c, then the corresponding value of y is within ε units of L.

We can write "x is within δ units of c" mathematically as

$$|x - c| < \delta, \qquad \text{which is equivalent to} \qquad c - \delta < x < c + \delta.$$

Letting the symbol "$\longrightarrow$" represent the word "implies," we can rewrite **3″** as

$$|x - c| < \delta \longrightarrow |y - L| < \varepsilon \qquad \text{or} \qquad c - \delta < x < c + \delta \longrightarrow L - \varepsilon < y < L + \varepsilon.$$

The point is that δ and ε, being tolerances, can be any positive (but typically small) values. Finally, we have the formal definition of the limit with the notation seen in the previous section.

Notes:

> **Definition 1   The Limit of a Function f**
>
> Let I be an open interval containing c, and let f be a function defined on I, except possibly at c. The **limit of $f(x)$, as x approaches c, is** L, denoted by
> $$\lim_{x \to c} f(x) = L,$$
> means that given any $\varepsilon > 0$, there exists $\delta > 0$ such that for all $x \neq c$, if $|x - c| < \delta$, then $|f(x) - L| < \varepsilon$.

(Mathematicians often enjoy writing ideas without using any words. Here is the wordless definition of the limit:

$$\lim_{x \to c} f(x) = L \iff \forall \varepsilon > 0, \exists \delta > 0 \ s.t. \ 0 < |x - c| < \delta \longrightarrow |f(x) - L| < \varepsilon.)$$

Note the order in which ε and δ are given. In the definition, the y-tolerance ε is given *first* and then the limit will exist **if** we can find an x-tolerance δ that works.

An example will help us understand this definition. Note that the explanation is long, but it will take one through all steps necessary to understand the ideas.

Example 6   Evaluating a limit using the definition
Show that $\lim\limits_{x \to 4} \sqrt{x} = 2$.

SOLUTION   Before we use the formal definition, let's try some numerical tolerances. What if the y tolerance is 0.5, or $\varepsilon = 0.5$? How close to 4 does x have to be so that y is within 0.5 units of 2, i.e., $1.5 < y < 2.5$? In this case, we can proceed as follows:

$$
\begin{aligned}
1.5 &< y < 2.5 \\
1.5 &< \sqrt{x} < 2.5 \\
1.5^2 &< x < 2.5^2 \\
2.25 &< x < 6.25.
\end{aligned}
$$

So, what is the desired x tolerance? Remember, we want to find a symmetric interval of x values, namely $4 - \delta < x < 4 + \delta$. The lower bound of 2.25 is 1.75 units from 4; the upper bound of 6.25 is 2.25 units from 4. We need the smaller of these two distances; we must have $\delta \leq 1.75$. See Figure 1.17.

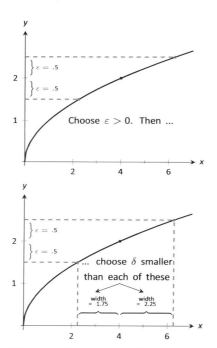

With $\varepsilon = 0.5$, we pick any $\delta < 1.75$.

Figure 1.17: Illustrating the $\varepsilon - \delta$ process.

Notes:

Given the y tolerance $\varepsilon = 0.5$, we have found an x tolerance, $\delta \leq 1.75$, such that whenever x is within δ units of 4, then y is within ε units of 2. That's what we were trying to find.

Let's try another value of ε.

What if the y tolerance is 0.01, i.e., $\varepsilon = 0.01$? How close to 4 does x have to be in order for y to be within 0.01 units of 2 (or $1.99 < y < 2.01$)? Again, we just square these values to get $1.99^2 < x < 2.01^2$, or

$$3.9601 < x < 4.0401.$$

What is the desired x tolerance? In this case we must have $\delta \leq 0.0399$, which is the minimum distance from 4 of the two bounds given above.

What we have so far: if $\varepsilon = 0.5$, then $\delta \leq 1.75$ and if $\varepsilon = 0.01$, then $\delta \leq 0.0399$. A pattern is not easy to see, so we switch to general ε try to determine δ symbolically. We start by assuming $y = \sqrt{x}$ is within ε units of 2:

$$
\begin{aligned}
|y - 2| &< \varepsilon \\
-\varepsilon < y - 2 &< \varepsilon && \text{(Definition of absolute value)} \\
-\varepsilon < \sqrt{x} - 2 &< \varepsilon && (y = \sqrt{x}) \\
2 - \varepsilon < \sqrt{x} &< 2 + \varepsilon && \text{(Add 2)} \\
(2 - \varepsilon)^2 < x &< (2 + \varepsilon)^2 && \text{(Square all)} \\
4 - 4\varepsilon + \varepsilon^2 < x &< 4 + 4\varepsilon + \varepsilon^2 && \text{(Expand)} \\
4 - (4\varepsilon - \varepsilon^2) < x &< 4 + (4\varepsilon + \varepsilon^2). && \text{(Rewrite in the desired form)}
\end{aligned}
$$

The "desired form" in the last step is "$4 - something < x < 4 + something$." Since we want this last interval to describe an x tolerance around 4, we have that either $\delta \leq 4\varepsilon - \varepsilon^2$ or $\delta \leq 4\varepsilon + \varepsilon^2$, whichever is smaller:

$$\delta \leq \min\{4\varepsilon - \varepsilon^2, 4\varepsilon + \varepsilon^2\}.$$

Since $\varepsilon > 0$, the minimum is $\delta \leq 4\varepsilon - \varepsilon^2$. That's the formula: given an ε, set $\delta \leq 4\varepsilon - \varepsilon^2$.

We can check this for our previous values. If $\varepsilon = 0.5$, the formula gives $\delta \leq 4(0.5) - (0.5)^2 = 1.75$ and when $\varepsilon = 0.01$, the formula gives $\delta \leq 4(0.01) - (0.01)^2 = 0.399$.

So given any $\varepsilon > 0$, set $\delta \leq 4\varepsilon - \varepsilon^2$. Then if $|x - 4| < \delta$ (and $x \neq 4$), then $|f(x) - 2| < \varepsilon$, satisfying the definition of the limit. We have shown formally (and finally!) that $\lim_{x \to 4} \sqrt{x} = 2$.

Notes:

The previous example was a little long in that we sampled a few specific cases of ε before handling the general case. Normally this is not done. The previous example is also a bit unsatisfying in that $\sqrt{4} = 2$; why work so hard to prove something so obvious? Many ε-δ proofs are long and difficult to do. In this section, we will focus on examples where the answer is, frankly, obvious, because the non–obvious examples are even harder. In the next section we will learn some theorems that allow us to evaluate limits *analytically*, that is, without using the ε-δ definition.

Example 7   **Evaluating a limit using the definition**

Show that $\lim\limits_{x \to 2} x^2 = 4$.

SOLUTION   Let's do this example symbolically from the start. Let $\varepsilon > 0$ be given; we want $|y - 4| < \varepsilon$, i.e., $|x^2 - 4| < \varepsilon$. How do we find δ such that when $|x - 2| < \delta$, we are guaranteed that $|x^2 - 4| < \varepsilon$?

This is a bit trickier than the previous example, but let's start by noticing that $|x^2 - 4| = |x - 2| \cdot |x + 2|$. Consider:

$$|x^2 - 4| < \varepsilon \longrightarrow |x - 2| \cdot |x + 2| < \varepsilon \longrightarrow |x - 2| < \frac{\varepsilon}{|x + 2|}. \tag{1.1}$$

Could we not set $\delta = \dfrac{\varepsilon}{|x + 2|}$?

We are close to an answer, but the catch is that δ must be a *constant* value (so it can't contain x). There is a way to work around this, but we do have to make an assumption. Remember that ε is supposed to be a small number, which implies that δ will also be a small value. In particular, we can (probably) assume that $\delta < 1$. If this is true, then $|x - 2| < \delta$ would imply that $|x - 2| < 1$, giving $1 < x < 3$.

Now, back to the fraction $\dfrac{\varepsilon}{|x + 2|}$. If $1 < x < 3$, then $3 < x + 2 < 5$ (add 2 to all terms in the inequality). Taking reciprocals, we have

$$\frac{1}{5} < \frac{1}{|x + 2|} < \frac{1}{3} \qquad \text{which implies}$$

$$\frac{1}{5} < \frac{1}{|x + 2|} \qquad \text{which implies}$$

$$\frac{\varepsilon}{5} < \frac{\varepsilon}{|x + 2|}. \tag{1.2}$$

This suggests that we set $\delta \le \dfrac{\varepsilon}{5}$. To see why, let consider what follows when we assume $|x - 2| < \delta$:

Notes:

$$|x - 2| < \delta$$

$$|x - 2| < \frac{\varepsilon}{5} \qquad \text{(Our choice of } \delta)$$

$$|x - 2| \cdot |x + 2| < |x + 2| \cdot \frac{\varepsilon}{5} \qquad \text{(Multiply by } |x + 2|)$$

$$|x^2 - 4| < |x + 2| \cdot \frac{\varepsilon}{5} \qquad \text{(Combine left side)}$$

$$|x^2 - 4| < |x + 2| \cdot \frac{\varepsilon}{5} < |x + 2| \cdot \frac{\varepsilon}{|x + 2|} = \varepsilon \qquad \text{(Using (1.2) as long as } \delta < 1)$$

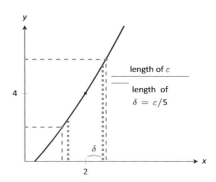

Figure 1.18: Choosing $\delta = \varepsilon/5$ in Example 7.

We have arrived at $|x^2 - 4| < \varepsilon$ as desired. Note again, in order to make this happen we needed δ to first be less than 1. That is a safe assumption; we want ε to be arbitrarily small, forcing δ to also be small.

We have also picked δ to be smaller than "necessary." We could get by with a slightly larger δ, as shown in Figure 1.18. The dashed outer lines show the boundaries defined by our choice of ε. The dotted inner lines show the boundaries defined by setting $\delta = \varepsilon/5$. Note how these dotted lines are within the dashed lines. That is perfectly fine; by choosing x within the dotted lines we are guaranteed that $f(x)$ will be within ε of 4.

In summary, given $\varepsilon > 0$, set $\delta = \leq \varepsilon/5$. Then $|x - 2| < \delta$ implies $|x^2 - 4| < \varepsilon$ (i.e. $|y - 4| < \varepsilon$) as desired. This shows that $\lim_{x \to 2} x^2 = 4$. Figure 1.18 gives a visualization of this; by restricting x to values within $\delta = \varepsilon/5$ of 2, we see that $f(x)$ is within ε of 4.

Make note of the general pattern exhibited in these last two examples. In some sense, each starts out "backwards." That is, while we want to

1. start with $|x - c| < \delta$ and conclude that

2. $|f(x) - L| < \varepsilon$,

we actually start by assuming

1. $|f(x) - L| < \varepsilon$, then perform some algebraic manipulations to give an inequality of the form

2. $|x - c| < something$.

When we have properly done this, the *something* on the "greater than" side of the inequality becomes our δ. We can refer to this as the "scratch–work" phase of our proof. Once we have δ, we can formally start with $|x - c| < \delta$ and use algebraic manipulations to conclude that $|f(x) - L| < \varepsilon$, usually by using the same steps of our "scratch–work" in reverse order.

Notes:

We highlight this process in the following example.

Example 8   **Evaluating a limit using the definition**
Prove that $\lim\limits_{x \to 1} x^3 - 2x = -1$.

SOLUTION   We start our scratch–work by considering $|f(x) - (-1)| < \varepsilon$:

$$|f(x) - (-1)| < \varepsilon$$
$$|x^3 - 2x + 1| < \varepsilon \qquad \text{(Now factor)}$$
$$|(x-1)(x^2 + x - 1)| < \varepsilon$$
$$|x - 1| < \frac{\varepsilon}{|x^2 + x - 1|}. \qquad (1.3)$$

We are at the phase of saying that $|x - 1| < something$, where $something = \varepsilon/|x^2 + x - 1|$. We want to turn that $something$ into δ.

Since x is approaching 1, we are safe to assume that x is between 0 and 2. So

$$0 < x < 2$$
$$0 < x^2 < 4. \qquad \text{(squared each term)}$$

Since $0 < x < 2$, we can add 0, x and 2, respectively, to each part of the inequality and maintain the inequality.

$$0 < x^2 + x < 6$$
$$-1 < x^2 + x - 1 < 5. \qquad \text{(subtracted 1 from each part)}$$

In Equation (1.3), we wanted $|x - 1| < \varepsilon/|x^2 + x - 1|$. The above shows that given any x in $[0, 2]$, we know that

$$x^2 + x - 1 < 5 \qquad \text{which implies that}$$
$$\frac{1}{5} < \frac{1}{x^2 + x - 1} \qquad \text{which implies that}$$
$$\frac{\varepsilon}{5} < \frac{\varepsilon}{x^2 + x - 1}. \qquad (1.4)$$

So we set $\delta \leq \varepsilon/5$. This ends our scratch–work, and we begin the formal proof (which also helps us understand why this was a good choice of δ).

Given ε, let $\delta \leq \varepsilon/5$. We want to show that when $|x - 1| < \delta$, then $|(x^3 -$

Notes:

$2x) - (-1)| < \varepsilon$. We start with $|x - 1| < \delta$:

$$|x - 1| < \delta$$
$$|x - 1| < \frac{\varepsilon}{5}$$
$$|x - 1| < \frac{\varepsilon}{5} < \frac{\varepsilon}{|x^2 + x - 1|} \qquad \text{(for } x \text{ near 1, from Equation (1.4))}$$
$$|x - 1| \cdot |x^2 + x - 1| < \varepsilon$$
$$|x^3 - 2x + 1| < \varepsilon$$
$$|(x^3 - 2x) - (-1)| < \varepsilon,$$

which is what we wanted to show. Thus $\lim\limits_{x \to 1} x^3 - 2x = -1$.

We illustrate evaluating limits once more.

Example 9   **Evaluating a limit using the definition**
Prove that $\lim\limits_{x \to 0} e^x = 1$.

SOLUTION   Symbolically, we want to take the equation $|e^x - 1| < \varepsilon$ and unravel it to the form $|x - 0| < \delta$. Here is our scratch–work:

$$|e^x - 1| < \varepsilon$$
$$-\varepsilon < e^x - 1 < \varepsilon \qquad \text{(Definition of absolute value)}$$
$$1 - \varepsilon < e^x < 1 + \varepsilon \qquad \text{(Add 1)}$$
$$\ln(1 - \varepsilon) < x < \ln(1 + \varepsilon) \qquad \text{(Take natural logs)}$$

Note: Recall $\ln 1 = 0$ and $\ln x < 0$ when $0 < x < 1$. So $\ln(1 - \varepsilon) < 0$, hence we consider its absolute value.

Making the safe assumption that $\varepsilon < 1$ ensures the last inequality is valid (i.e., so that $\ln(1 - \varepsilon)$ is defined). We can then set δ to be the minimum of $|\ln(1 - \varepsilon)|$ and $\ln(1 + \varepsilon)$; i.e.,

$$\delta = \min\{|\ln(1 - \varepsilon)|, \ln(1 + \varepsilon)\} = \ln(1 + \varepsilon).$$

Now, we work through the actual the proof:

$$|x - 0| < \delta$$
$$-\delta < x < \delta \qquad \text{(Definition of absolute value)}$$
$$-\ln(1 + \varepsilon) < x < \ln(1 + \varepsilon).$$
$$\ln(1 - \varepsilon) < x < \ln(1 + \varepsilon). \qquad \text{(since } \ln(1 - \varepsilon) < -\ln(1 + \varepsilon))$$

Notes:

The above line is true by our choice of δ and by the fact that since $|\ln(1-\varepsilon)| > \ln(1+\varepsilon)$ and $\ln(1-\varepsilon) < 0$, we know $\ln(1-\varepsilon) < -\ln(1+\varepsilon)$.

$$1 - \varepsilon < e^x < 1 + \varepsilon \qquad \text{(Exponentiate)}$$
$$-\varepsilon < e^x - 1 < \varepsilon \qquad \text{(Subtract 1)}$$

In summary, given $\varepsilon > 0$, let $\delta = \ln(1+\varepsilon)$. Then $|x - 0| < \delta$ implies $|e^x - 1| < \varepsilon$ as desired. We have shown that $\lim_{x \to 0} e^x = 1$.

We note that we could actually show that $\lim_{x \to c} e^x = e^c$ for any constant c. We do this by factoring out e^c from both sides, leaving us to show $\lim_{x \to c} e^{x-c} = 1$ instead. By using the substitution $u = x - c$, this reduces to showing $\lim_{u \to 0} e^u = 1$ which we just did in the last example. As an added benefit, this shows that in fact the function $f(x) = e^x$ is *continuous* at all values of x, an important concept we will define in Section 1.5.

This formal definition of the limit is not an easy concept grasp. Our examples are actually "easy" examples, using "simple" functions like polynomials, square–roots and exponentials. It is very difficult to prove, using the techniques given above, that $\lim_{x \to 0} (\sin x)/x = 1$, as we approximated in the previous section.

There is hope. The next section shows how one can evaluate complicated limits using certain basic limits as building blocks. While limits are an incredibly important part of calculus (and hence much of higher mathematics), rarely are limits evaluated using the definition. Rather, the techniques of the following section are employed.

Notes:

Exercises 1.2

Terms and Concepts

1. What is wrong with the following "definition" of a limit?

   "The limit of $f(x)$, as x approaches a, is K" means that given any $\delta > 0$ there exists $\varepsilon > 0$ such that whenever $|f(x) - K| < \varepsilon$, we have $|x - a| < \delta$.

2. Which is given first in establishing a limit, the x–tolerance or the y–tolerance?

3. T/F: ε must always be positive.

4. T/F: δ must always be positive.

Problems

In Exercises 5 – 11, prove the given limit using an $\varepsilon - \delta$ proof.

5. $\lim\limits_{x \to 5} 3 - x = -2$

6. $\lim\limits_{x \to 3} x^2 - 3 = 6$

7. $\lim\limits_{x \to 4} x^2 + x - 5 = 15$

8. $\lim\limits_{x \to 2} x^3 - 1 = 7$

9. $\lim\limits_{x \to 2} 5 = 5$

10. $\lim\limits_{x \to 0} e^{2x} - 1 = 0$

11. $\lim\limits_{x \to 0} \sin x = 0$ (Hint: use the fact that $|\sin x| \le |x|$, with equality only when $x = 0$.)

1.3 Finding Limits Analytically

In Section 1.1 we explored the concept of the limit without a strict definition, meaning we could only make approximations. In the previous section we gave the definition of the limit and demonstrated how to use it to verify our approximations were correct. Thus far, our method of finding a limit is 1) make a really good approximation either graphically or numerically, and 2) verify our approximation is correct using a ε-δ proof.

Recognizing that ε-δ proofs are cumbersome, this section gives a series of theorems which allow us to find limits much more quickly and intuitively.

Suppose that $\lim_{x \to 2} f(x) = 2$ and $\lim_{x \to 2} g(x) = 3$. What is $\lim_{x \to 2}(f(x) + g(x))$? Intuition tells us that the limit should be 5, as we expect limits to behave in a nice way. The following theorem states that already established limits do behave nicely.

Theorem 1 Basic Limit Properties

Let b, c, L and K be real numbers, let n be a positive integer, and let f and g be functions with the following limits:

$$\lim_{x \to c} f(x) = L \text{ and } \lim_{x \to c} g(x) = K.$$

The following limits hold.

1. Constants: $\lim_{x \to c} b = b$

2. Identity: $\lim_{x \to c} x = c$

3. Sums/Differences: $\lim_{x \to c}(f(x) \pm g(x)) = L \pm K$

4. Scalar Multiples: $\lim_{x \to c} b \cdot f(x) = bL$

5. Products: $\lim_{x \to c} f(x) \cdot g(x) = LK$

6. Quotients: $\lim_{x \to c} f(x)/g(x) = L/K, (K \neq 0)$

7. Powers: $\lim_{x \to c} f(x)^n = L^n$

8. Roots: $\lim_{x \to c} \sqrt[n]{f(x)} = \sqrt[n]{L}$

9. Compositions: Adjust our previously given limit situation to:

$$\lim_{x \to c} f(x) = L \text{ and } \lim_{x \to L} g(x) = K.$$

Then $\lim_{x \to c} g(f(x)) = K$.

Notes:

We make a note about Property #8: when n is even, L must be greater than 0. If n is odd, then the statement is true for all L.

We apply the theorem to an example.

Example 10   Using basic limit properties

Let

$$\lim_{x\to 2} f(x) = 2, \quad \lim_{x\to 2} g(x) = 3 \quad \text{and} \quad p(x) = 3x^2 - 5x + 7.$$

Find the following limits:

1. $\lim_{x\to 2} \left(f(x) + g(x)\right)$

2. $\lim_{x\to 2} \left(5f(x) + g(x)^2\right)$

3. $\lim_{x\to 2} p(x)$

SOLUTION

1. Using the Sum/Difference rule, we know that $\lim_{x\to 2} \left(f(x) + g(x)\right) = 2 + 3 = 5$.

2. Using the Scalar Multiple and Sum/Difference rules, we find that $\lim_{x\to 2} \left(5f(x) + g(x)^2\right) = 5 \cdot 2 + 3^2 = 19$.

3. Here we combine the Power, Scalar Multiple, Sum/Difference and Constant Rules. We show quite a few steps, but in general these can be omitted:

$$\begin{aligned}
\lim_{x\to 2} p(x) &= \lim_{x\to 2}(3x^2 - 5x + 7) \\
&= \lim_{x\to 2} 3x^2 - \lim_{x\to 2} 5x + \lim_{x\to 2} 7 \\
&= 3 \cdot 2^2 - 5 \cdot 2 + 7 \\
&= 9
\end{aligned}$$

Part 3 of the previous example demonstrates how the limit of a quadratic polynomial can be determined using the properties of Theorem 1. Not only that, recognize that

$$\lim_{x\to 2} p(x) = 9 = p(2);$$

i.e., the limit at 2 was found just by plugging 2 into the function. This holds true for all polynomials, and also for rational functions (which are quotients of polynomials), as stated in the following theorem.

Notes:

Theorem 2   **Limits of Polynomial and Rational Functions**

Let $p(x)$ and $q(x)$ be polynomials and c a real number. Then:

1. $\lim\limits_{x \to c} p(x) = p(c)$

2. $\lim\limits_{x \to c} \dfrac{p(x)}{q(x)} = \dfrac{p(c)}{q(c)}$, where $q(c) \neq 0$.

Example 11   **Finding a limit of a rational function**

Using Theorem 2, find

$$\lim_{x \to -1} \frac{3x^2 - 5x + 1}{x^4 - x^2 + 3}.$$

SOLUTION   Using Theorem 2, we can quickly state that

$$\lim_{x \to -1} \frac{3x^2 - 5x + 1}{x^4 - x^2 + 3} = \frac{3(-1)^2 - 5(-1) + 1}{(-1)^4 - (-1)^2 + 3}$$

$$= \frac{9}{3} = 3.$$

It was likely frustrating in Section 1.2 to do a lot of work to prove that

$$\lim_{x \to 2} x^2 = 4$$

as it seemed fairly obvious. The previous theorems state that many functions behave in such an "obvious" fashion, as demonstrated by the rational function in Example 11.

Polynomial and rational functions are not the only functions to behave in such a predictable way. The following theorem gives a list of functions whose behavior is particularly "nice" in terms of limits. In the next section, we will give a formal name to these functions that behave "nicely."

Theorem 3   **Special Limits**

Let c be a real number in the domain of the given function and let n be a positive integer. The following limits hold:

1. $\lim\limits_{x \to c} \sin x = \sin c$

2. $\lim\limits_{x \to c} \cos x = \cos c$

3. $\lim\limits_{x \to c} \tan x = \tan c$

4. $\lim\limits_{x \to c} \csc x = \csc c$

5. $\lim\limits_{x \to c} \sec x = \sec c$

6. $\lim\limits_{x \to c} \cot x = \cot c$

7. $\lim\limits_{x \to c} a^x = a^c \ (a > 0)$

8. $\lim\limits_{x \to c} \ln x = \ln c$

9. $\lim\limits_{x \to c} \sqrt[n]{x} = \sqrt[n]{c}$

Notes:

Example 12 **Evaluating limits analytically**

Evaluate the following limits.

1. $\lim\limits_{x \to \pi} \cos x$

2. $\lim\limits_{x \to 3} (\sec^2 x - \tan^2 x)$

3. $\lim\limits_{x \to \pi/2} \cos x \sin x$

4. $\lim\limits_{x \to 1} e^{\ln x}$

5. $\lim\limits_{x \to 0} \dfrac{\sin x}{x}$

SOLUTION

1. This is a straightforward application of Theorem 3. $\lim\limits_{x \to \pi} \cos x = \cos \pi = -1$.

2. We can approach this in at least two ways. First, by directly applying Theorem 3, we have:

$$\lim\limits_{x \to 3} (\sec^2 x - \tan^2 x) = \sec^2 3 - \tan^2 3.$$

Using the Pythagorean Theorem, this last expression is 1; therefore

$$\lim\limits_{x \to 3} (\sec^2 x - \tan^2 x) = 1.$$

We can also use the Pythagorean Theorem from the start.

$$\lim\limits_{x \to 3} (\sec^2 x - \tan^2 x) = \lim\limits_{x \to 3} 1 = 1,$$

using the Constant limit rule. Either way, we find the limit is 1.

3. Applying the Product limit rule of Theorem 1 and Theorem 3 gives

$$\lim\limits_{x \to \pi/2} \cos x \sin x = \cos(\pi/2) \sin(\pi/2) = 0 \cdot 1 = 0.$$

4. Again, we can approach this in two ways. First, we can use the exponential/logarithmic identity that $e^{\ln x} = x$ and evaluate $\lim\limits_{x \to 1} e^{\ln x} = \lim\limits_{x \to 1} x = 1$.

We can also use the Composition limit rule of Theorem 1. Using Theorem 3, we have $\lim\limits_{x \to 1} \ln x = \ln 1 = 0$. Applying the Composition rule,

$$\lim\limits_{x \to 1} e^{\ln x} = \lim\limits_{x \to 0} e^x = e^0 = 1.$$

Both approaches are valid, giving the same result.

Notes:

5. We encountered this limit in Section 1.1. Applying our theorems, we attempt to find the limit as

$$\lim_{x \to 0} \frac{\sin x}{x} \to \frac{\sin 0}{0} \to \frac{\text{``} 0 \text{''}}{0}.$$

This, of course, violates a condition of Theorem 1, as the limit of the denominator is not allowed to be 0. Therefore, we are still unable to evaluate this limit with tools we currently have at hand.

The section could have been titled "Using Known Limits to Find Unknown Limits." By knowing certain limits of functions, we can find limits involving sums, products, powers, etc., of these functions. We further the development of such comparative tools with the Squeeze Theorem, a clever and intuitive way to find the value of some limits.

Before stating this theorem formally, suppose we have functions f, g and h where g always takes on values between f and h; that is, for all x in an interval,

$$f(x) \le g(x) \le h(x).$$

If f and h have the same limit at c, and g is always "squeezed" between them, then g must have the same limit as well. That is what the Squeeze Theorem states.

Theorem 4   Squeeze Theorem

Let f, g and h be functions on an open interval I containing c such that for all x in I,

$$f(x) \le g(x) \le h(x).$$

If

$$\lim_{x \to c} f(x) = L = \lim_{x \to c} h(x),$$

then

$$\lim_{x \to c} g(x) = L.$$

It can take some work to figure out appropriate functions by which to "squeeze" the given function of which you are trying to evaluate a limit. However, that is generally the only place work is necessary; the theorem makes the "evaluating the limit part" very simple.

We use the Squeeze Theorem in the following example to finally prove that $\lim_{x \to 0} \frac{\sin x}{x} = 1$.

Notes:

Example 13   Using the Squeeze Theorem

Use the Squeeze Theorem to show that

$$\lim_{x \to 0} \frac{\sin x}{x} = 1.$$

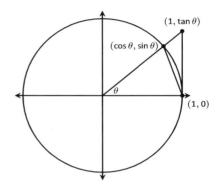

Figure 1.19: The unit circle and related triangles.

SOLUTION   We begin by considering the unit circle. Each point on the unit circle has coordinates $(\cos\theta, \sin\theta)$ for some angle θ as shown in Figure 1.19. Using similar triangles, we can extend the line from the origin through the point to the point $(1, \tan\theta)$, as shown. (Here we are assuming that $0 \le \theta \le \pi/2$. Later we will show that we can also consider $\theta \le 0$.)

Figure 1.19 shows three regions have been constructed in the first quadrant, two triangles and a sector of a circle, which are also drawn below. The area of the large triangle is $\frac{1}{2}\tan\theta$; the area of the sector is $\theta/2$; the area of the triangle contained inside the sector is $\frac{1}{2}\sin\theta$. It is then clear from the diagram that

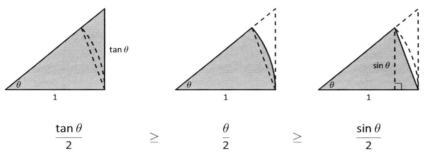

$$\frac{\tan\theta}{2} \qquad \ge \qquad \frac{\theta}{2} \qquad \ge \qquad \frac{\sin\theta}{2}$$

Multiply all terms by $\dfrac{2}{\sin\theta}$, giving

$$\frac{1}{\cos\theta} \ge \frac{\theta}{\sin\theta} \ge 1.$$

Taking reciprocals reverses the inequalities, giving

$$\cos\theta \le \frac{\sin\theta}{\theta} \le 1.$$

(These inequalities hold for all values of θ near 0, even negative values, since $\cos(-\theta) = \cos\theta$ and $\sin(-\theta) = -\sin\theta$.)

Now take limits.

$$\lim_{\theta \to 0} \cos\theta \le \lim_{\theta \to 0} \frac{\sin\theta}{\theta} \le \lim_{\theta \to 0} 1$$

Notes:

$$\cos 0 \leq \lim_{\theta \to 0} \frac{\sin \theta}{\theta} \leq 1$$

$$1 \leq \lim_{\theta \to 0} \frac{\sin \theta}{\theta} \leq 1$$

Clearly this means that $\lim_{\theta \to 0} \frac{\sin \theta}{\theta} = 1$.

Two notes about the previous example are worth mentioning. First, one might be discouraged by this application, thinking "I would *never* have come up with that on my own. This is too hard!" Don't be discouraged; within this text we will guide you in your use of the Squeeze Theorem. As one gains mathematical maturity, clever proofs like this are easier and easier to create.

Second, this limit tells us more than just that as x approaches 0, $\sin(x)/x$ approaches 1. Both x and $\sin x$ are approaching 0, but the *ratio* of x and $\sin x$ approaches 1, meaning that they are approaching 0 in essentially the same way. Another way of viewing this is: for small x, the functions $y = x$ and $y = \sin x$ are essentially indistinguishable.

We include this special limit, along with three others, in the following theorem.

Theorem 5   Special Limits

1. $\lim_{x \to 0} \dfrac{\sin x}{x} = 1$

2. $\lim_{x \to 0} \dfrac{\cos x - 1}{x} = 0$

3. $\lim_{x \to 0} (1 + x)^{\frac{1}{x}} = e$

4. $\lim_{x \to 0} \dfrac{e^x - 1}{x} = 1$

A short word on how to interpret the latter three limits. We know that as x goes to 0, $\cos x$ goes to 1. So, in the second limit, both the numerator and denominator are approaching 0. However, since the limit is 0, we can interpret this as saying that "$\cos x$ is approaching 1 faster than x is approaching 0."

In the third limit, inside the parentheses we have an expression that is approaching 1 (though never equaling 1), and we know that 1 raised to any power is still 1. At the same time, the power is growing toward infinity. What happens to a number near 1 raised to a very large power? In this particular case, the result approaches Euler's number, e, approximately 2.718.

In the fourth limit, we see that as $x \to 0$, e^x approaches 1 "just as fast" as $x \to 0$, resulting in a limit of 1.

Notes:

Our final theorem for this section will be motivated by the following example.

Example 14   **Using algebra to evaluate a limit**
Evaluate the following limit:

$$\lim_{x \to 1} \frac{x^2 - 1}{x - 1}.$$

SOLUTION   We begin by attempting to apply Theorem 3 and substituting 1 for x in the quotient. This gives:

$$\lim_{x \to 1} \frac{x^2 - 1}{x - 1} = \frac{1^2 - 1}{1 - 1} = \frac{\text{``} 0 \text{''}}{0},$$

and indeterminate form. We cannot apply the theorem.

By graphing the function, as in Figure 1.20, we see that the function seems to be linear, implying that the limit should be easy to evaluate. Recognize that the numerator of our quotient can be factored:

$$\frac{x^2 - 1}{x - 1} = \frac{(x - 1)(x + 1)}{x - 1}.$$

The function is not defined when $x = 1$, but for all other x,

$$\frac{x^2 - 1}{x - 1} = \frac{(x - 1)(x + 1)}{x - 1} = \frac{\cancel{(x - 1)}(x + 1)}{\cancel{x - 1}} = x + 1.$$

Clearly $\lim_{x \to 1} x + 1 = 2$. Recall that when considering limits, we are not concerned with the value of the function at 1, only the value the function approaches as x approaches 1. Since $(x^2 - 1)/(x - 1)$ and $x + 1$ are the same at all points except $x = 1$, they both approach the same value as x approaches 1. Therefore we can conclude that

$$\lim_{x \to 1} \frac{x^2 - 1}{x - 1} = 2.$$

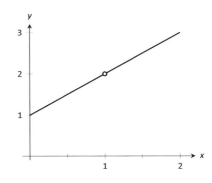

Figure 1.20: Graphing f in Example 14 to understand a limit.

The key to the above example is that the functions $y = (x^2 - 1)/(x - 1)$ and $y = x + 1$ are identical except at $x = 1$. Since limits describe a value the function is approaching, not the value the function actually attains, the limits of the two functions are always equal.

Notes:

Theorem 6   Limits of Functions Equal At All But One Point

Let $g(x) = f(x)$ for all x in an open interval, except possibly at c, and let $\lim_{x \to c} g(x) = L$ for some real number L. Then

$$\lim_{x \to c} f(x) = L.$$

The Fundamental Theorem of Algebra tells us that when dealing with a rational function of the form $g(x)/f(x)$ and directly evaluating the limit $\lim_{x \to c} \dfrac{g(x)}{f(x)}$ returns "0/0", then $(x - c)$ is a factor of both $g(x)$ and $f(x)$. One can then use algebra to factor this term out, cancel, then apply Theorem 6. We demonstrate this once more.

Example 15   Evaluating a limit using Theorem 6

Evaluate $\lim_{x \to 3} \dfrac{x^3 - 2x^2 - 5x + 6}{2x^3 + 3x^2 - 32x + 15}$.

SOLUTION   We begin by applying Theorem 3 and substituting 3 for x. This returns the familiar indeterminate form of "0/0". Since the numerator and denominator are each polynomials, we know that $(x-3)$ is factor of each. Using whatever method is most comfortable to you, factor out $(x-3)$ from each (using polynomial division, synthetic division, a computer algebra system, etc.). We find that

$$\frac{x^3 - 2x^2 - 5x + 6}{2x^3 + 3x^2 - 32x + 15} = \frac{(x - 3)(x^2 + x - 2)}{(x - 3)(2x^2 + 9x - 5)}.$$

We can cancel the $(x-3)$ terms as long as $x \neq 3$. Using Theorem 6 we conclude:

$$\lim_{x \to 3} \frac{x^3 - 2x^2 - 5x + 6}{2x^3 + 3x^2 - 32x + 15} = \lim_{x \to 3} \frac{(x - 3)(x^2 + x - 2)}{(x - 3)(2x^2 + 9x - 5)}$$
$$= \lim_{x \to 3} \frac{(x^2 + x - 2)}{(2x^2 + 9x - 5)}$$
$$= \frac{10}{40} = \frac{1}{4}.$$

We end this section by revisiting a limit first seen in Section 1.1, a limit of a difference quotient. Let $f(x) = -1.5x^2 + 11.5x$; we approximated the limit $\lim_{h \to 0} \dfrac{f(1 + h) - f(1)}{h} \approx 8.5$. We formally evaluate this limit in the following example.

Notes:

Example 16   **Evaluating the limit of a difference quotient**

Let $f(x) = -1.5x^2 + 11.5x$; find $\lim\limits_{h \to 0} \dfrac{f(1+h) - f(1)}{h}$.

SOLUTION   Since f is a polynomial, our first attempt should be to employ Theorem 3 and substitute 0 for h. However, we see that this gives us "0/0." Knowing that we have a rational function hints that some algebra will help. Consider the following steps:

$$\lim_{h \to 0} \frac{f(1+h) - f(1)}{h} = \lim_{h \to 0} \frac{-1.5(1+h)^2 + 11.5(1+h) - \left(-1.5(1)^2 + 11.5(1)\right)}{h}$$

$$= \lim_{h \to 0} \frac{-1.5(1 + 2h + h^2) + 11.5 + 11.5h - 10}{h}$$

$$= \lim_{h \to 0} \frac{-1.5h^2 + 8.5h}{h}$$

$$= \lim_{h \to 0} \frac{h(-1.5h + 8.5)}{h}$$

$$= \lim_{h \to 0} (-1.5h + 8.5) \quad \left(\text{using Theorem 6, as } h \neq 0\right)$$

$$= 8.5 \quad \left(\text{using Theorem 3}\right)$$

This matches our previous approximation.

This section contains several valuable tools for evaluating limits. One of the main results of this section is Theorem 3; it states that many functions that we use regularly behave in a very nice, predictable way. In the next section we give a name to this nice behavior; we label such functions as *continuous*. Defining that term will require us to look again at what a limit is and what causes limits to not exist.

Notes:

Exercises 1.3

Terms and Concepts

1. Explain in your own words, without using ε-δ formality, why $\lim\limits_{x \to c} b = b$.

2. Explain in your own words, without using ε-δ formality, why $\lim\limits_{x \to c} x = c$.

3. What does the text mean when it says that certain functions' "behavior is 'nice' in terms of limits"? What, in particular, is "nice"?

4. Sketch a graph that visually demonstrates the Squeeze Theorem.

5. You are given the following information:

   (a) $\lim\limits_{x \to 1} f(x) = 0$

   (b) $\lim\limits_{x \to 1} g(x) = 0$

   (c) $\lim\limits_{x \to 1} f(x)/g(x) = 2$

   What can be said about the relative sizes of $f(x)$ and $g(x)$ as x approaches 1?

Problems

Using:
$$\lim_{x \to 9} f(x) = 6 \qquad \lim_{x \to 6} f(x) = 9$$
$$\lim_{x \to 9} g(x) = 3 \qquad \lim_{x \to 6} g(x) = 3$$
evaluate the limits given in Exercises 6 – 13, where possible. If it is not possible to know, state so.

6. $\lim\limits_{x \to 9} (f(x) + g(x))$

7. $\lim\limits_{x \to 9} (3f(x)/g(x))$

8. $\lim\limits_{x \to 9} \left(\dfrac{f(x) - 2g(x)}{g(x)} \right)$

9. $\lim\limits_{x \to 6} \left(\dfrac{f(x)}{3 - g(x)} \right)$

10. $\lim\limits_{x \to 9} g(f(x))$

11. $\lim\limits_{x \to 6} f(g(x))$

12. $\lim\limits_{x \to 6} g(f(f(x)))$

13. $\lim\limits_{x \to 6} f(x)g(x) - f^2(x) + g^2(x)$

Using:
$$\lim_{x \to 1} f(x) = 2 \qquad \lim_{x \to 10} f(x) = 1$$
$$\lim_{x \to 1} g(x) = 0 \qquad \lim_{x \to 10} g(x) = \pi$$
evaluate the limits given in Exercises 14 – 17, where possible. If it is not possible to know, state so.

14. $\lim\limits_{x \to 1} f(x)^{g(x)}$

15. $\lim\limits_{x \to 10} \cos(g(x))$

16. $\lim\limits_{x \to 1} f(x)g(x)$

17. $\lim\limits_{x \to 1} g(5f(x))$

In Exercises 18 – 32, evaluate the given limit.

18. $\lim\limits_{x \to 3} x^2 - 3x + 7$

19. $\lim\limits_{x \to \pi} \left(\dfrac{x - 3}{x - 5} \right)^7$

20. $\lim\limits_{x \to \pi/4} \cos x \sin x$

21. $\lim\limits_{x \to 0} \ln x$

22. $\lim\limits_{x \to 3} 4^{x^3 - 8x}$

23. $\lim\limits_{x \to \pi/6} \csc x$

24. $\lim\limits_{x \to 0} \ln(1 + x)$

25. $\lim\limits_{x \to \pi} \dfrac{x^2 + 3x + 5}{5x^2 - 2x - 3}$

26. $\lim\limits_{x \to \pi} \dfrac{3x + 1}{1 - x}$

27. $\lim\limits_{x \to 6} \dfrac{x^2 - 4x - 12}{x^2 - 13x + 42}$

28. $\lim\limits_{x \to 0} \dfrac{x^2 + 2x}{x^2 - 2x}$

29. $\lim\limits_{x \to 2} \dfrac{x^2 + 6x - 16}{x^2 - 3x + 2}$

30. $\lim\limits_{x \to 2} \dfrac{x^2 - 10x + 16}{x^2 - x - 2}$

31. $\lim\limits_{x \to -2} \dfrac{x^2 - 5x - 14}{x^2 + 10x + 16}$

32. $\lim\limits_{x \to -1} \dfrac{x^2 + 9x + 8}{x^2 - 6x - 7}$

Use the Squeeze Theorem in Exercises 33 – 36, where appropriate, to evaluate the given limit.

33. $\lim\limits_{x \to 0} x \sin\left(\dfrac{1}{x}\right)$

34. $\lim\limits_{x \to 0} \sin x \cos\left(\dfrac{1}{x^2}\right)$

35. $\lim\limits_{x \to 1} f(x)$, where $3x - 2 \leq f(x) \leq x^3$.

36. $\lim\limits_{x \to 3^+} f(x)$, where $6x - 9 \leq f(x) \leq x^2$ on $[0, 3]$.

Exercises 37 – 40 challenge your understanding of limits but can be evaluated using the knowledge gained in this section.

37. $\lim\limits_{x \to 0} \dfrac{\sin 3x}{x}$

38. $\lim\limits_{x \to 0} \dfrac{\sin 5x}{8x}$

39. $\lim\limits_{x \to 0} \dfrac{\ln(1 + x)}{x}$

40. $\lim\limits_{x \to 0} \dfrac{\sin x}{x}$, where x is measured in degrees, not radians.

1.4 One Sided Limits

We introduced the concept of a limit gently, approximating their values graphically and numerically. Next came the rigorous definition of the limit, along with an admittedly tedious method for evaluating them. The previous section gave us tools (which we call theorems) that allow us to compute limits with greater ease. Chief among the results were the facts that polynomials and rational, trigonometric, exponential and logarithmic functions (and their sums, products, etc.) all behave "nicely." In this section we rigorously define what we mean by "nicely."

In Section 1.1 we explored the three ways in which limits of functions failed to exist:

1. The function approached different values from the left and right,

2. The function grows without bound, and

3. The function oscillates.

In this section we explore in depth the concepts behind #1 by introducing the *one-sided limit*. We begin with formal definitions that are very similar to the definition of the limit given in Section 1.2, but the notation is slightly different and "$x \neq c$" is replaced with either "$x < c$" or "$x > c$."

Definition 2   One Sided Limits

Left-Hand Limit
Let I be an open interval containing c, and let f be a function defined on I, except possibly at c. The **limit of** $f(x)$**, as** x **approaches** c **from the left, is** L, or, **the left–hand limit of** f **at** c **is** L, denoted by

$$\lim_{x \to c^-} f(x) = L,$$

means that given any $\varepsilon > 0$, there exists $\delta > 0$ such that for all $x < c$, if $|x - c| < \delta$, then $|f(x) - L| < \varepsilon$.

Right-Hand Limit
Let I be an open interval containing c, and let f be a function defined on I, except possibly at c. The **limit of** $f(x)$**, as** x **approaches** c **from the right, is** L, or, **the right–hand limit of** f **at** c **is** L, denoted by

$$\lim_{x \to c^+} f(x) = L,$$

means that given any $\varepsilon > 0$, there exists $\delta > 0$ such that for all $x > c$, if $|x - c| < \delta$, then $|f(x) - L| < \varepsilon$.

Notes:

Practically speaking, when evaluating a left-hand limit, we consider only values of x "to the left of c," i.e., where $x < c$. The admittedly imperfect notation $x \to c^-$ is used to imply that we look at values of x to the left of c. The notation has nothing to do with positive or negative values of either x or c. A similar statement holds for evaluating right-hand limits; there we consider only values of x to the right of c, i.e., $x > c$. We can use the theorems from previous sections to help us evaluate these limits; we just restrict our view to one side of c.

We practice evaluating left and right-hand limits through a series of examples.

Example 17   **Evaluating one sided limits**

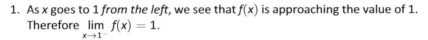 Let $f(x) = \begin{cases} x & 0 \le x \le 1 \\ 3-x & 1 < x < 2 \end{cases}$, as shown in Figure 1.21. Find each of the following:

1. $\lim\limits_{x \to 1^-} f(x)$

2. $\lim\limits_{x \to 1^+} f(x)$

3. $\lim\limits_{x \to 1} f(x)$

4. $f(1)$

5. $\lim\limits_{x \to 0^+} f(x)$

6. $f(0)$

7. $\lim\limits_{x \to 2^-} f(x)$

8. $f(2)$

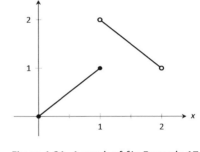

Figure 1.21: A graph of f in Example 17.

SOLUTION   For these problems, the visual aid of the graph is likely more effective in evaluating the limits than using f itself. Therefore we will refer often to the graph.

1. As x goes to 1 *from the left*, we see that $f(x)$ is approaching the value of 1. Therefore $\lim\limits_{x \to 1^-} f(x) = 1$.

2. As x goes to 1 *from the right*, we see that $f(x)$ is approaching the value of 2. Recall that it does not matter that there is an "open circle" there; we are evaluating a limit, not the value of the function. Therefore $\lim\limits_{x \to 1^+} f(x) = 2$.

3. *The* limit of f as x approaches 1 does not exist, as discussed in the first section. The function does not approach one particular value, but two different values from the left and the right.

4. Using the definition and by looking at the graph we see that $f(1) = 1$.

5. As x goes to 0 from the right, we see that $f(x)$ is also approaching 0. Therefore $\lim\limits_{x \to 0^+} f(x) = 0$. Note we cannot consider a left-hand limit at 0 as f is not defined for values of $x < 0$.

Notes:

6. Using the definition and the graph, $f(0) = 0$.

7. As x goes to 2 from the left, we see that $f(x)$ is approaching the value of 1. Therefore $\lim\limits_{x \to 2^-} f(x) = 1$.

8. The graph and the definition of the function show that $f(2)$ is not defined.

Note how the left and right-hand limits were different at $x = 1$. This, of course, causes *the* limit to not exist. The following theorem states what is fairly intuitive: *the* limit exists precisely when the left and right-hand limits are equal.

Theorem 7   Limits and One Sided Limits

Let f be a function defined on an open interval I containing c. Then

$$\lim_{x \to c} f(x) = L$$

if, and only if,

$$\lim_{x \to c^-} f(x) = L \quad \text{and} \quad \lim_{x \to c^+} f(x) = L.$$

The phrase "if, and only if" means the two statements are *equivalent*: they are either both true or both false. If the limit equals L, then the left and right hand limits both equal L. If the limit is not equal to L, then at least one of the left and right-hand limits is not equal to L (it may not even exist).

One thing to consider in Examples 17 – 20 is that the value of the function may/may not be equal to the value(s) of its left/right-hand limits, even when these limits agree.

Example 18   Evaluating limits of a piecewise–defined function

Let $f(x) = \begin{cases} 2 - x & 0 < x < 1 \\ (x-2)^2 & 1 < x < 2 \end{cases}$, as shown in Figure 1.22. Evaluate the following.

1. $\lim\limits_{x \to 1^-} f(x)$

2. $\lim\limits_{x \to 1^+} f(x)$

3. $\lim\limits_{x \to 1} f(x)$

4. $f(1)$

5. $\lim\limits_{x \to 0^+} f(x)$

6. $f(0)$

7. $\lim\limits_{x \to 2^-} f(x)$

8. $f(2)$

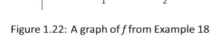

Figure 1.22: A graph of f from Example 18

Notes:

SOLUTION Again we will evaluate each using both the definition of f and its graph.

1. As x approaches 1 from the left, we see that $f(x)$ approaches 1. Therefore $\lim\limits_{x \to 1^-} f(x) = 1$.

2. As x approaches 1 from the right, we see that again $f(x)$ approaches 1. Therefore $\lim\limits_{x \to 1+} f(x) = 1$.

3. *The* limit of f as x approaches 1 exists and is 1, as f approaches 1 from both the right and left. Therefore $\lim\limits_{x \to 1} f(x) = 1$.

4. $f(1)$ is not defined. Note that 1 is not in the domain of f as defined by the problem, which is indicated on the graph by an open circle when $x = 1$.

5. As x goes to 0 from the right, $f(x)$ approaches 2. So $\lim\limits_{x \to 0^+} f(x) = 2$.

6. $f(0)$ is not defined as 0 is not in the domain of f.

7. As x goes to 2 from the left, $f(x)$ approaches 0. So $\lim\limits_{x \to 2^-} f(x) = 0$.

8. $f(2)$ is not defined as 2 is not in the domain of f.

Example 19 **Evaluating limits of a piecewise–defined function**

Let $f(x) = \begin{cases} (x-1)^2 & 0 \le x \le 2, x \ne 1 \\ 1 & x = 1 \end{cases}$, as shown in Figure 1.23. Evaluate the following.

1. $\lim\limits_{x \to 1^-} f(x)$

2. $\lim\limits_{x \to 1+} f(x)$

3. $\lim\limits_{x \to 1} f(x)$

4. $f(1)$

SOLUTION It is clear by looking at the graph that both the left and right-hand limits of f, as x approaches 1, is 0. Thus it is also clear that *the* limit is 0; i.e., $\lim\limits_{x \to 1} f(x) = 0$. It is also clearly stated that $f(1) = 1$.

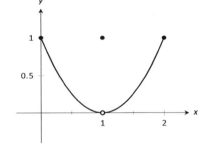

Figure 1.23: Graphing f in Example 19

Example 20 **Evaluating limits of a piecewise–defined function**

Let $f(x) = \begin{cases} x^2 & 0 \le x \le 1 \\ 2-x & 1 < x \le 2 \end{cases}$, as shown in Figure 1.24. Evaluate the following.

Notes:

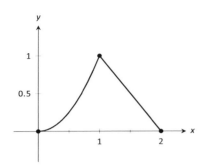

Figure 1.24: Graphing f in Example 20

1. $\lim\limits_{x \to 1^-} f(x)$

2. $\lim\limits_{x \to 1^+} f(x)$

3. $\lim\limits_{x \to 1} f(x)$

4. $f(1)$

SOLUTION It is clear from the definition of the function and its graph that all of the following are equal:

$$\lim_{x \to 1^-} f(x) = \lim_{x \to 1^+} f(x) = \lim_{x \to 1} f(x) = f(1) = 1.$$

In Examples 17 – 20 we were asked to find both $\lim\limits_{x \to 1} f(x)$ and $f(1)$. Consider the following table:

	$\lim\limits_{x \to 1} f(x)$	$f(1)$
Example 17	does not exist	1
Example 18	1	not defined
Example 19	0	1
Example 20	1	1

Only in Example 20 do both the function and the limit exist and agree. This seems "nice;" in fact, it seems "normal." This is in fact an important situation which we explore in the next section, entitled "Continuity." In short, a *continuous function* is one in which when a function approaches a value as $x \to c$ (i.e., when $\lim\limits_{x \to c} f(x) = L$), it actually *attains* that value at c. Such functions behave nicely as they are very predictable.

Notes:

Exercises 1.4

Terms and Concepts

1. What are the three ways in which a limit may fail to exist?

2. T/F: If $\lim_{x \to 1^-} f(x) = 5$, then $\lim_{x \to 1} f(x) = 5$

3. T/F: If $\lim_{x \to 1^-} f(x) = 5$, then $\lim_{x \to 1^+} f(x) = 5$

4. T/F: If $\lim_{x \to 1} f(x) = 5$, then $\lim_{x \to 1^-} f(x) = 5$

Problems

In Exercises 5 – 12, evaluate each expression using the given graph of $f(x)$.

5.

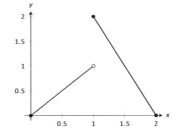

  (a) $\lim_{x \to 1^-} f(x)$ (d) $f(1)$

  (b) $\lim_{x \to 1^+} f(x)$ (e) $\lim_{x \to 0^-} f(x)$

  (c) $\lim_{x \to 1} f(x)$ (f) $\lim_{x \to 0^+} f(x)$

6.

  (a) $\lim_{x \to 1^-} f(x)$ (d) $f(1)$

  (b) $\lim_{x \to 1^+} f(x)$ (e) $\lim_{x \to 2^-} f(x)$

  (c) $\lim_{x \to 1} f(x)$ (f) $\lim_{x \to 2^+} f(x)$

7.

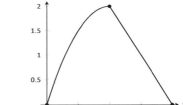

  (a) $\lim_{x \to 1^-} f(x)$ (d) $f(1)$

  (b) $\lim_{x \to 1^+} f(x)$ (e) $\lim_{x \to 2^-} f(x)$

  (c) $\lim_{x \to 1} f(x)$ (f) $\lim_{x \to 0^+} f(x)$

8.

  (a) $\lim_{x \to 1^-} f(x)$ (c) $\lim_{x \to 1} f(x)$

  (b) $\lim_{x \to 1^+} f(x)$ (d) $f(1)$

9.

  (a) $\lim_{x \to 1^-} f(x)$ (c) $\lim_{x \to 1} f(x)$

  (b) $\lim_{x \to 1^+} f(x)$ (d) $f(1)$

10.

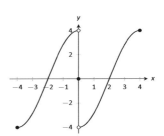

  (a) $\lim_{x \to 0^-} f(x)$ (c) $\lim_{x \to 0} f(x)$

  (b) $\lim_{x \to 0^+} f(x)$ (d) $f(0)$

11.

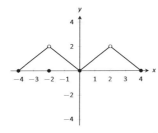

(a) $\lim\limits_{x \to -2^-} f(x)$ (e) $\lim\limits_{x \to 2^-} f(x)$

(b) $\lim\limits_{x \to -2^+} f(x)$ (f) $\lim\limits_{x \to 2^+} f(x)$

(c) $\lim\limits_{x \to -2} f(x)$ (g) $\lim\limits_{x \to 2} f(x)$

(d) $f(-2)$ (h) $f(2)$

12.

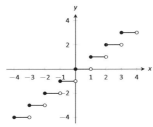

Let $-3 \le a \le 3$ be an integer.

(a) $\lim\limits_{x \to a^-} f(x)$ (c) $\lim\limits_{x \to a} f(x)$

(b) $\lim\limits_{x \to a^+} f(x)$ (d) $f(a)$

In Exercises 13 – 21, evaluate the given limits of the piecewise defined functions f.

13. $f(x) = \begin{cases} x+1 & x \le 1 \\ x^2 - 5 & x > 1 \end{cases}$

(a) $\lim\limits_{x \to 1^-} f(x)$ (c) $\lim\limits_{x \to 1} f(x)$

(b) $\lim\limits_{x \to 1^+} f(x)$ (d) $f(1)$

14. $f(x) = \begin{cases} 2x^2 + 5x - 1 & x < 0 \\ \sin x & x \ge 0 \end{cases}$

(a) $\lim\limits_{x \to 0^-} f(x)$ (c) $\lim\limits_{x \to 0} f(x)$

(b) $\lim\limits_{x \to 0^+} f(x)$ (d) $f(0)$

15. $f(x) = \begin{cases} x^2 - 1 & x < -1 \\ x^3 + 1 & -1 \le x \le 1 \\ x^2 + 1 & x > 1 \end{cases}$

(a) $\lim\limits_{x \to -1^-} f(x)$ (e) $\lim\limits_{x \to 1^-} f(x)$

(b) $\lim\limits_{x \to -1^+} f(x)$ (f) $\lim\limits_{x \to 1^+} f(x)$

(c) $\lim\limits_{x \to -1} f(x)$ (g) $\lim\limits_{x \to 1} f(x)$

(d) $f(-1)$ (h) $f(1)$

16. $f(x) = \begin{cases} \cos x & x < \pi \\ \sin x & x \ge \pi \end{cases}$

(a) $\lim\limits_{x \to \pi^-} f(x)$ (c) $\lim\limits_{x \to \pi} f(x)$

(b) $\lim\limits_{x \to \pi^+} f(x)$ (d) $f(\pi)$

17. $f(x) = \begin{cases} 1 - \cos^2 x & x < a \\ \sin^2 x & x \ge a \end{cases}$,

where a is a real number.

(a) $\lim\limits_{x \to a^-} f(x)$ (c) $\lim\limits_{x \to a} f(x)$

(b) $\lim\limits_{x \to a^+} f(x)$ (d) $f(a)$

18. $f(x) = \begin{cases} x+1 & x < 1 \\ 1 & x = 1 \\ x-1 & x > 1 \end{cases}$

(a) $\lim\limits_{x \to 1^-} f(x)$ (c) $\lim\limits_{x \to 1} f(x)$

(b) $\lim\limits_{x \to 1^+} f(x)$ (d) $f(1)$

19. $f(x) = \begin{cases} x^2 & x < 2 \\ x+1 & x = 2 \\ -x^2 + 2x + 4 & x > 2 \end{cases}$

(a) $\lim\limits_{x \to 2^-} f(x)$ (c) $\lim\limits_{x \to 2} f(x)$

(b) $\lim\limits_{x \to 2^+} f(x)$ (d) $f(2)$

20. $f(x) = \begin{cases} a(x-b)^2 + c & x < b \\ a(x-b) + c & x \ge b \end{cases}$,

where a, b and c are real numbers.

(a) $\lim\limits_{x \to b^-} f(x)$ (c) $\lim\limits_{x \to b} f(x)$

(b) $\lim\limits_{x \to b^+} f(x)$ (d) $f(b)$

21. $f(x) = \begin{cases} \frac{|x|}{x} & x \ne 0 \\ 0 & x = 0 \end{cases}$

(a) $\lim\limits_{x \to 0^-} f(x)$ (c) $\lim\limits_{x \to 0} f(x)$

(b) $\lim\limits_{x \to 0^+} f(x)$ (d) $f(0)$

Review

22. Evaluate the limit: $\lim\limits_{x \to -1} \dfrac{x^2 + 5x + 4}{x^2 - 3x - 4}$.

23. Evaluate the limit: $\lim\limits_{x \to -4} \dfrac{x^2 - 16}{x^2 - 4x - 32}$.

24. Evaluate the limit: $\lim\limits_{x \to -6} \dfrac{x^2 - 15x + 54}{x^2 - 6x}$.

25. Approximate the limit numerically: $\lim\limits_{x \to 0.4} \dfrac{x^2 - 4.4x + 1.6}{x^2 - 0.4x}$.

26. Approximate the limit numerically: $\lim\limits_{x \to 0.2} \dfrac{x^2 + 5.8x - 1.2}{x^2 - 4.2x + 0.8}$.

1.5   Continuity

As we have studied limits, we have gained the intuition that limits measure "where a function is heading." That is, if $\lim\limits_{x\to 1} f(x) = 3$, then as x is close to 1, $f(x)$ is close to 3. We have seen, though, that this is not necessarily a good indicator of what $f(1)$ actually this. This can be problematic; functions can tend to one value but attain another. This section focuses on functions that *do not* exhibit such behavior.

Definition 3   Continuous Function

Let f be a function defined on an open interval I containing c.

1. f is **continuous at** c if $\lim\limits_{x\to c} f(x) = f(c)$.

2. f is **continuous on** I if f is continuous at c for all values of c in I. If f is continuous on $(-\infty, \infty)$, we say f is **continuous everywhere**.

A useful way to establish whether or not a function f is continuous at c is to verify the following three things:

1. $\lim\limits_{x\to c} f(x)$ exists,

2. $f(c)$ is defined, and

3. $\lim\limits_{x\to c} f(x) = f(c)$.

Example 21   Finding intervals of continuity
Let f be defined as shown in Figure 1.25. Give the interval(s) on which f is continuous.

SOLUTION   We proceed by examining the three criteria for continuity.

1. The limits $\lim\limits_{x\to c} f(x)$ exists for all c between 0 and 3.

2. $f(c)$ is defined for all c between 0 and 3, *except for* $c = 1$. We know immediately that f cannot be continuous at $x = 1$.

3. The limit $\lim\limits_{x\to c} f(x) = f(c)$ for all c between 0 and 3, except, of course, for $c = 1$.

We conclude that f is continuous at every point of $(0, 3)$ except at $x = 1$. Therefore f is continuous on $(0, 1) \cup (1, 3)$.

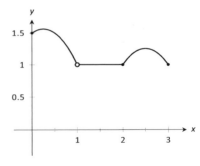

Figure 1.25: A graph of f in Example 21.

Notes:

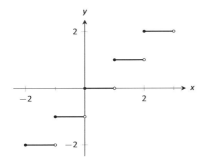

Figure 1.26: A graph of the step function in Example 22.

Example 22   **Finding intervals of continuity**

The *floor function*, $f(x) = \lfloor x \rfloor$, returns the largest integer smaller than the input x. (For example, $f(\pi) = \lfloor \pi \rfloor = 3$.) The graph of f in Figure 1.26 demonstrates why this is often called a "step function."

Give the intervals on which f is continuous.

> **SOLUTION**   We examine the three criteria for continuity.
>
> 1. The limits $\lim_{x \to c} f(x)$ do not exist at the jumps from one "step" to the next, which occur at all integer values of c. Therefore the limits exist for all c except when c is an integer.
>
> 2. The function is defined for all values of c.
>
> 3. The limit $\lim_{x \to c} f(x) = f(c)$ for all values of c where the limit exist, since each step consists of just a line.
>
> We conclude that f is continuous everywhere except at integer values of c. So the intervals on which f is continuous are
>
> $$\ldots, (-2, -1), (-1, 0), (0, 1), (1, 2), \ldots.$$

Our definition of continuity on an interval specifies the interval is an open interval. We can extend the definition of continuity to closed intervals by considering the appropriate one-sided limits at the endpoints.

Definition 4   **Continuity on Closed Intervals**

Let f be defined on the closed interval $[a, b]$ for some real numbers a, b. f is **continuous on** $[a, b]$ if:

1. f is continuous on (a, b),

2. $\lim_{x \to a^+} f(x) = f(a)$ and

3. $\lim_{x \to b^-} f(x) = f(b)$.

We can make the appropriate adjustments to talk about continuity on half–open intervals such as $[a, b)$ or $(a, b]$ if necessary.

Notes:

Example 23 **Determining intervals on which a function is continuous**

For each of the following functions, give the domain of the function and the interval(s) on which it is continuous.

1. $f(x) = 1/x$

2. $f(x) = \sin x$

3. $f(x) = \sqrt{x}$

4. $f(x) = \sqrt{1 - x^2}$

5. $f(x) = |x|$

SOLUTION We examine each in turn.

1. The domain of $f(x) = 1/x$ is $(-\infty, 0) \cup (0, \infty)$. As it is a rational function, we apply Theorem 2 to recognize that f is continuous on all of its domain.

2. The domain of $f(x) = \sin x$ is all real numbers, or $(-\infty, \infty)$. Applying Theorem 3 shows that $\sin x$ is continuous everywhere.

3. The domain of $f(x) = \sqrt{x}$ is $[0, \infty)$. Applying Theorem 3 shows that $f(x) = \sqrt{x}$ is continuous on its domain of $[0, \infty)$.

4. The domain of $f(x) = \sqrt{1 - x^2}$ is $[-1, 1]$. Applying Theorems 1 and 3 shows that f is continuous on all of its domain, $[-1, 1]$.

5. The domain of $f(x) = |x|$ is $(-\infty, \infty)$. We can define the absolute value function as $f(x) = \begin{cases} -x & x < 0 \\ x & x \geq 0 \end{cases}$. Each "piece" of this piecewise defined function is continuous on all of its domain, giving that f is continuous on $(-\infty, 0)$ and $[0, \infty)$. We cannot assume this implies that f is continuous on $(-\infty, \infty)$; we need to check that $\lim\limits_{x \to 0} f(x) = f(0)$, as $x = 0$ is the point where f transitions from one "piece" of its definition to the other. It is easy to verify that this is indeed true, hence we conclude that $f(x) = |x|$ is continuous everywhere.

Continuity is inherently tied to the properties of limits. Because of this, the properties of limits found in Theorems 1 and 2 apply to continuity as well. Further, now knowing the definition of continuity we can re–read Theorem 3 as giving a list of functions that are continuous on their domains. The following theorem states how continuous functions can be combined to form other continuous functions, followed by a theorem which formally lists functions that we know are continuous on their domains.

Notes:

Theorem 8   Properties of Continuous Functions

Let f and g be continuous functions on an interval I, let c be a real number and let n be a positive integer. The following functions are continuous on I.

1. Sums/Differences: $f \pm g$

2. Constant Multiples: $c \cdot f$

3. Products: $f \cdot g$

4. Quotients: f/g (as long as $g \neq 0$ on I)

5. Powers: f^n

6. Roots: $\sqrt[n]{f}$ (if n is even then $f \geq 0$ on I; if n is odd, then true for all values of f on I.)

7. Compositions: Adjust the definitions of f and g to: Let f be continuous on I, where the range of f on I is J, and let g be continuous on J. Then $g \circ f$, i.e., $g(f(x))$, is continuous on I.

Theorem 9   Continuous Functions

The following functions are continuous on their domains.

1. $f(x) = \sin x$ 2. $f(x) = \cos x$

3. $f(x) = \tan x$ 4. $f(x) = \cot x$

5. $f(x) = \sec x$ 6. $f(x) = \csc x$

7. $f(x) = \ln x$ 8. $f(x) = \sqrt[n]{x}$,

9. $f(x) = a^x \ (a > 0)$ (where n is a positive integer)

We apply these theorems in the following Example.

Notes:

Example 24   **Determining intervals on which a function is continuous**

State the interval(s) on which each of the following functions is continuous.

1. $f(x) = \sqrt{x-1} + \sqrt{5-x}$

2. $f(x) = x \sin x$

3. $f(x) = \tan x$

4. $f(x) = \sqrt{\ln x}$

SOLUTION   We examine each in turn, applying Theorems 8 and 9 as appropriate.

1. The square–root terms are continuous on the intervals $[1, \infty)$ and $(-\infty, 5]$, respectively. As f is continuous only where each term is continuous, f is continuous on $[1, 5]$, the intersection of these two intervals. A graph of f is given in Figure 1.27.

2. The functions $y = x$ and $y = \sin x$ are each continuous everywhere, hence their product is, too.

3. Theorem 9 states that $f(x) = \tan x$ is continuous "on its domain." Its domain includes all real numbers except odd multiples of $\pi/2$. Thus $f(x) = \tan x$ is continuous on

$$\ldots \left(-\frac{3\pi}{2}, -\frac{\pi}{2} \right), \left(-\frac{\pi}{2}, \frac{\pi}{2} \right), \left(\frac{\pi}{2}, \frac{3\pi}{2} \right), \ldots,$$

or, equivalently, on $D = \{ x \in \mathbb{R} \mid x \neq n \cdot \frac{\pi}{2}, \text{n is an odd integer} \}$.

4. The domain of $y = \sqrt{x}$ is $[0, \infty)$. The range of $y = \ln x$ is $(-\infty, \infty)$, but if we restrict its domain to $[1, \infty)$ its range is $[0, \infty)$. So restricting $y = \ln x$ to the domain of $[1, \infty)$ restricts its output is $[0, \infty)$, on which $y = \sqrt{x}$ is defined. Thus the domain of $f(x) = \sqrt{\ln x}$ is $[1, \infty)$.

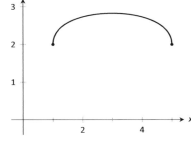

Figure 1.27: A graph of f in Example 24(a).

A common way of thinking of a continuous function is that "its graph can be sketched without lifting your pencil." That is, its graph forms a "continuous" curve, without holes, breaks or jumps. While beyond the scope of this text, this pseudo–definition glosses over some of the finer points of continuity. Very strange functions are continuous that one would be hard pressed to actually sketch by hand.

This intuitive notion of continuity does help us understand another important concept as follows. Suppose f is defined on $[1, 2]$ and $f(1) = -10$ and $f(2) = 5$. If f is continuous on $[1, 2]$ (i.e., its graph can be sketched as a continuous curve from $(1, -10)$ to $(2, 5)$) then we know intuitively that somewhere on $[1, 2]$ f must be equal to -9, and -8, and -7, -6, $\ldots$, 0, $1/2$, etc. In short, f

Notes:

takes on all *intermediate* values between -10 and 5. It may take on more values; f may actually equal 6 at some time, for instance, but we are guaranteed all values between -10 and 5.

While this notion seems intuitive, it is not trivial to prove and its importance is profound. Therefore the concept is stated in the form of a theorem.

Theorem 10   Intermediate Value Theorem

Let f be a continuous function on $[a, b]$ and, without loss of generality, let $f(a) < f(b)$. Then for every value y, where $f(a) < y < f(b)$, there is a value c in $[a, b]$ such that $f(c) = y$.

One important application of the Intermediate Value Theorem is root finding. Given a function f, we are often interested in finding values of x where $f(x) = 0$. These roots may be very difficult to find exactly. Good approximations can be found through successive applications of this theorem. Suppose through direct computation we find that $f(a) < 0$ and $f(b) > 0$, where $a < b$. The Intermediate Value Theorem states that there is a c in $[a, b]$ such that $f(c) = 0$. The theorem does not give us any clue as to where that value is in the interval $[a, b]$, just that it exists.

There is a technique that produces a good approximation of c. Let d be the midpoint of the interval $[a, b]$ and consider $f(d)$. There are three possibilities:

1. $f(d) = 0$ – we got lucky and stumbled on the actual value. We stop as we found a root.

2. $f(d) < 0$ Then we know there is a root of f on the interval $[d, b]$ – we have halved the size of our interval, hence are closer to a good approximation of the root.

3. $f(d) > 0$ Then we know there is a root of f on the interval $[a, d]$ – again, we have halved the size of our interval, hence are closer to a good approximation of the root.

Successively applying this technique is called the **Bisection Method** of root finding. We continue until the interval is sufficiently small. We demonstrate this in the following example.

Example 25   Using the Bisection Method
Approximate the root of $f(x) = x - \cos x$, accurate to three places after the decimal.

SOLUTION   Consider the graph of $f(x) = x - \cos x$, shown in Figure 1.28. It is clear that the graph crosses the x-axis somewhere near $x = 0.8$. To start the

---

Notes:

Bisection Method, pick an interval that contains 0.8. We choose $[0.7, 0.9]$. Note that all we care about are signs of $f(x)$, not their actual value, so this is all we display.

Iteration 1: $f(0.7) < 0, f(0.9) > 0$, and $f(0.8) > 0$. So replace 0.9 with 0.8 and repeat.

Iteration 2: $f(0.7) < 0, f(0.8) > 0$, and at the midpoint, 0.75, we have $f(0.75) > 0$. So replace 0.8 with 0.75 and repeat. Note that we don't need to continue to check the endpoints, just the midpoint. Thus we put the rest of the iterations in Table 1.29.

Notice that in the 12th iteration we have the endpoints of the interval each starting with 0.739. Thus we have narrowed the zero down to an accuracy of the first three places after the decimal. Using a computer, we have

$$f(0.7390) = -0.00014, \quad f(0.7391) = 0.000024.$$

Either endpoint of the interval gives a good approximation of where f is 0. The Intermediate Value Theorem states that the actual zero is still within this interval. While we do not know its exact value, we know it starts with 0.739.

This type of exercise is rarely done by hand. Rather, it is simple to program a computer to run such an algorithm and stop when the endpoints differ by a preset small amount. One of the authors did write such a program and found the zero of f, accurate to 10 places after the decimal, to be 0.7390851332. While it took a few minutes to write the program, it took less than a thousandth of a second for the program to run the necessary 35 iterations. In less than 8 hundredths of a second, the zero was calculated to 100 decimal places (with less than 200 iterations).

It is a simple matter to extend the Bisection Method to solve problems similar to "Find x, where $f(x) = 0$." For instance, we can find x, where $f(x) = 1$. It actually works very well to define a new function g where $g(x) = f(x) - 1$. Then use the Bisection Method to solve $g(x) = 0$.

Similarly, given two functions f and g, we can use the Bisection Method to solve $f(x) = g(x)$. Once again, create a new function h where $h(x) = f(x) - g(x)$ and solve $h(x) = 0$.

In Section 4.1 another equation solving method will be introduced, called Newton's Method. In many cases, Newton's Method is much faster. It relies on more advanced mathematics, though, so we will wait before introducing it.

This section formally defined what it means to be a continuous function. "Most" functions that we deal with are continuous, so often it feels odd to have to formally define this concept. Regardless, it is important, and forms the basis of the next chapter.

In the next section we examine one more aspect of limits: limits that involve infinity.

Notes:

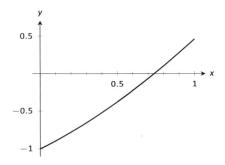

Figure 1.28: Graphing a root of $f(x) = x - \cos x$.

Iteration #	Interval	Midpoint Sign
1	$[0.7, 0.9]$	$f(0.8) > 0$
2	$[0.7, 0.8]$	$f(0.75) > 0$
3	$[0.7, 0.75]$	$f(0.725) < 0$
4	$[0.725, 0.75]$	$f(0.7375) < 0$
5	$[0.7375, 0.75]$	$f(0.7438) > 0$
6	$[0.7375, 0.7438]$	$f(0.7407) > 0$
7	$[0.7375, 0.7407]$	$f(0.7391) > 0$
8	$[0.7375, 0.7391]$	$f(0.7383) < 0$
9	$[0.7383, 0.7391]$	$f(0.7387) < 0$
10	$[0.7387, 0.7391]$	$f(0.7389) < 0$
11	$[0.7389, 0.7391]$	$f(0.7390) < 0$
12	$[0.7390, 0.7391]$	

Figure 1.29: Iterations of the Bisection Method of Root Finding

Exercises 1.5

Terms and Concepts

1. In your own words, describe what it means for a function to be continuous.

2. In your own words, describe what the Intermediate Value Theorem states.

3. What is a "root" of a function?

4. Given functions f and g on an interval I, how can the Bisection Method be used to find a value c where $f(c) = g(c)$?

5. T/F: If f is defined on an open interval containing c, and $\lim\limits_{x \to c} f(x)$ exists, then f is continuous at c.

6. T/F: If f is continuous at c, then $\lim\limits_{x \to c} f(x)$ exists.

7. T/F: If f is continuous at c, then $\lim\limits_{x \to c^+} f(x) = f(c)$.

8. T/F: If f is continuous on $[a, b]$, then $\lim\limits_{x \to a^-} f(x) = f(a)$.

9. T/F: If f is continuous on $[0, 1)$ and $[1, 2)$, then f is continuous on $[0, 2)$.

10. T/F: The sum of continuous functions is also continuous.

Problems

In Exercises 11 – 17, a graph of a function f is given along with a value a. Determine if f is continuous at a; if it is not, state why it is not.

11. $a = 1$

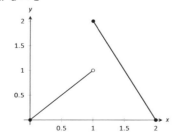

12. $a = 1$

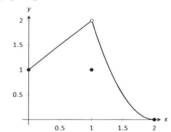

13. $a = 1$

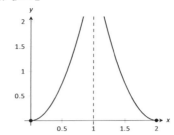

14. $a = 0$

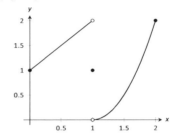

15. $a = 1$

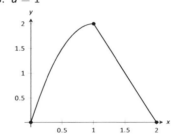

16. $a = 4$

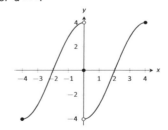

17.   (a) $a = -2$

    (b) $a = 0$

    (c) $a = 2$

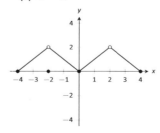

In Exercises 18 – 21, determine if f is continuous at the indicated values. If not, explain why.

18. $f(x) = \begin{cases} 1 & x = 0 \\ \frac{\sin x}{x} & x > 0 \end{cases}$

   (a) $x = 0$

   (b) $x = \pi$

19. $f(x) = \begin{cases} x^3 - x & x < 1 \\ x - 2 & x \geq 1 \end{cases}$

   (a) $x = 0$

   (b) $x = 1$

20. $f(x) = \begin{cases} \frac{x^2 + 5x + 4}{x^2 + 3x + 2} & x \neq -1 \\ 3 & x = -1 \end{cases}$

   (a) $x = -1$

   (b) $x = 10$

21. $f(x) = \begin{cases} \frac{x^2 - 64}{x^2 - 11x + 24} & x \neq 8 \\ 5 & x = 8 \end{cases}$

   (a) $x = 0$

   (b) $x = 8$

In Exercises 22 – 32, give the intervals on which the given function is continuous.

22. $f(x) = x^2 - 3x + 9$

23. $g(x) = \sqrt{x^2 - 4}$

24. $h(k) = \sqrt{1 - k} + \sqrt{k + 1}$

25. $f(t) = \sqrt{5t^2 - 30}$

26. $g(t) = \dfrac{1}{\sqrt{1 - t^2}}$

27. $g(x) = \dfrac{1}{1 + x^2}$

28. $f(x) = e^x$

29. $g(s) = \ln s$

30. $h(t) = \cos t$

31. $f(k) = \sqrt{1 - e^k}$

32. $f(x) = \sin(e^x + x^2)$

33. Let f be continuous on $[1, 5]$ where $f(1) = -2$ and $f(5) = -10$. Does a value $1 < c < 5$ exist such that $f(c) = -9$? Why/why not?

34. Let g be continuous on $[-3, 7]$ where $g(0) = 0$ and $g(2) = 25$. Does a value $-3 < c < 7$ exist such that $g(c) = 15$? Why/why not?

35. Let f be continuous on $[-1, 1]$ where $f(-1) = -10$ and $f(1) = 10$. Does a value $-1 < c < 1$ exist such that $f(c) = 11$? Why/why not?

36. Let h be a function on $[-1, 1]$ where $h(-1) = -10$ and $h(1) = 10$. Does a value $-1 < c < 1$ exist such that $h(c) = 0$? Why/why not?

In Exercises 37 – 40, use the Bisection Method to approximate, accurate to two decimal places, the value of the root of the given function in the given interval.

37. $f(x) = x^2 + 2x - 4$ on $[1, 1.5]$.

38. $f(x) = \sin x - 1/2$ on $[0.5, 0.55]$

39. $f(x) = e^x - 2$ on $[0.65, 0.7]$.

40. $f(x) = \cos x - \sin x$ on $[0.7, 0.8]$.

Review

41. Let $f(x) = \begin{cases} x^2 - 5 & x < 5 \\ 5x & x \geq 5 \end{cases}$.

   (a) $\lim\limits_{x \to 5^-} f(x)$ (c) $\lim\limits_{x \to 5} f(x)$

   (b) $\lim\limits_{x \to 5^+} f(x)$ (d) $f(5)$

42. Numerically approximate the following limits:

   (a) $\lim\limits_{x \to -4/5^+} \dfrac{x^2 - 8.2x - 7.2}{x^2 + 5.8x + 4}$

   (b) $\lim\limits_{x \to -4/5^-} \dfrac{x^2 - 8.2x - 7.2}{x^2 + 5.8x + 4}$

43. Give an example of function $f(x)$ for which $\lim\limits_{x \to 0} f(x)$ does not exist.

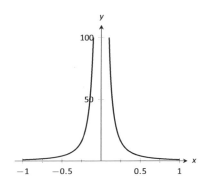

Figure 1.30: Graphing $f(x) = 1/x^2$ for values of x near 0.

1.6 Limits Involving Infinity

In Definition 1 we stated that in the equation $\lim_{x \to c} f(x) = L$, both c and L were numbers. In this section we relax that definition a bit by considering situations when it makes sense to let c and/or L be "infinity."

As a motivating example, consider $f(x) = 1/x^2$, as shown in Figure 1.30. Note how, as x approaches 0, $f(x)$ grows very, very large. It seems appropriate, and descriptive, to state that

$$\lim_{x \to 0} \frac{1}{x^2} = \infty.$$

Also note that as x gets very large, $f(x)$ gets very, very small. We could represent this concept with notation such as

$$\lim_{x \to \infty} \frac{1}{x^2} = 0.$$

We explore both types of use of ∞ in turn.

Definition 5   **Limit of Infinity, ∞**

We say $\lim_{x \to c} f(x) = \infty$ if for every $M > 0$ there exists $\delta > 0$ such that for all $x \neq c$, if $|x - c| < \delta$, then $f(x) \geq M$.

This is just like the ε–δ definition from Section 1.2. In that definition, given any (small) value ε, if we let x get close enough to c (within δ units of c) then $f(x)$ is guaranteed to be within ε of $f(c)$. Here, given any (large) value M, if we let x get close enough to c (within δ units of c), then $f(x)$ will be at least as large as M. In other words, if we get close enough to c, then we can make $f(x)$ as large as we want. We can define limits equal to $-\infty$ in a similar way.

It is important to note that by saying $\lim_{x \to c} f(x) = \infty$ we are implicitly stating that *the* limit of $f(x)$, as x approaches c, *does not exist*. A limit only exists when $f(x)$ approaches an actual numeric value. We use the concept of limits that approach infinity because it is helpful and descriptive.

Example 26   **Evaluating limits involving infinity**
Find $\lim_{x \to 1} \dfrac{1}{(x - 1)^2}$ as shown in Figure 1.31.

SOLUTION   In Example 4 of Section 1.1, by inspecting values of x close to 1 we concluded that this limit does not exist. That is, it cannot equal any real

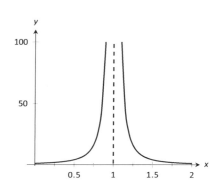

Figure 1.31: Observing infinite limit as $x \to 1$ in Example 26.

Notes:

number. But the limit could be infinite. And in fact, we see that the function does appear to be growing larger and larger, as $f(.99) = 10^4$, $f(.999) = 10^6$, $f(.9999) = 10^8$. A similar thing happens on the other side of 1. In general, let a "large" value M be given. Let $\delta = 1/\sqrt{M}$. If x is within δ of 1, i.e., if $|x - 1| < 1/\sqrt{M}$, then:

$$|x - 1| < \frac{1}{\sqrt{M}}$$
$$(x - 1)^2 < \frac{1}{M}$$
$$\frac{1}{(x - 1)^2} > M,$$

which is what we wanted to show. So we may say $\lim\limits_{x \to 1} 1/(x - 1)^2 = \infty$.

Example 27 **Evaluating limits involving infinity**
Find $\lim\limits_{x \to 0} \dfrac{1}{x}$, as shown in Figure 1.32.

SOLUTION It is easy to see that the function grows without bound near 0, but it does so in different ways on different sides of 0. Since its behavior is not consistent, we cannot say that $\lim\limits_{x \to 0} \dfrac{1}{x} = \infty$. However, we can make a statement about one–sided limits. We can state that $\lim\limits_{x \to 0^+} \dfrac{1}{x} = \infty$ and $\lim\limits_{x \to 0^-} \dfrac{1}{x} = -\infty$.

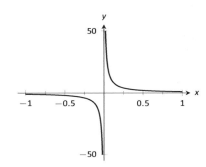

Figure 1.32: Evaluating $\lim\limits_{x \to 0} \dfrac{1}{x}$.

Vertical asymptotes

If the limit of $f(x)$ as x approaches c from either the left or right (or both) is ∞ or $-\infty$, we say the function has a **vertical asymptote** at c.

Example 28 **Finding vertical asymptotes**
Find the vertical asymptotes of $f(x) = \dfrac{3x}{x^2 - 4}$.

SOLUTION Vertical asymptotes occur where the function grows without bound; this can occur at values of c where the denominator is 0. When x is near c, the denominator is small, which in turn can make the function take on large values. In the case of the given function, the denominator is 0 at $x = \pm 2$. Substituting in values of x close to 2 and -2 seems to indicate that the function tends toward ∞ or $-\infty$ at those points. We can graphically confirm this by looking at Figure 1.33. Thus the vertical asymptotes are at $x = \pm 2$.

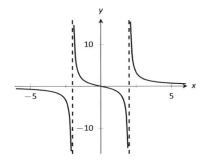

Figure 1.33: Graphing $f(x) = \dfrac{3x}{x^2 - 4}$.

Notes:

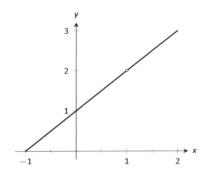

Figure 1.34: Graphically showing that $f(x) = \dfrac{x^2 - 1}{x - 1}$ does not have an asymptote at $x = 1$.

When a rational function has a vertical asymptote at $x = c$, we can conclude that the denominator is 0 at $x = c$. However, just because the denominator is 0 at a certain point does not mean there is a vertical asymptote there. For instance, $f(x) = (x^2 - 1)/(x - 1)$ does not have a vertical asymptote at $x = 1$, as shown in Figure 1.34. While the denominator does get small near $x = 1$, the numerator gets small too, matching the denominator step for step. In fact, factoring the numerator, we get

$$f(x) = \frac{(x - 1)(x + 1)}{x - 1}.$$

Canceling the common term, we get that $f(x) = x + 1$ for $x \neq 1$. So there is clearly no asymptote, rather a hole exists in the graph at $x = 1$.

The above example may seem a little contrived. Another example demonstrating this important concept is $f(x) = (\sin x)/x$. We have considered this function several times in the previous sections. We found that $\lim\limits_{x \to 0} \dfrac{\sin x}{x} = 1$; i.e., there is no vertical asymptote. No simple algebraic cancellation makes this fact obvious; we used the Squeeze Theorem in Section 1.3 to prove this.

If the denominator is 0 at a certain point but the numerator is not, then there will usually be a vertical asymptote at that point. On the other hand, if the numerator and denominator are both zero at that point, then there may or may not be a vertical asymptote at that point. This case where the numerator and denominator are both zero returns us to an important topic.

Indeterminate Forms

We have seen how the limits

$$\lim_{x \to 0} \frac{\sin x}{x} \quad \text{and} \quad \lim_{x \to 1} \frac{x^2 - 1}{x - 1}$$

each return the indeterminate form "0/0" when we blindly plug in $x = 0$ and $x = 1$, respectively. However, 0/0 is not a valid arithmetical expression. It gives no indication that the respective limits are 1 and 2.

With a little cleverness, one can come up 0/0 expressions which have a limit of ∞, 0, or any other real number. That is why this expression is called *indeterminate*.

A key concept to understand is that such limits do not really return 0/0. Rather, keep in mind that we are taking *limits*. What is really happening is that the numerator is shrinking to 0 while the denominator is also shrinking to 0. The respective rates at which they do this are very important and determine the actual value of the limit.

Notes:

An indeterminate form indicates that one needs to do more work in order to compute the limit. That work may be algebraic (such as factoring and canceling) or it may require a tool such as the Squeeze Theorem. In a later section we will learn a technique called l'Hospital's Rule that provides another way to handle indeterminate forms.

Some other common indeterminate forms are $\infty - \infty$, $\infty \cdot 0$, ∞/∞, 0^0, ∞^0 and 1^∞. Again, keep in mind that these are the "blind" results of evaluating a limit, and each, in and of itself, has no meaning. The expression $\infty - \infty$ does not really mean "subtract infinity from infinity." Rather, it means "One quantity is subtracted from the other, but both are growing without bound." What is the result? It is possible to get every value between $-\infty$ and ∞

Note that $1/0$ and $\infty/0$ are not indeterminate forms, though they are not exactly valid mathematical expressions, either. In each, the function is growing without bound, indicating that the limit will be ∞, $-\infty$, or simply not exist if the left- and right-hand limits do not match.

Limits at Infinity and Horizontal Asymptotes

At the beginning of this section we briefly considered what happens to $f(x) = 1/x^2$ as x grew very large. Graphically, it concerns the behavior of the function to the "far right" of the graph. We make this notion more explicit in the following definition.

Definition 6   Limits at Infinity and Horizontal Asymptote

1. We say $\lim_{x \to \infty} f(x) = L$ if for every $\varepsilon > 0$ there exists $M > 0$ such that if $x \geq M$, then $|f(x) - L| < \varepsilon$.

2. We say $\lim_{x \to -\infty} f(x) = L$ if for every $\varepsilon > 0$ there exists $M < 0$ such that if $x \leq M$, then $|f(x) - L| < \varepsilon$.

3. If $\lim_{x \to \infty} f(x) = L$ or $\lim_{x \to -\infty} f(x) = L$, we say that $y = L$ is a **horizontal asymptote** of f.

We can also define limits such as $\lim_{x \to \infty} f(x) = \infty$ by combining this definition with Definition 5.

Notes:

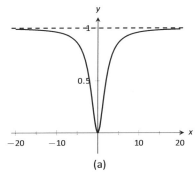

Figure 1.35: Using a graph and a table to approximate a horizontal asymptote in Example 29.

x	f(x)
10	0.9615
100	0.9996
10000	0.999996
−10	0.9615
−100	0.9996
−10000	0.999996

(b)

Example 29   **Approximating horizontal asymptotes**

Approximate the horizontal asymptote(s) of $f(x) = \dfrac{x^2}{x^2 + 4}$.

SOLUTION   We will approximate the horizontal asymptotes by approximating the limits

$$\lim_{x \to -\infty} \frac{x^2}{x^2 + 4} \quad \text{and} \quad \lim_{x \to \infty} \frac{x^2}{x^2 + 4}.$$

Figure 1.35(a) shows a sketch of f, and part (b) gives values of $f(x)$ for large magnitude values of x. It seems reasonable to conclude from both of these sources that f has a horizontal asymptote at $y = 1$.

Later, we will show how to determine this analytically.

Horizontal asymptotes can take on a variety of forms. Figure 1.36(a) shows that $f(x) = x/(x^2 + 1)$ has a horizontal asymptote of $y = 0$, where 0 is approached from both above and below.

Figure 1.36(b) shows that $f(x) = x/\sqrt{x^2 + 1}$ has two horizontal asymptotes; one at $y = 1$ and the other at $y = -1$.

Figure 1.36(c) shows that $f(x) = (\sin x)/x$ has even more interesting behavior than at just $x = 0$; as x approaches $\pm\infty$, $f(x)$ approaches 0, but oscillates as it does this.

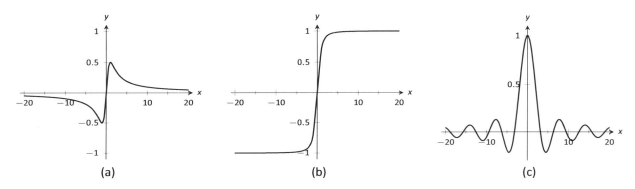

Figure 1.36: Considering different types of horizontal asymptotes.

We can analytically evaluate limits at infinity for rational functions once we understand $\lim\limits_{x \to \infty} 1/x$. As x gets larger and larger, the $1/x$ gets smaller and smaller, approaching 0. We can, in fact, make $1/x$ as small as we want by choosing a large

Notes:

enough value of x. Given ε, we can make $1/x < \varepsilon$ by choosing $x > 1/\varepsilon$. Thus we have $\lim_{x \to \infty} 1/x = 0$.

It is now not much of a jump to conclude the following:

$$\lim_{x \to \infty} \frac{1}{x^n} = 0 \quad \text{and} \quad \lim_{x \to -\infty} \frac{1}{x^n} = 0$$

Now suppose we need to compute the following limit:

$$\lim_{x \to \infty} \frac{x^3 + 2x + 1}{4x^3 - 2x^2 + 9}.$$

A good way of approaching this is to divide through the numerator and denominator by x^3 (hence dividing by 1), which is the largest power of x to appear in the function. Doing this, we get

$$\lim_{x \to \infty} \frac{x^3 + 2x + 1}{4x^3 - 2x^2 + 9} = \lim_{x \to \infty} \frac{1/x^3}{1/x^3} \cdot \frac{x^3 + 2x + 1}{4x^3 - 2x^2 + 9}$$

$$= \lim_{x \to \infty} \frac{x^3/x^3 + 2x/x^3 + 1/x^3}{4x^3/x^3 - 2x^2/x^3 + 9/x^3}$$

$$= \lim_{x \to \infty} \frac{1 + 2/x^2 + 1/x^3}{4 - 2/x + 9/x^3}.$$

Then using the rules for limits (which also hold for limits at infinity), as well as the fact about limits of $1/x^n$, we see that the limit becomes

$$\frac{1 + 0 + 0}{4 - 0 + 0} = \frac{1}{4}.$$

This procedure works for any rational function. In fact, it gives us the following theorem.

Theorem 11   Limits of Rational Functions at Infinity

Let $f(x)$ be a rational function of the following form:

$$f(x) = \frac{a_n x^n + a_{n-1} x^{n-1} + \cdots + a_1 x + a_0}{b_m x^m + b_{m-1} x^{m-1} + \cdots + b_1 x + b_0},$$

where any of the coefficients may be 0 except for a_n and b_m.

1. If $n = m$, then $\lim_{x \to \infty} f(x) = \lim_{x \to -\infty} f(x) = \dfrac{a_n}{b_m}$.

2. If $n < m$, then $\lim_{x \to \infty} f(x) = \lim_{x \to -\infty} f(x) = 0$.

3. If $n > m$, then $\lim_{x \to \infty} f(x)$ and $\lim_{x \to -\infty} f(x)$ are both infinite.

Notes:

We can see why this is true. If the highest power of x is the same in both the numerator and denominator (i.e. $n = m$), we will be in a situation like the example above, where we will divide by x^n and in the limit all the terms will approach 0 except for $a_n x^n / x^n$ and $b_m x^m / x^n$. Since $n = m$, this will leave us with the limit a_n / b_m. If $n < m$, then after dividing through by x^m, all the terms in the numerator will approach 0 in the limit, leaving us with $0 / b_m$ or 0. If $n > m$, and we try dividing through by x^n, we end up with all the terms in the denominator tending toward 0, while the x^n term in the numerator does not approach 0. This is indicative of some sort of infinite limit.

Intuitively, as x gets very large, all the terms in the numerator are small in comparison to $a_n x^n$, and likewise all the terms in the denominator are small compared to $b_n x^m$. If $n = m$, looking only at these two important terms, we have $(a_n x^n)/(b_n x^m)$. This reduces to a_n / b_m. If $n < m$, the function behaves like $a_n / (b_m x^{m-n})$, which tends toward 0. If $n > m$, the function behaves like $a_n x^{n-m} / b_m$, which will tend to either ∞ or $-\infty$ depending on the values of n, m, a_n, b_m and whether you are looking for $\lim_{x \to \infty} f(x)$ or $\lim_{x \to -\infty} f(x)$.

With care, we can quickly evaluate limits at infinity for a large number of functions by considering the largest powers of x. For instance, consider again $\lim\limits_{x \to \pm\infty} \dfrac{x}{\sqrt{x^2 + 1}}$, graphed in Figure 1.36(b). When x is very large, $x^2 + 1 \approx x^2$. Thus

$$\sqrt{x^2 + 1} \approx \sqrt{x^2} = |x|, \quad \text{and} \quad \frac{x}{\sqrt{x^2 + 1}} \approx \frac{x}{|x|}.$$

This expression is 1 when x is positive and -1 when x is negative. Hence we get asymptotes of $y = 1$ and $y = -1$, respectively.

Example 30   **Finding a limit of a rational function**

Confirm analytically that $y = 1$ is the horizontal asymptote of $f(x) = \dfrac{x^2}{x^2 + 4}$, as approximated in Example 29.

SOLUTION   Before using Theorem 11, let's use the technique of evaluating limits at infinity of rational functions that led to that theorem. The largest power of x in f is 2, so divide the numerator and denominator of f by x^2, then

---

Notes:

take limits.

$$\lim_{x\to\infty} \frac{x^2}{x^2+4} = \lim_{x\to\infty} \frac{x^2/x^2}{x^2/x^2+4/x^2}$$
$$= \lim_{x\to\infty} \frac{1}{1+4/x^2}$$
$$= \frac{1}{1+0}$$
$$= 1.$$

We can also use Theorem 11 directly; in this case $n = m$ so the limit is the ratio of the leading coefficients of the numerator and denominator, i.e., 1/1 = 1.

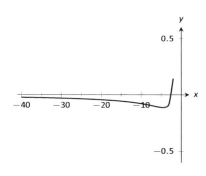

(a)

Example 31 Finding limits of rational functions

Use Theorem 11 to evaluate each of the following limits.

1. $\displaystyle\lim_{x\to-\infty} \frac{x^2+2x-1}{x^3+1}$

2. $\displaystyle\lim_{x\to\infty} \frac{x^2+2x-1}{1-x-3x^2}$

3. $\displaystyle\lim_{x\to\infty} \frac{x^2-1}{3-x}$

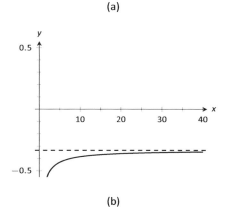

(b)

SOLUTION

1. The highest power of x is in the denominator. Therefore, the limit is 0; see Figure 1.37(a).

2. The highest power of x is x^2, which occurs in both the numerator and denominator. The limit is therefore the ratio of the coefficients of x^2, which is $-1/3$. See Figure 1.37(b).

3. The highest power of x is in the numerator so the limit will be ∞ or $-\infty$. To see which, consider only the dominant terms from the numerator and denominator, which are x^2 and $-x$. The expression in the limit will behave like $x^2/(-x) = -x$ for large values of x. Therefore, the limit is $-\infty$. See Figure 1.37(c).

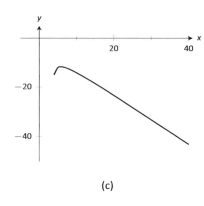

(c)

Figure 1.37: Visualizing the functions in Example 31.

Notes:

Chapter Summary

In this chapter we:

- defined the limit,

- found accessible ways to approximate their values numerically and graphically,

- developed a not–so–easy method of proving the value of a limit (ε-δ proofs),

- explored when limits do not exist,

- defined continuity and explored properties of continuous functions, and

- considered limits that involved infinity.

Why? Mathematics is famous for building on itself and calculus proves to be no exception. In the next chapter we will be interested in "dividing by 0." That is, we will want to divide a quantity by a smaller and smaller number and see what value the quotient approaches. In other words, we will want to find a limit. These limits will enable us to, among other things, determine *exactly* how fast something is moving when we are only given position information.

Later, we will want to add up an infinite list of numbers. We will do so by first adding up a finite list of numbers, then take a limit as the number of things we are adding approaches infinity. Surprisingly, this sum often is finite; that is, we can add up an infinite list of numbers and get, for instance, 42.

These are just two quick examples of why we are interested in limits. Many students dislike this topic when they are first introduced to it, but over time an appreciation is often formed based on the scope of its applicability.

Notes:

Exercises 1.6

Terms and Concepts

1. T/F: If $\lim\limits_{x\to 5} f(x) = \infty$, then we are implicitly stating that the limit exists.

2. T/F: If $\lim\limits_{x\to\infty} f(x) = 5$, then we are implicitly stating that the limit exists.

3. T/F: If $\lim\limits_{x\to 1^-} f(x) = -\infty$, then $\lim\limits_{x\to 1^+} f(x) = \infty$

4. T/F: If $\lim\limits_{x\to 5} f(x) = \infty$, then f has a vertical asymptote at $x = 5$.

5. T/F: $\infty/0$ is not an indeterminate form.

6. List 5 indeterminate forms.

7. Construct a function with a vertical asymptote at $x = 5$ and a horizontal asymptote at $y = 5$.

8. Let $\lim\limits_{x\to 7} f(x) = \infty$. Explain how we know that f is/is not continuous at $x = 7$.

Problems

In Exercises 9 – 14, evaluate the given limits using the graph of the function.

9. $f(x) = \dfrac{1}{(x+1)^2}$

   (a) $\lim\limits_{x\to -1^-} f(x)$

   (b) $\lim\limits_{x\to -1^+} f(x)$

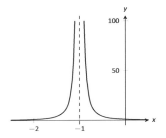

10. $f(x) = \dfrac{1}{(x-3)(x-5)^2}$.

   (a) $\lim\limits_{x\to 3^-} f(x)$ (d) $\lim\limits_{x\to 5^-} f(x)$

   (b) $\lim\limits_{x\to 3^+} f(x)$ (e) $\lim\limits_{x\to 5^+} f(x)$

   (c) $\lim\limits_{x\to 3} f(x)$ (f) $\lim\limits_{x\to 5} f(x)$

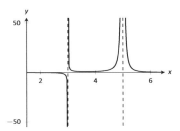

11. $f(x) = \dfrac{1}{e^x + 1}$

   (a) $\lim\limits_{x\to -\infty} f(x)$ (c) $\lim\limits_{x\to 0^-} f(x)$

   (b) $\lim\limits_{x\to\infty} f(x)$ (d) $\lim\limits_{x\to 0^+} f(x)$

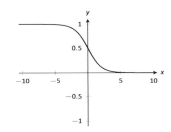

12. $f(x) = x^2 \sin(\pi x)$

   (a) $\lim\limits_{x\to -\infty} f(x)$

   (b) $\lim\limits_{x\to\infty} f(x)$

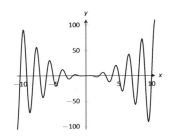

13. $f(x) = \cos(x)$

    (a) $\lim\limits_{x \to -\infty} f(x)$

    (b) $\lim\limits_{x \to \infty} f(x)$

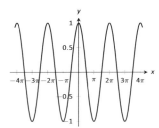

14. $f(x) = 2^x + 10$

    (a) $\lim\limits_{x \to -\infty} f(x)$

    (b) $\lim\limits_{x \to \infty} f(x)$

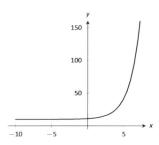

In Exercises 15 – 18, numerically approximate the following limits:

    (a) $\lim\limits_{x \to 3^-} f(x)$

    (b) $\lim\limits_{x \to 3^+} f(x)$

    (c) $\lim\limits_{x \to 3} f(x)$

15. $f(x) = \dfrac{x^2 - 1}{x^2 - x - 6}$

16. $f(x) = \dfrac{x^2 + 5x - 36}{x^3 - 5x^2 + 3x + 9}$

17. $f(x) = \dfrac{x^2 - 11x + 30}{x^3 - 4x^2 - 3x + 18}$

18. $f(x) = \dfrac{x^2 - 9x + 18}{x^2 - x - 6}$

In Exercises 19 – 24, identify the horizontal and vertical asymptotes, if any, of the given function.

19. $f(x) = \dfrac{2x^2 - 2x - 4}{x^2 + x - 20}$

20. $f(x) = \dfrac{-3x^2 - 9x - 6}{5x^2 - 10x - 15}$

21. $f(x) = \dfrac{x^2 + x - 12}{7x^3 - 14x^2 - 21x}$

22. $f(x) = \dfrac{x^2 - 9}{9x - 9}$

23. $f(x) = \dfrac{x^2 - 9}{9x + 27}$

24. $f(x) = \dfrac{x^2 - 1}{-x^2 - 1}$

In Exercises 25 – 28, evaluate the given limit.

25. $\lim\limits_{x \to \infty} \dfrac{x^3 + 2x^2 + 1}{x - 5}$

26. $\lim\limits_{x \to \infty} \dfrac{x^3 + 2x^2 + 1}{5 - x}$

27. $\lim\limits_{x \to -\infty} \dfrac{x^3 + 2x^2 + 1}{x^2 - 5}$

28. $\lim\limits_{x \to -\infty} \dfrac{x^3 + 2x^2 + 1}{5 - x^2}$

Review

29. Use an $\varepsilon - \delta$ proof to show that
$$\lim\limits_{x \to 1} 5x - 2 = 3.$$

30. Let $\lim\limits_{x \to 2} f(x) = 3$ and $\lim\limits_{x \to 2} g(x) = -1$. Evaluate the following limits.

    (a) $\lim\limits_{x \to 2} (f + g)(x)$ (c) $\lim\limits_{x \to 2} (f/g)(x)$

    (b) $\lim\limits_{x \to 2} (fg)(x)$ (d) $\lim\limits_{x \to 2} f(x)^{g(x)}$

31. Let $f(x) = \begin{cases} x^2 - 1 & x < 3 \\ x + 5 & x \geq 3 \end{cases}$.

    Is f continuous everywhere?

32. Evaluate the limit: $\lim\limits_{x \to e} \ln x$.

2: DERIVATIVES

The previous chapter introduced the most fundamental of calculus topics: the limit. This chapter introduces the second most fundamental of calculus topics: the derivative. Limits describe *where* a function is going; derivatives describe *how fast* the function is going.

2.1 Instantaneous Rates of Change: The Derivative

A common amusement park ride lifts riders to a height then allows them to freefall a certain distance before safely stopping them. Suppose such a ride drops riders from a height of 150 feet. Student of physics may recall that the height (in feet) of the riders, t seconds after freefall (and ignoring air resistance, etc.) can be accurately modeled by $f(t) = -16t^2 + 150$.

Using this formula, it is easy to verify that, without intervention, the riders will hit the ground at $t = 2.5\sqrt{1.5} \approx 3.06$ seconds. Suppose the designers of the ride decide to begin slowing the riders' fall after 2 seconds (corresponding to a height of 86 ft.). How fast will the riders be traveling at that time?

We have been given a *position* function, but what we want to compute is a velocity at a specific point in time, i.e., we want an *instantaneous velocity*. We do not currently know how to calculate this.

However, we do know from common experience how to calculate an *average velocity*. (If we travel 60 miles in 2 hours, we know we had an average velocity of 30 mph.) We looked at this concept in Section 1.1 when we introduced the difference quotient. We have

$$\frac{\text{change in distance}}{\text{change in time}} = \frac{\text{`` rise ''}}{\text{run}} = \text{average velocity}.$$

We can approximate the instantaneous velocity at $t = 2$ by considering the average velocity over some time period containing $t = 2$. If we make the time interval small, we will get a good approximation. (This fact is commonly used. For instance, high speed cameras are used to track fast moving objects. Distances are measured over a fixed number of frames to generate an accurate approximation of the velocity.)

Consider the interval from $t = 2$ to $t = 3$ (just before the riders hit the ground). On that interval, the average velocity is

$$\frac{f(3) - f(2)}{3 - 2} = \frac{f(3) - f(2)}{1} = -80 \text{ ft/s},$$

where the minus sign indicates that the riders are moving *down*. By narrowing the interval we consider, we will likely get a better approximation of the instantaneous velocity. On $[2, 2.5]$ we have

$$\frac{f(2.5) - f(2)}{2.5 - 2} = \frac{f(2.5) - f(2)}{0.5} = -72 \text{ ft/s}.$$

We can do this for smaller and smaller intervals of time. For instance, over a time span of $1/10^{\text{th}}$ of a second, i.e., on $[2, 2.1]$, we have

$$\frac{f(2.1) - f(2)}{2.1 - 2} = \frac{f(2.1) - f(2)}{0.1} = -65.6 \text{ ft/s}.$$

Over a time span of $1/100^{\text{th}}$ of a second, on $[2, 2.01]$, the average velocity is

$$\frac{f(2.01) - f(2)}{2.01 - 2} = \frac{f(2.01) - f(2)}{0.01} = -64.16 \text{ ft/s}.$$

What we are really computing is the average velocity on the interval $[2, 2+h]$ for small values of h. That is, we are computing

$$\frac{f(2 + h) - f(2)}{h}$$

where h is small.

What we really want is for $h = 0$, but this, of course, returns the familiar "0/0" indeterminate form. So we employ a limit, as we did in Section 1.1.

We can approximate the value of this limit numerically with small values of h as seen in Figure 2.1. It looks as though the velocity is approaching -64 ft/s. Computing the limit directly gives

$$\lim_{h \to 0} \frac{f(2 + h) - f(2)}{h} = \lim_{h \to 0} \frac{-16(2 + h)^2 + 150 - (-16(2)^2 + 150)}{h}$$
$$= \lim_{h \to 0} \frac{-64h - 16h^2}{h}$$
$$= \lim_{h \to 0} -64 - 16h$$
$$= -64.$$

	Average Velocity
h	ft/s
1	−80
0.5	−72
0.1	−65.6
0.01	−64.16
0.001	−64.016

Figure 2.1: Approximating the instantaneous velocity with average velocities over a small time period h.

Graphically, we can view the average velocities we computed numerically as the slopes of secant lines on the graph of f going through the points $(2, f(2))$ and $(2+h, f(2+h))$. In Figure 2.2, the secant line corresponding to $h = 1$ is shown in three contexts. Figure 2.2(a) shows a "zoomed out" version of f with its secant line. In (b), we zoom in around the points of intersection between f and the secant line. Notice how well this secant line approximates f between those two points – it is a common practice to approximate functions with straight lines.

Notes:

As $h \rightarrow 0$, these secant lines approach the *tangent line*, a line that goes through the point $(2, f(2))$ with the special slope of -64. In parts (c) and (d) of Figure 2.2, we zoom in around the point $(2, 86)$. In (c) we see the secant line, which approximates f well, but not as well the tangent line shown in (d).

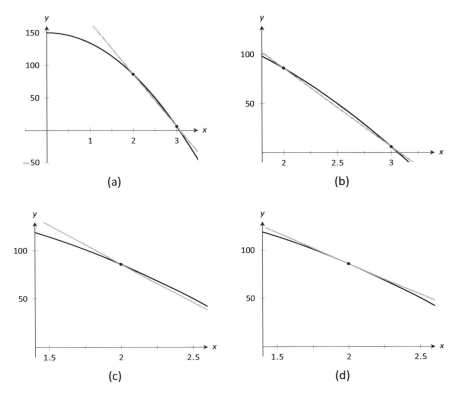

(a)   (b)

(c)   (d)

Figure 2.2: Parts (a), (b) and (c) show the secant line to $f(x)$ with $h = 1$, zoomed in different amounts. Part (d) shows the tangent line to f at $x = 2$.

We have just introduced a number of important concepts that we will flesh out more within this section. First, we formally define two of them.

Notes:

Definition 7   Derivative at a Point

Let f be a continuous function on an open interval I and let c be in I. The **derivative of f at c**, denoted $f'(c)$, is

$$\lim_{h \to 0} \frac{f(c+h) - f(c)}{h},$$

provided the limit exists. If the limit exists, we say that f **is differentiable at** c; if the limit does not exist, then f **is not differentiable at** c. If f is differentiable at every point in I, then f **is differentiable on** I.

Definition 8   Tangent Line

Let f be continuous on an open interval I and differentiable at c, for some c in I. The line with equation $\ell(x) = f'(c)(x-c) + f(c)$ is the **tangent line** to the graph of f at c; that is, it is the line through $(c, f(c))$ whose slope is the derivative of f at c.

Some examples will help us understand these definitions.

Example 32   Finding derivatives and tangent lines

Let $f(x) = 3x^2 + 5x - 7$. Find:

1. $f'(1)$

2. The equation of the tangent line to the graph of f at $x = 1$.

3. $f'(3)$

4. The equation of the tangent line to the graph f at $x = 3$.

SOLUTION

1. We compute this directly using Definition 7.

$$
\begin{aligned}
f'(1) &= \lim_{h \to 0} \frac{f(1+h) - f(1)}{h} \\
&= \lim_{h \to 0} \frac{3(1+h)^2 + 5(1+h) - 7 - (3(1)^2 + 5(1) - 7)}{h} \\
&= \lim_{h \to 0} \frac{3h^2 + 11h}{h} \\
&= \lim_{h \to 0} 3h + 11 = 11.
\end{aligned}
$$

Notes:

2. The tangent line at $x = 1$ has slope $f'(1)$ and goes through the point $(1, f(1)) = (1, 1)$. Thus the tangent line has equation, in point-slope form, $y = 11(x - 1) + 1$. In slope-intercept form we have $y = 11x - 10$.

3. Again, using the definition,

$$f'(3) = \lim_{h \to 0} \frac{f(3 + h) - f(3)}{h}$$

$$= \lim_{h \to 0} \frac{3(3 + h)^2 + 5(3 + h) - 7 - (3(3)^2 + 5(3) - 7)}{h}$$

$$= \lim_{h \to 0} \frac{3h^2 + 23h}{h}$$

$$= \lim_{h \to 0} 3h + 23$$

$$= 23.$$

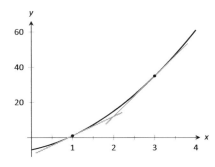

Figure 2.3: A graph of $f(x) = 3x^2 + 5x - 7$ and its tangent lines at $x = 1$ and $x = 3$.

4. The tangent line at $x = 3$ has slope 23 and goes through the point $(3, f(3)) = (3, 35)$. Thus the tangent line has equation $y = 23(x - 3) + 35 = 23x - 34$.

A graph of f is given in Figure 2.3 along with the tangent lines at $x = 1$ and $x = 3$.

Another important line that can be created using information from the derivative is the **normal line.** It is perpendicular to the tangent line, hence its slope is the opposite–reciprocal of the tangent line's slope.

Definition 9   Normal Line

Let f be continuous on an open interval I and differentiable at c, for some c in I. The **normal line** to the graph of f at c is the line with equation

$$n(x) = \frac{-1}{f'(c)}(x - c) + f(c),$$

where $f'(c) \neq 0$. When $f'(c) = 0$, the normal line is the vertical line through $(c, f(c))$; that is, $x = c$.

Example 33   Finding equations of normal lines
Let $f(x) = 3x^2 + 5x - 7$, as in Example 32. Find the equations of the normal lines to the graph of f at $x = 1$ and $x = 3$.

SOLUTION   In Example 32, we found that $f'(1) = 11$. Hence at $x = 1$,

---

Notes:

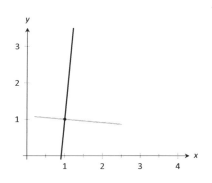

Figure 2.4: A graph of $f(x) = 3x^2 + 5x - 7$, along with its normal line at $x = 1$.

the normal line will have slope $-1/11$. An equation for the normal line is

$$n(x) = \frac{-1}{11}(x - 1) + 1.$$

The normal line is plotted with $y = f(x)$ in Figure 2.4. Note how the line looks perpendicular to f. (A key word here is "looks." Mathematically, we say that the normal line *is* perpendicular to f at $x = 1$ as the slope of the normal line is the opposite–reciprocal of the slope of the tangent line. However, normal lines may not always *look* perpendicular. The aspect ratio of the picture of the graph plays a big role in this.)

We also found that $f'(3) = 23$, so the normal line to the graph of f at $x = 3$ will have slope $-1/23$. An equation for the normal line is

$$n(x) = \frac{-1}{23}(x - 3) + 35.$$

Linear functions are easy to work with; many functions that arise in the course of solving real problems are not easy to work with. A common practice in mathematical problem solving is to approximate difficult functions with not–so–difficult functions. Lines are a common choice. It turns out that at any given point on the graph of a differentiable function f, the best linear approximation to f is its tangent line. That is one reason we'll spend considerable time finding tangent lines to functions.

One type of function that does not benefit from a tangent–line approximation is a line; it is rather simple to recognize that the tangent line to a line is the line itself. We look at this in the following example.

Example 34   Finding the Derivative of a Line

Consider $f(x) = 3x + 5$. Find the equation of the tangent line to f at $x = 1$ and $x = 7$.

SOLUTION   We find the slope of the tangent line by using Definition 7.

$$
\begin{aligned}
f'(1) &= \lim_{h \to 0} \frac{f(1 + h) - f(1)}{h} \\
&= \lim_{h \to 0} \frac{3(1 + h) + 5 - (3 + 5)}{h} \\
&= \lim_{h \to 0} \frac{3h}{h} \\
&= \lim_{h \to 0} 3 \\
&= 3.
\end{aligned}
$$

Notes:

We just found that $f'(1) = 3$. That is, we found the *instantaneous rate of change* of $f(x) = 3x + 5$ is 3. This is not surprising; lines are characterized by being the *only* functions with a *constant rate of change.* That rate of change is called the *slope* of the line. Since their rates of change are constant, their *instantaneous* rates of change are always the same; they are all the slope.

So given a line $f(x) = ax + b$, the derivative at any point x will be a; that is, $f'(x) = a$.

It is now easy to see that the tangent line to the graph of f at $x = 1$ is just f, with the same being true for $x = 7$.

We often desire to find the tangent line to the graph of a function without knowing the actual derivative of the function. In these cases, the best we may be able to do is approximate the tangent line. We demonstrate this in the next example.

Example 35   Numerical Approximation of the Tangent Line
Approximate the equation of the tangent line to the graph of $f(x) = \sin x$ at $x = 0$.

SOLUTION   In order to find the equation of the tangent line, we need a slope and a point. The point is given to us: $(0, \sin 0) = (0, 0)$. To compute the slope, we need the derivative. This is where we will make an approximation. Recall that

$$f'(0) \approx \frac{\sin(0 + h) - \sin 0}{h}$$

for a small value of h. We choose (somewhat arbitrarily) to let $h = 0.1$. Thus

$$f'(0) \approx \frac{\sin(0.1) - \sin 0}{0.1} \approx 0.9983.$$

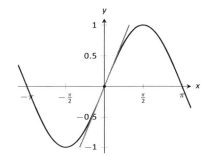

Figure 2.5: $f(x) = \sin x$ graphed with an approximation to its tangent line at $x = 0$.

Thus our approximation of the equation of the tangent line is $y = 0.9983(x - 0) + 0 = 0.9983x$; it is graphed in Figure 2.5. The graph seems to imply the approximation is rather good.

Recall from Section 1.3 that $\lim\limits_{x \to 0} \dfrac{\sin x}{x} = 1$, meaning for values of x near 0, $\sin x \approx x$. Since the slope of the line $y = x$ is 1 at $x = 0$, it should seem reasonable that "the slope of $f(x) = \sin x$" is near 1 at $x = 0$. In fact, since we *approximated* the value of the slope to be 0.9983, we might guess the *actual value* is 1. We'll come back to this later.

Consider again Example 32. To find the derivative of f at $x = 1$, we needed to evaluate a limit. To find the derivative of f at $x = 3$, we needed to again evaluate a limit. We have this process:

Notes:

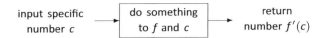

This process describes a *function*; given one input (the value of *c*), we return exactly one output (the value of $f'(c)$). The "do something" box is where the tedious work (taking limits) of this function occurs.

Instead of applying this function repeatedly for different values of *c*, let us apply it just once to the variable *x*. We then take a limit just once. The process now looks like:

input variable *x* → do something to *f* and *x* → return function $f'(x)$

The output is the "derivative function," $f'(x)$. The $f'(x)$ function will take a number *c* as input and return the derivative of *f* at *c*. This calls for a definition.

Definition 10   Derivative Function

Let *f* be a differentiable function on an open interval *I*. The function

$$f'(x) = \lim_{h \to 0} \frac{f(x+h) - f(x)}{h}$$

is **the derivative of** *f*.

Notation:
Let $y = f(x)$. The following notations all represent the derivative:

$$f'(x) = y' = \frac{dy}{dx} = \frac{df}{dx} = \frac{d}{dx}(f) = \frac{d}{dx}(y).$$

Important: The notation $\dfrac{dy}{dx}$ is one symbol; it is **not** the fraction "*dy/dx*". The notation, while somewhat confusing at first, was chosen with care. A fraction–looking symbol was chosen because the derivative has many fraction–like properties. Among other places, we see these properties at work when we talk about the units of the derivative, when we discuss the Chain Rule, and when we learn about integration (topics that appear in later sections and chapters).

Examples will help us understand this definition.

Example 36   Finding the derivative of a function
Let $f(x) = 3x^2 + 5x - 7$ as in Example 32. Find $f'(x)$.

Notes:

SOLUTION We apply Definition 10.

$$f'(x) = \lim_{h \to 0} \frac{f(x+h) - f(x)}{h}$$
$$= \lim_{x \to 0} \frac{3(x+h)^2 + 5(x+h) - 7 - (3x^2 + 5x - 7)}{h}$$
$$= \lim_{x \to 0} \frac{3h^2 + 6xh + 5h}{h}$$
$$= \lim_{x \to 0} 3h + 6x + 5$$
$$= 6x + 5$$

So $f'(x) = 6x + 5$. Recall earlier we found that $f'(1) = 11$ and $f'(3) = 23$. Note our new computation of $f'(x)$ affirm these facts.

Example 37   **Finding the derivative of a function**

Let $f(x) = \dfrac{1}{x+1}$. Find $f'(x)$.

SOLUTION We apply Definition 10.

$$f'(x) = \lim_{h \to 0} \frac{f(x+h) - f(x)}{h}$$
$$= \lim_{h \to 0} \frac{\frac{1}{x+h+1} - \frac{1}{x+1}}{h}$$

Now find common denominator then subtract; pull $1/h$ out front to facilitate reading.

$$= \lim_{h \to 0} \frac{1}{h} \cdot \left(\frac{x+1}{(x+1)(x+h+1)} - \frac{x+h+1}{(x+1)(x+h+1)} \right)$$
$$= \lim_{h \to 0} \frac{1}{h} \cdot \left(\frac{x+1 - (x+h+1)}{(x+1)(x+h+1)} \right)$$
$$= \lim_{h \to 0} \frac{1}{h} \cdot \left(\frac{-h}{(x+1)(x+h+1)} \right)$$
$$= \lim_{h \to 0} \frac{-1}{(x+1)(x+h+1)}$$
$$= \frac{-1}{(x+1)(x+1)}$$
$$= \frac{-1}{(x+1)^2}$$

Notes:

So $f'(x) = \dfrac{-1}{(x+1)^2}$. To practice using our notation, we could also state

$$\frac{d}{dx}\left(\frac{1}{x+1}\right) = \frac{-1}{(x+1)^2}.$$

Example 38 Finding the derivative of a function

Find the derivative of $f(x) = \sin x$.

SOLUTION Before applying Definition 10, note that once this is found, we can find the actual tangent line to $f(x) = \sin x$ at $x = 0$, whereas we settled for an approximation in Example 35.

$$\begin{aligned}
f'(x) &= \lim_{h\to 0} \frac{\sin(x+h) - \sin x}{h} && \left(\begin{array}{c}\text{Use trig identity}\\ \sin(x+h) = \sin x \cos h + \cos x \sin h\end{array}\right)\\
&= \lim_{h\to 0} \frac{\sin x \cos h + \cos x \sin h - \sin x}{h} && (\text{regroup})\\
&= \lim_{h\to 0} \frac{\sin x(\cos h - 1) + \cos x \sin h}{h} && (\text{split into two fractions})\\
&= \lim_{h\to 0} \left(\frac{\sin x(\cos h - 1)}{h} + \frac{\cos x \sin h}{h}\right) && \left(\text{use } \lim_{h\to 0}\frac{\cos h - 1}{h} = 0 \text{ and } \lim_{h\to 0}\frac{\sin h}{h} = 1\right)\\
&= \sin x \cdot 0 + \cos x \cdot 1\\
&= \cos x\ !
\end{aligned}$$

We have found that when $f(x) = \sin x$, $f'(x) = \cos x$. This should be somewhat surprising; the result of a tedious limit process and the sine function is a nice function. Then again, perhaps this is not entirely surprising. The sine function is periodic – it repeats itself on regular intervals. Therefore its rate of change also repeats itself on the same regular intervals. We should have known the derivative would be periodic; we now know exactly which periodic function it is.

Thinking back to Example 35, we can find the slope of the tangent line to $f(x) = \sin x$ at $x = 0$ using our derivative. We approximated the slope as 0.9983; we now know the slope is *exactly* $\cos 0 = 1$.

Example 39 Finding the derivative of a piecewise defined function

Find the derivative of the absolute value function,

$$f(x) = |x| = \begin{cases} -x & x < 0 \\ x & x \ge 0 \end{cases}.$$

See Figure 2.6.

SOLUTION We need to evaluate $\lim_{h\to 0}\dfrac{f(x+h) - f(x)}{h}$. As f is piecewise–defined, we need to consider separately the limits when $x < 0$ and when $x > 0$.

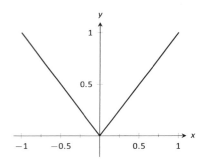

Figure 2.6: The absolute value function, $f(x) = |x|$. Notice how the slope of the lines (and hence the tangent lines) abruptly changes at $x = 0$.

Notes:

When $x < 0$:

$$\frac{d}{dx}(-x) = \lim_{h \to 0} \frac{-(x+h) - (-x)}{h}$$
$$= \lim_{h \to 0} \frac{-h}{h}$$
$$= \lim_{h \to 0} -1$$
$$= -1.$$

When $x > 0$, a similar computation shows that $\frac{d}{dx}(x) = 1$.

We need to also find the derivative at $x = 0$. By the definition of the derivative at a point, we have

$$f'(0) = \lim_{h \to 0} \frac{f(0+h) - f(0)}{h}.$$

Since $x = 0$ is the point where our function's definition switches from one piece to other, we need to consider left and right-hand limits. Consider the following, where we compute the left and right hand limits side by side.

$$\lim_{h \to 0^-} \frac{f(0+h) - f(0)}{h} = \qquad\qquad \lim_{h \to 0^+} \frac{f(0+h) - f(0)}{h} =$$
$$\lim_{h \to 0^-} \frac{-h - 0}{h} = \qquad\qquad \lim_{h \to 0^+} \frac{h - 0}{h} =$$
$$\lim_{h \to 0^-} -1 = -1 \qquad\qquad \lim_{h \to 0^+} 1 = 1$$

The last lines of each column tell the story: the left and right hand limits are not equal. Therefore the limit does not exist at 0, and f is not differentiable at 0. So we have

$$f'(x) = \begin{cases} -1 & x < 0 \\ 1 & x > 0 \end{cases}.$$

At $x = 0$, $f'(x)$ does not exist; there is a jump discontinuity at 0; see Figure 2.7. So $f(x) = |x|$ is differentiable everywhere except at 0.

The point of non-differentiability came where the piecewise defined function switched from one piece to the other. Our next example shows that this does not always cause trouble.

Example 40   **Finding the derivative of a piecewise defined function**

Find the derivative of $f(x)$, where $f(x) = \begin{cases} \sin x & x \le \pi/2 \\ 1 & x > \pi/2 \end{cases}$. See Figure 2.8.

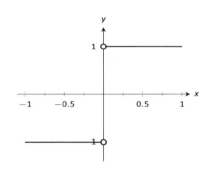

Figure 2.7: A graph of the derivative of $f(x) = |x|$.

Notes:

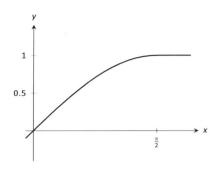

Figure 2.8: A graph of $f(x)$ as defined in Example 40.

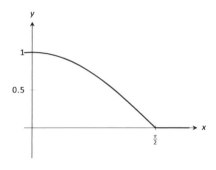

Figure 2.9: A graph of $f'(x)$ in Example 40.

SOLUTION   Using Example 38, we know that when $x < \pi/2, f'(x) = \cos x$. It is easy to verify that when $x > \pi/2, f'(x) = 0$; consider:

$$\lim_{x \to 0} \frac{f(x+h) - f(x)}{h} = \lim_{x \to 0} \frac{1-1}{h} = \lim_{h \to 0} 0 = 0.$$

So far we have

$$f'(x) = \begin{cases} \cos x & x < \pi/2 \\ 0 & x > \pi/2 \end{cases}.$$

We still need to find $f'(\pi/2)$. Notice at $x = \pi/2$ that both pieces of f' are 0, meaning we can state that $f'(\pi/2) = 0$.

Being more rigorous, we can again evaluate the difference quotient limit at $x = \pi/2$, utilizing again left and right–hand limits:

$$\lim_{h \to 0^-} \frac{f(\pi/2 + h) - f(\pi/2)}{h} = \qquad \lim_{h \to 0^+} \frac{f(\pi/2 + h) - f(\pi/2)}{h} =$$

$$\lim_{h \to 0^-} \frac{\sin(\pi/2 + h) - \sin(\pi/2)}{h} = \qquad \lim_{h \to 0^+} \frac{1-1}{h} =$$

$$\lim_{h \to 0^-} \frac{\sin(\frac{\pi}{2})\cos(h) + \sin(h)\cos(\frac{\pi}{2}) - \sin(\frac{\pi}{2})}{h} = \qquad \lim_{h \to 0^+} \frac{0}{h} =$$

$$\lim_{h \to 0^-} \frac{1 \cdot \cos(h) + \sin(h) \cdot 0 - 1}{h} = \qquad 0$$

$$0$$

Since both the left and right hand limits are 0 at $x = \pi/2$, the limit exists and $f'(\pi/2)$ exists (and is 0). Therefore we can fully write f' as

$$f'(x) = \begin{cases} \cos x & x \le \pi/2 \\ 0 & x > \pi/2 \end{cases}.$$

See Figure 2.9 for a graph of this function.

Recall we pseudo–defined a continuous function as one in which we could sketch its graph without lifting our pencil. We can give a pseudo–definition for differentiability as well: it is a continuous function that does not have any "sharp corners." One such sharp corner is shown in Figure 2.6. Even though the function f in Example 40 is piecewise–defined, the transition is "smooth" hence it is differentiable. Note how in the graph of f in Figure 2.8 it is difficult to tell when f switches from one piece to the other; there is no "corner."

This section defined the derivative; in some sense, it answers the question of "What *is* the derivative?" The next section addresses the question "What does the derivative *mean*?"

Notes:

Exercises 2.1

Terms and Concepts

1. T/F: Let f be a position function. The average rate of change on $[a, b]$ is the slope of the line through the points $(a, f(a))$ and $(b, f(b))$.

2. T/F: The definition of the derivative of a function at a point involves taking a limit.

3. In your own words, explain the difference between the average rate of change and instantaneous rate of change.

4. In your own words, explain the difference between Definitions 7 and 10.

5. Let $y = f(x)$. Give three different notations equivalent to "$f'(x)$."

Problems

In Exercises 6 – 12, use the definition of the derivative to compute the derivative of the given function.

6. $f(x) = 6$

7. $f(x) = 2x$

8. $f(t) = 4 - 3t$

9. $g(x) = x^2$

10. $f(x) = 3x^2 - x + 4$

11. $r(x) = \dfrac{1}{x}$

12. $r(s) = \dfrac{1}{s - 2}$

In Exercises 13 – 19, a function and an x–value c are given. (Note: these functions are the same as those given in Exercises 6 through 12.)

   (a) **Find the tangent line to the graph of the function at c.**

   (b) **Find the normal line to the graph of the function at c.**

13. $f(x) = 6$, at $x = -2$.

14. $f(x) = 2x$, at $x = 3$.

15. $f(x) = 4 - 3x$, at $x = 7$.

16. $g(x) = x^2$, at $x = 2$.

17. $f(x) = 3x^2 - x + 4$, at $x = -1$.

18. $r(x) = \dfrac{1}{x}$, at $x = -2$.

19. $r(x) = \dfrac{1}{x - 2}$, at $x = 3$.

In Exercises 20 – 23, a function f and an x–value a are given. Approximate the equation of the tangent line to the graph of f at $x = a$ by numerically approximating $f'(a)$, using $h = 0.1$.

20. $f(x) = x^2 + 2x + 1$, $x = 3$

21. $f(x) = \dfrac{10}{x + 1}$, $x = 9$

22. $f(x) = e^x$, $x = 2$

23. $f(x) = \cos x$, $x = 0$

24. The graph of $f(x) = x^2 - 1$ is shown.

   (a) Use the graph to approximate the slope of the tangent line to f at the following points: $(-1, 0)$, $(0, -1)$ and $(2, 3)$.

   (b) Using the definition, find $f'(x)$.

   (c) Find the slope of the tangent line at the points $(-1, 0)$, $(0, -1)$ and $(2, 3)$.

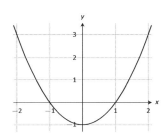

25. The graph of $f(x) = \dfrac{1}{x + 1}$ is shown.

   (a) Use the graph to approximate the slope of the tangent line to f at the following points: $(0, 1)$ and $(1, 0.5)$.

   (b) Using the definition, find $f'(x)$.

   (c) Find the slope of the tangent line at the points $(0, 1)$ and $(1, 0.5)$.

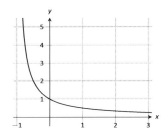

In Exercises 26 – 29, a graph of a function $f(x)$ is given. Using the graph, sketch $f'(x)$.

26.

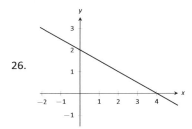

27.

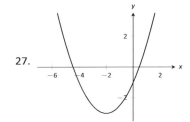

28.

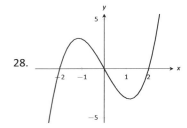

29.

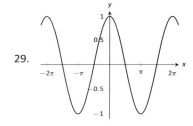

30. Using the graph of $g(x)$ below, answer the following questions.

(a) Where is $g(x) > 0$?   (c) Where is $g'(x) < 0$?

(b) Where is $g(x) < 0$?   (d) Where is $g'(x) > 0$?

(c) Where is $g(x) = 0$?   (e) Where is $g'(x) = 0$?

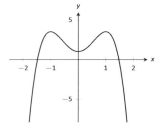

Review

31. Approximate $\lim\limits_{x \to 5} \dfrac{x^2 + 2x - 35}{x^2 - 10.5x + 27.5}$.

32. Use the Bisection Method to approximate, accurate to two decimal places, the root of $g(x) = x^3 + x^2 + x - 1$ on $[0.5, 0.6]$.

33. Give intervals on which each of the following functions are continuous.

(a) $\dfrac{1}{e^x + 1}$ (c) $\sqrt{5 - x}$

(b) $\dfrac{1}{x^2 - 1}$ (d) $\sqrt{5 - x^2}$

34. Use the graph of $f(x)$ provided to answer the following.

(a) $\lim\limits_{x \to -3^-} f(x) = ?$ (c) $\lim\limits_{x \to -3} f(x) = ?$

(b) $\lim\limits_{x \to -3^+} f(x) = ?$ (d) Where is f continuous?

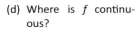

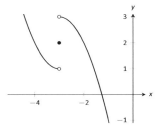

2.2   Interpretations of the Derivative

The previous section defined the derivative of a function and gave examples of how to compute it using its definition (i.e., using limits). The section also started with a brief motivation for this definition, that is, finding the instantaneous velocity of a falling object given its position function. The next section will give us more accessible tools for computing the derivative, tools that are easier to use than repeated use of limits.

This section falls in between the "What is the definition of the derivative?" and "How do I compute the derivative?" sections. Here we are concerned with "What does the derivative mean?", or perhaps, when read with the right emphasis, "What *is* the derivative?" We offer two interconnected interpretations of the derivative, hopefully explaining why we care about it and why it is worthy of study.

Interpretation of the Derivative #1: Instantaneous Rate of Change

The previous section started with an example of using the position of an object (in this case, a falling amusement–park rider) to find the object's velocity. This type of example is often used when introducing the derivative because we tend to readily recognize that velocity is the *instantaneous rate of change of position*. In general, if f is a function of x, then $f'(x)$ measures the instantaneous rate of change of f with respect to x. Put another way, the derivative answers "When x changes, at what rate does f change?" Thinking back to the amusement–park ride, we asked "When time changed, at what rate did the height change?" and found the answer to be "By -64 feet per second."

Now imagine driving a car and looking at the speedometer, which reads "60 mph." Five minutes later, you wonder how far you have traveled. Certainly, lots of things could have happened in those 5 minutes; you could have intentionally sped up significantly, you might have come to a complete stop, you might have slowed to 20 mph as you passed through construction. But suppose that you know, as the driver, none of these things happened. You know you maintained a fairly consistent speed over those 5 minutes. What is a good approximation of the distance traveled?

One could argue the *only* good approximation, given the information provided, would be based on "distance = rate $\times$ time." In this case, we assume a constant rate of 60 mph with a time of $5/60$ hours. Hence we would approximate the distance traveled as 5 miles.

Referring back to the falling amusement–park ride, knowing that at $t = 2$ the velocity was -64 ft/s, we could reasonably assume that 1 second later the rid-

---

Notes:

ers' height would have dropped by about 64 feet. Knowing that the riders were *accelerating* as they fell would inform us that this is an *under–approximation*. If all we knew was that $f(2) = 86$ and $f'(2) = -64$, we'd know that we'd have to stop the riders quickly otherwise they would hit the ground!

Units of the Derivative

It is useful to recognize the *units* of the derivative function. If y is a function of x, i.e., $y = f(x)$ for some function f, and y is measured in feet and x in seconds, then the units of $y' = f'$ are "feet per second," commonly written as "ft/s." In general, if y is measured in units P and x is measured in units Q, then y' will be measured in units "P per Q," or "P/Q." Here we see the fraction–like behavior of the derivative in the notation:

$$\text{the units of} \quad \frac{dy}{dx} \quad \text{are} \quad \frac{\text{units of } y}{\text{units of } x}.$$

Example 41   The meaning of the derivative: World Population

Let $P(t)$ represent the world population t minutes after 12:00 a.m., January 1, 2012. It is fairly accurate to say that $P(0) = 7,028,734,178$ (www.prb.org). It is also fairly accurate to state that $P'(0) = 156$; that is, at midnight on January 1, 2012, the population of the world was growing by about 156 *people per minute* (note the units). Twenty days later (or, 28,800 minutes later) we could reasonably assume the population grew by about $28,800 \cdot 156 = 4,492,800$ people.

Example 42   The meaning of the derivative: Manufacturing

The term *widget* is an economic term for a generic unit of manufacturing output. Suppose a company produces widgets and knows that the market supports a price of $10 per widget. Let $P(n)$ give the profit, in dollars, earned by manufacturing and selling n widgets. The company likely cannot make a (positive) profit making just one widget; the start–up costs will likely exceed $10. Mathematically, we would write this as $P(1) < 0$.

What do $P(1000) = 500$ and $P'(1000) = 0.25$ mean? Approximate $P(1100)$.

SOLUTION   The equation $P(1000) = 500$ means that selling 1,000 widgets returns a profit of $500. We interpret $P'(1000) = 0.25$ as meaning that the profit is increasing at rate of $0.25 per widget (the units are "dollars per widget.") Since we have no other information to use, our best approximation for $P(1100)$ is:

$$P(1100) \approx P(1000) + P'(1000) \times 100 = \$500 + 100 \cdot 0.25 = \$525.$$

We approximate that selling 1,100 widgets returns a profit of $525.

Notes:

The previous examples made use of an important approximation tool that we first used in our previous "driving a car at 60 mph" example at the beginning of this section. Five minutes after looking at the speedometer, our best approximation for distance traveled assumed the rate of change was constant. In Examples 41 and 42 we made similar approximations. We were given rate of change information which we used to approximate total change. Notationally, we would say that

$$f(c + h) \approx f(c) + f'(c) \cdot h.$$

This approximation is best when h is "small." "Small" is a relative term; when dealing with the world population, $h = 22$ days = 28,800 minutes is small in comparison to years. When manufacturing widgets, 100 widgets is small when one plans to manufacture thousands.

The Derivative and Motion

One of the most fundamental applications of the derivative is the study of motion. Let $s(t)$ be a position function, where t is time and $s(t)$ is distance. For instance, s could measure the height of a projectile or the distance an object has traveled.

Let's let $s(t)$ measure the distance traveled, in feet, of an object after t seconds of travel. Then $s'(t)$ has units "feet per second," and $s'(t)$ measures the *instantaneous rate of distance change* — it measures **velocity**.

Now consider $v(t)$, a velocity function. That is, at time t, $v(t)$ gives the velocity of an object. The derivative of v, $v'(t)$, gives the *instantaneous rate of velocity change* — **acceleration**. (We often think of acceleration in terms of cars: a car may "go from 0 to 60 in 4.8 seconds." This is an *average* acceleration, a measurement of how quickly the velocity changed.) If velocity is measured in feet per second, and time is measured in seconds, then the units of acceleration (i.e., the units of $v'(t)$) are "feet per second per second," or (ft/s)/s. We often shorten this to "feet per second squared," or ft/s^2, but this tends to obscure the meaning of the units.

Perhaps the most well known acceleration is that of gravity. In this text, we use $g = 32$ft/s^2 or $g = 9.8$m/s^2. What do these numbers mean?

A constant acceleration of 32(ft/s)/s means that the velocity changes by 32ft/s each second. For instance, let $v(t)$ measures the velocity of a ball thrown straight up into the air, where v has units ft/s and t is measured in seconds. The ball will have a positive velocity while traveling upwards and a negative velocity while falling down. The acceleration is thus -32ft/s^2. If $v(1) = 20$ft/s, then when $t = 2$, the velocity will have decreased by 32ft/s; that is, $v(2) = -12$ft/s. We can continue: $v(3) = -44$ft/s, and we can also figure that $v(0) = 42$ft/s.

These ideas are so important we write them out as a Key Idea.

Notes:

Key Idea 1   The Derivative and Motion

1. Let $s(t)$ be the position function of an object. Then $s'(t)$ is the velocity function of the object.

2. Let $v(t)$ be the velocity function of an object. Then $v'(t)$ is the acceleration function of the object.

We now consider the second interpretation of the derivative given in this section. This interpretation is not independent from the first by any means; many of the same concepts will be stressed, just from a slightly different perspective.

Interpretation of the Derivative #2: The Slope of the Tangent Line

Given a function $y = f(x)$, the difference quotient $\dfrac{f(c + h) - f(c)}{h}$ gives a change in y values divided by a change in x values; i.e., it is a measure of the "rise over run," or "slope," of the line that goes through two points on the graph of f: $(c, f(c))$ and $(c+h, f(c+h))$. As h shrinks to 0, these two points come close together; in the limit we find $f'(c)$, the slope of a special line called the tangent line that intersects f only once near $x = c$.

Lines have a constant rate of change, their slope. Nonlinear functions do not have a constant rate of change, but we can measure their *instantaneous rate of change* at a given x value c by computing $f'(c)$. We can get an idea of how f is behaving by looking at the slopes of its tangent lines. We explore this idea in the following example.

Example 43   Understanding the derivative: the rate of change
Consider $f(x) = x^2$ as shown in Figure 2.10. It is clear that at $x = 3$ the function is growing faster than at $x = 1$, as it is steeper at $x = 3$. How much faster is it growing?

SOLUTION   We can answer this directly after the following section, where we learn to quickly compute derivatives. For now, we will answer graphically, by considering the slopes of the respective tangent lines.

With practice, one can fairly effectively sketch tangent lines to a curve at a particular point. In Figure 2.11, we have sketched the tangent lines to f at $x = 1$ and $x = 3$, along with a grid to help us measure the slopes of these lines. At

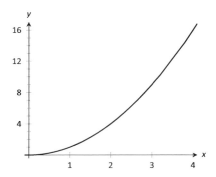

Figure 2.10: A graph of $f(x) = x^2$.

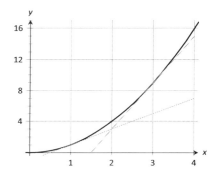

Figure 2.11: A graph of $f(x) = x^2$ and tangent lines.

Notes:

$x = 1$, the slope is 2; at $x = 3$, the slope is 6. Thus we can say not only is f growing faster at $x = 3$ than at $x = 1$, it is growing *three times as fast*.

Example 44   Understanding the graph of the derivative

Consider the graph of $f(x)$ and its derivative, $f'(x)$, in Figure 2.12(a). Use these graphs to find the slopes of the tangent lines to the graph of f at $x = 1$, $x = 2$, and $x = 3$.

    SOLUTION   To find the appropriate slopes of tangent lines to the graph of f, we need to look at the corresponding values of f'.

    The slope of the tangent line to f at $x = 1$ is $f'(1)$; this looks to be about -1.

    The slope of the tangent line to f at $x = 2$ is $f'(2)$; this looks to be about 4.

    The slope of the tangent line to f at $x = 3$ is $f'(3)$; this looks to be about 3.

    Using these slopes, the tangent lines to f are sketched in Figure 2.12(b). Included on the graph of f' in this figure are filled circles where $x = 1$, $x = 2$ and $x = 3$ to help better visualize the y value of f' at those points.

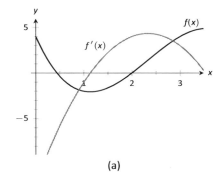

(a)

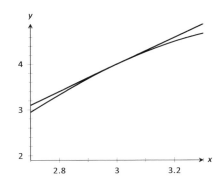

(b)

Figure 2.12: Graphs of f and f' in Example 44, along with tangent lines in (b).

Example 45   Approximation with the derivative

Consider again the graph of $f(x)$ and its derivative $f'(x)$ in Example 44. Use the tangent line to f at $x = 3$ to approximate the value of $f(3.1)$.

    SOLUTION   Figure 2.13 shows the graph of f along with its tangent line, zoomed in at $x = 3$. Notice that near $x = 3$, the tangent line makes an excellent approximation of f. Since lines are easy to deal with, often it works well to approximate a function with its tangent line. (This is especially true when you don't actually know much about the function at hand, as we don't in this example.)

    While the tangent line to f was drawn in Example 44, it was not explicitly computed. Recall that the tangent line to f at $x = c$ is $y = f'(c)(x - c) + f(c)$. While f is not explicitly given, by the graph it looks like $f(3) = 4$. Recalling that $f'(3) = 3$, we can compute the tangent line to be approximately $y = 3(x-3)+4$. It is often useful to leave the tangent line in point–slope form.

    To use the tangent line to approximate $f(3.1)$, we simply evaluate y at 3.1 instead of f.

$$f(3.1) \approx y(3.1) = 3(3.1 - 3) + 4 = .1 * 3 + 4 = 4.3.$$

    We approximate $f(3.1) \approx 4.3$.

    To demonstrate the accuracy of the tangent line approximation, we now state that in Example 45, $f(x) = -x^3 + 7x^2 - 12x + 4$. We can evaluate $f(3.1) = 4.279$. Had we known f all along, certainly we could have just made this computation. In reality, we often only know two things:

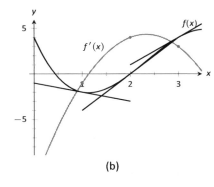

Figure 2.13: Zooming in on f at $x = 3$ for the function given in Examples 44 and 45.

Notes:

1. What $f(c)$ is, for some value of c, and

2. what $f'(c)$ is.

For instance, we can easily observe the location of an object and its instantaneous velocity at a particular point in time. We do not have a "function f" for the location, just an observation. This is enough to create an approximating function for f.

This last example has a direct connection to our approximation method explained above after Example 42. We stated there that

$$f(c + h) \approx f(c) + f'(c) \cdot h.$$

If we know $f(c)$ and $f'(c)$ for some value $x = c$, then computing the tangent line at $(c, f(c))$ is easy: $y(x) = f'(c)(x - c) + f(c)$. In Example 45, we used the tangent line to approximate a value of f. Let's use the tangent line at $x = c$ to approximate a value of f near $x = c$; i.e., compute $y(c + h)$ to approximate $f(c + h)$, assuming again that h is "small." Note:

$$y(c + h) = f'(c)\big((c + h) - c\big) + f(c) = f'(c) \cdot h + f(c).$$

This is the exact same approximation method used above! Not only does it make intuitive sense, as explained above, it makes analytical sense, as this approximation method is simply using a tangent line to approximate a function's value.

The importance of understanding the derivative cannot be understated. When f is a function of x, $f'(x)$ measures the instantaneous rate of change of f with respect to x and gives the slope of the tangent line to f at x.

Notes:

Exercises 2.2

Terms and Concepts

1. What is the instantaneous rate of change of position called?

2. Given a function $y = f(x)$, in your own words describe how to find the units of $f'(x)$.

3. What functions have a constant rate of change?

Problems

4. Given $f(5) = 10$ and $f'(5) = 2$, approximate $f(6)$.

5. Given $P(100) = -67$ and $P'(100) = 5$, approximate $P(110)$.

6. Given $z(25) = 187$ and $z'(25) = 17$, approximate $z(20)$.

7. Knowing $f(10) = 25$ and $f'(10) = 5$ and the methods described in this section, which approximation is likely to be most accurate: $f(10.1)$, $f(11)$, or $f(20)$? Explain your reasoning.

8. Given $f(7) = 26$ and $f(8) = 22$, approximate $f'(7)$.

9. Given $H(0) = 17$ and $H(2) = 29$, approximate $H'(2)$.

10. Let $V(x)$ measure the volume, in decibels, measured inside a restaurant with x customers. What are the units of $V'(x)$?

11. Let $v(t)$ measure the velocity, in ft/s, of a car moving in a straight line t seconds after starting. What are the units of $v'(t)$?

12. The height H, in feet, of a river is recorded t hours after midnight, April 1. What are the units of $H'(t)$?

13. P is the profit, in thousands of dollars, of producing and selling c cars.

    (a) What are the units of $P'(c)$?

    (b) What is likely true of $P(0)$?

14. T is the temperature in degrees Fahrenheit, h hours after midnight on July 4 in Sidney, NE.

    (a) What are the units of $T'(h)$?

    (b) Is $T'(8)$ likely greater than or less than 0? Why?

    (c) Is $T(8)$ likely greater than or less than 0? Why?

In Exercises 15 – 18, graphs of functions $f(x)$ and $g(x)$ are given. Identify which function is the derivative of the other.)

15.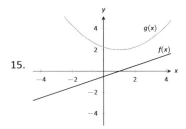

16.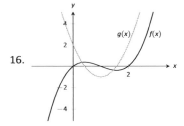

17.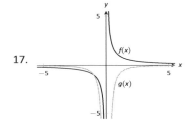

18.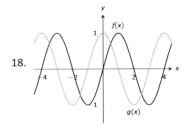

Review

In Exercises 19 – 20, use the definition to compute the derivatives of the following functions.

19. $f(x) = 5x^2$

20. $f(x) = (x - 2)^3$

In Exercises 21 – 22, numerically approximate the value of $f'(x)$ at the indicated x value.

21. $f(x) = \cos x$ at $x = \pi$.

22. $f(x) = \sqrt{x}$ at $x = 9$.

2.3   Basic Differentiation Rules

The derivative is a powerful tool but is admittedly awkward given its reliance on limits. Fortunately, one thing mathematicians are good at is *abstraction*. For instance, instead of continually finding derivatives at a point, we abstracted and found the derivative function.

Let's practice abstraction on linear functions, $y = mx + b$. What is y'? Without limits, recognize that linear function are characterized by being functions with a constant rate of change (the slope). The derivative, y', gives the instantaneous rate of change; with a linear function, this is constant, m. Thus $y' = m$.

Let's abstract once more. Let's find the derivative of the general quadratic function, $f(x) = ax^2 + bx + c$. Using the definition of the derivative, we have:

$$\begin{aligned} f'(x) &= \lim_{h \to 0} \frac{a(x+h)^2 + b(x+h) + c - (ax^2 + bx + c)}{h} \\ &= \lim_{h \to 0} \frac{ah^2 + 2ahx + bh}{h} \\ &= \lim_{h \to 0} ah + 2ax + b \\ &= 2ax + b. \end{aligned}$$

So if $y = 6x^2 + 11x - 13$, we can immediately compute $y' = 12x + 11$.

In this section (and in some sections to follow) we will learn some of what mathematicians have already discovered about the derivatives of certain functions and how derivatives interact with arithmetic operations. We start with a theorem.

Theorem 12   Derivatives of Common Functions

1. **Constant Rule:**

   $\dfrac{d}{dx}(c) = 0$, where c is a constant.

2. **Power Rule:**

   $\dfrac{d}{dx}(x^n) = nx^{n-1}$, where n is an integer, $n > 0$.

3. $\dfrac{d}{dx}(\sin x) = \cos x$

4. $\dfrac{d}{dx}(\cos x) = -\sin x$

5. $\dfrac{d}{dx}(e^x) = e^x$

6. $\dfrac{d}{dx}(\ln x) = \dfrac{1}{x}$

This theorem starts by stating an intuitive fact: constant functions have no rate of change as they are *constant*. Therefore their derivative is 0 (they change

---

Notes:

at the rate of 0). The theorem then states some fairly amazing things. The Power Rule states that the derivatives of Power Functions (of the form $y = x^n$) are very straightforward: multiply by the power, then subtract 1 from the power. We see something incredible about the function $y = e^x$: it is its own derivative. We also see a new connection between the sine and cosine functions.

One special case of the Power Rule is when $n = 1$, i.e., when $f(x) = x$. What is $f'(x)$? According to the Power Rule,

$$f'(x) = \frac{d}{dx}(x) = \frac{d}{dx}(x^1) = 1 \cdot x^0 = 1.$$

In words, we are asking "At what rate does f change with respect to x?" Since f is x, we are asking "At what rate does x change with respect to x?" The answer is: 1. They change at the same rate.

Let's practice using this theorem.

Example 46   **Using Theorem 12 to find, and use, derivatives**
Let $f(x) = x^3$.

1. Find $f'(x)$.

2. Find the equation of the line tangent to the graph of f at $x = -1$.

3. Use the tangent line to approximate $(-1.1)^3$.

4. Sketch f, f' and the found tangent line on the same axis.

SOLUTION

1. The Power Rule states that if $f(x) = x^3$, then $f'(x) = 3x^2$.

2. To find the equation of the line tangent to the graph of f at $x = -1$, we need a point and the slope. The point is $(-1, f(-1)) = (-1, -1)$. The slope is $f'(-1) = 3$. Thus the tangent line has equation $y = 3(x-(-1))+(-1) = 3x + 2$.

3. We can use the tangent line to approximate $(-1.1)^3$ as -1.1 is close to -1. We have
$$(-1.1)^3 \approx 3(-1.1) + 2 = -1.3.$$
We can easily find the actual answer; $(-1.1)^3 = -1.331$.

4. See Figure 2.14.

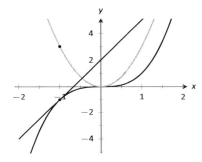

Figure 2.14: A graph of $f(x) = x^3$, along with its derivative $f'(x) = 3x^2$ and its tangent line at $x = -1$.

Notes:

Theorem 12 gives useful information, but we will need much more. For instance, using the theorem, we can easily find the derivative of $y = x^3$, but it does not tell how to compute the derivative of $y = 2x^3$, $y = x^3 + \sin x$ nor $y = x^3 \sin x$. The following theorem helps with the first two of these examples (the third is answered in the next section).

Theorem 13   Properties of the Derivative

Let f and g be differentiable on an open interval I and let c be a real number. Then:

1. **Sum/Difference Rule:**
$$\frac{d}{dx}\Big(f(x) \pm g(x)\Big) = \frac{d}{dx}\Big(f(x)\Big) \pm \frac{d}{dx}\Big(g(x)\Big) = f'(x) \pm g'(x)$$

2. **Constant Multiple Rule:**
$$\frac{d}{dx}\Big(c \cdot f(x)\Big) = c \cdot \frac{d}{dx}\Big(f(x)\Big) = c \cdot f'(x).$$

Theorem 13 allows us to find the derivatives of a wide variety of functions. It can be used in conjunction with the Power Rule to find the derivatives of any polynomial. Recall in Example 36 that we found, using the limit definition, the derivative of $f(x) = 3x^2 + 5x - 7$. We can now find its derivative without expressly using limits:

$$\frac{d}{dx}\Big(3x^2 + 5x + 7\Big) = 3\frac{d}{dx}\Big(x^2\Big) + 5\frac{d}{dx}\Big(x\Big) + \frac{d}{dx}\Big(7\Big)$$
$$= 3 \cdot 2x + 5 \cdot 1 + 0$$
$$= 6x + 5.$$

We were a bit pedantic here, showing every step. Normally we would do all the arithmetic and steps in our head and readily find $\frac{d}{dx}\Big(3x^2 + 5x + 7\Big) = 6x + 5$.

Example 47   Using the tangent line to approximate a function value
Let $f(x) = \sin x + 2x + 1$. Approximate $f(3)$ using an appropriate tangent line.

SOLUTION   This problem is intentionally ambiguous; we are to *approximate* using an *appropriate* tangent line. How good of an approximation are we seeking? What does appropriate mean?

In the "real world," people solving problems deal with these issues all time. One must make a judgment using whatever seems reasonable. In this example, the actual answer is $f(3) = \sin 3 + 7$, where the real problem spot is $\sin 3$. What is $\sin 3$?

Notes:

Since 3 is close to π, we can assume $\sin 3 \approx \sin \pi = 0$. Thus one guess is $f(3) \approx 7$. Can we do better? Let's use a tangent line as instructed and examine the results; it seems best to find the tangent line at $x = \pi$.

Using Theorem 12 we find $f'(x) = \cos x + 2$. The slope of the tangent line is thus $f'(\pi) = \cos \pi + 2 = 1$. Also, $f(\pi) = 2\pi + 1 \approx 7.28$. So the tangent line to the graph of f at $x = \pi$ is $y = 1(x - \pi) + 2\pi + 1 = x + \pi + 1 \approx x + 4.14$. Evaluated at $x = 3$, our tangent line gives $y = 3 + 4.14 = 7.14$. Using the tangent line, our final approximation is that $f(3) \approx 7.14$.

Using a calculator, we get an answer accurate to 4 places after the decimal: $f(3) = 7.1411$. Our initial guess was 7; our tangent line approximation was more accurate, at 7.14.

The point is *not* "Here's a cool way to do some math without a calculator." Sure, that might be handy sometime, but your phone could probably give you the answer. Rather, the point is to say that tangent lines are a good way of approximating, and many scientists, engineers and mathematicians often face problems too hard to solve directly. So they approximate.

Higher Order Derivatives

The derivative of a function f is itself a function, therefore we can take its derivative. The following definition gives a name to this concept and introduces its notation.

Definition 11   Higher Order Derivatives

Let $y = f(x)$ be a differentiable function on I.

1. The *second derivative* of f is:

$$f''(x) = \frac{d}{dx}\left(f'(x)\right) = \frac{d}{dx}\left(\frac{dy}{dx}\right) = \frac{d^2 y}{dx^2} = y''.$$

2. The *third derivative* of f is:

$$f'''(x) = \frac{d}{dx}\left(f''(x)\right) = \frac{d}{dx}\left(\frac{d^2 y}{dx^2}\right) = \frac{d^3 y}{dx^3} = y'''.$$

3. The nth *derivative* of f is:

$$f^{(n)}(x) = \frac{d}{dx}\left(f^{(n-1)}(x)\right) = \frac{d}{dx}\left(\frac{d^{n-1} y}{dx^{n-1}}\right) = \frac{d^n y}{dx^n} = y^{(n)}.$$

Note: Definition 11 comes with the caveat "Where the corresponding limits exist." With f differentiable on I, it is possible that f' is *not* differentiable on all of I, and so on.

Notes:

In general, when finding the fourth derivative and on, we resort to the $f^{(4)}(x)$ notation, not $f''''(x)$; after a while, too many ticks is too confusing.

Let's practice using this new concept.

Example 48   Finding higher order derivatives
Find the first four derivatives of the following functions:

1. $f(x) = 4x^2$ 3. $f(x) = 5e^x$

2. $f(x) = \sin x$

SOLUTION

1. Using the Power and Constant Multiple Rules, we have: $f'(x) = 8x$. Continuing on, we have

$$f''(x) = \frac{d}{dx}(8x) = 8; \qquad f'''(x) = 0; \qquad f^{(4)}(x) = 0.$$

Notice how all successive derivatives will also be 0.

2. We employ Theorem 12 repeatedly.

$$f'(x) = \cos x; \qquad f''(x) = -\sin x; \qquad f'''(x) = -\cos x; \qquad f^{(4)}(x) = \sin x.$$

Note how we have come right back to $f(x)$ again. (Can you quickly figure what $f^{(23)}(x)$ is?)

3. Employing Theorem 12 and the Constant Multiple Rule, we can see that

$$f'(x) = f''(x) = f'''(x) = f^{(4)}(x) = 5e^x.$$

Interpreting Higher Order Derivatives

What do higher order derivatives *mean*? What is the practical interpretation?

Our first answer is a bit wordy, but is technically correct and beneficial to understand. That is,

The second derivative of a function f is the rate of change of the rate of change of f.

Notes:

One way to grasp this concept is to let f describe a position function. Then, as stated in Key Idea 1, f' describes the rate of position change: velocity. We now consider f'', which describes the rate of velocity change. Sports car enthusiasts talk of how fast a car can go from 0 to 60 mph; they are bragging about the *acceleration* of the car.

We started this chapter with amusement–park riders free–falling with position function $f(t) = -16t^2 + 150$. It is easy to compute $f'(t) = -32t$ ft/s and $f''(t) = -32$ (ft/s)/s. We may recognize this latter constant; it is the acceleration due to gravity. In keeping with the unit notation introduced in the previous section, we say the units are "feet per second per second." This is usually shortened to "feet per second squared," written as "ft/s^2."

It can be difficult to consider the meaning of the third, and higher order, derivatives. The third derivative is "the rate of change of the rate of change of the rate of change of f." That is essentially meaningless to the uninitiated. In the context of our position/velocity/acceleration example, the third derivative is the "rate of change of acceleration," commonly referred to as "jerk."

Make no mistake: higher order derivatives have great importance even if their practical interpretations are hard (or "impossible") to understand. The mathematical topic of *series* makes extensive use of higher order derivatives.

Notes:

Exercises 2.3

Terms and Concepts

1. What is the name of the rule which states that $\frac{d}{dx}(x^n) = nx^{n-1}$, where $n > 0$ is an integer?

2. What is $\frac{d}{dx}(\ln x)$?

3. Give an example of a function $f(x)$ where $f'(x) = f(x)$.

4. Give an example of a function $f(x)$ where $f'(x) = 0$.

5. The derivative rules introduced in this section explain how to compute the derivative of which of the following functions?
   - $f(x) = \frac{3}{x^2}$
   - $j(x) = \sin x \cos x$
   - $g(x) = 3x^2 - x + 17$
   - $k(x) = e^{x^2}$
   - $h(x) = 5 \ln x$
   - $m(x) = \sqrt{x}$

6. Explain in your own words how to find the third derivative of a function $f(x)$.

7. Give an example of a function where $f'(x) \neq 0$ and $f''(x) = 0$.

8. Explain in your own words what the second derivative "means."

9. If $f(x)$ describes a position function, then $f'(x)$ describes what kind of function? What kind of function is $f''(x)$?

10. Let $f(x)$ be a function measured in pounds, where x is measured in feet. What are the units of $f''(x)$?

Problems

In Exercises 11 – 25, compute the derivative of the given function.

11. $f(x) = 7x^2 - 5x + 7$

12. $g(x) = 14x^3 + 7x^2 + 11x - 29$

13. $m(t) = 9t^5 - \frac{1}{8}t^3 + 3t - 8$

14. $f(\theta) = 9\sin\theta + 10\cos\theta$

15. $f(r) = 6e^r$

16. $g(t) = 10t^4 - \cos t + 7\sin t$

17. $f(x) = 2\ln x - x$

18. $p(s) = \frac{1}{4}s^4 + \frac{1}{3}s^3 + \frac{1}{2}s^2 + s + 1$

19. $h(t) = e^t - \sin t - \cos t$

20. $f(x) = \ln(5x^2)$

21. $f(t) = \ln(17) + e^2 + \sin \pi/2$

22. $g(t) = (1 + 3t)^2$

23. $g(x) = (2x - 5)^3$

24. $f(x) = (1 - x)^3$

25. $f(x) = (2 - 3x)^2$

26. A property of logarithms is that $\log_a x = \frac{\log_b x}{\log_b a}$, for all bases $a, b > 0, \neq 1$.
    (a) Rewrite this identity when $b = e$, i.e., using $\log_e x = \ln x$.
    (b) Use part (a) to find the derivative of $y = \log_a x$.
    (c) Give the derivative of $y = \log_{10} x$.

In Exercises 27 – 32, compute the first four derivatives of the given function.

27. $f(x) = x^6$

28. $g(x) = 2\cos x$

29. $h(t) = t^2 - e^t$

30. $p(\theta) = \theta^4 - \theta^3$

31. $f(\theta) = \sin\theta - \cos\theta$

32. $f(x) = 1,100$

In Exercises 33 – 38, find the equations of the tangent and normal lines to the graph of the function at the given point.

33. $f(x) = x^3 - x$ at $x = 1$

34. $f(t) = e^t + 3$ at $t = 0$

35. $g(x) = \ln x$ at $x = 1$

36. $f(x) = 4\sin x$ at $x = \pi/2$

37. $f(x) = -2\cos x$ at $x = \pi/4$

38. $f(x) = 2x + 3$ at $x = 5$

Review

39. Given that $e^0 = 1$, approximate the value of $e^{0.1}$ using the tangent line to $f(x) = e^x$ at $x = 0$.

40. Approximate the value of $(3.01)^4$ using the tangent line to $f(x) = x^4$ at $x = 3$.

2.4 The Product and Quotient Rules

The previous section showed that, in some ways, derivatives behave nicely. The Constant Multiple and Sum/Difference Rules established that the derivative of $f(x) = 5x^2 + \sin x$ was not complicated. We neglected computing the derivative of things like $g(x) = 5x^2 \sin x$ and $h(x) = \frac{5x^2}{\sin x}$ on purpose; their derivatives are *not* as straightforward. (If you had to guess what their respective derivatives are, you would probably guess wrong.) For these, we need the Product and Quotient Rules, respectively, which are defined in this section.

We begin with the Product Rule.

Theorem 14   Product Rule

Let f and g be differentiable functions on an open interval I. Then fg is a differentiable function on I, and

$$\frac{d}{dx}\Big(f(x)g(x)\Big) = f(x)g'(x) + f'(x)g(x).$$

Important: $\frac{d}{dx}\Big(f(x)g(x)\Big) \neq f'(x)g'(x)$! While this answer is simpler than the Product Rule, it is wrong.

We practice using this new rule in an example, followed by an example that demonstrates why this theorem is true.

Example 49   Using the Product Rule
Use the Product Rule to compute the derivative of $y = 5x^2 \sin x$. Evaluate the derivative at $x = \pi/2$.

SOLUTION   To make our use of the Product Rule explicit, let's set $f(x) = 5x^2$ and $g(x) = \sin x$. We easily compute/recall that $f'(x) = 10x$ and $g'(x) = \cos x$. Employing the rule, we have

$$\frac{d}{dx}\Big(5x^2 \sin x\Big) = 5x^2 \cos x + 10x \sin x.$$

At $x = \pi/2$, we have

$$y'(\pi/2) = 5\left(\frac{\pi}{2}\right)^2 \cos\left(\frac{\pi}{2}\right) + 10\frac{\pi}{2}\sin\left(\frac{\pi}{2}\right) = 5\pi.$$

We graph y and its tangent line at $x = \pi/2$, which has a slope of 5π, in Figure 2.15. While this does not *prove* that the Produce Rule is the correct way to handle derivatives of products, it helps validate its truth.

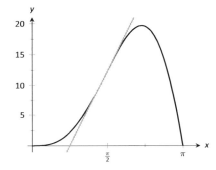

Figure 2.15: A graph of $y = 5x^2 \sin x$ and its tangent line at $x = \pi/2$.

Notes:

We now investigate why the Product Rule is true.

Example 50   A proof of the Product Rule
Use the definition of the derivative to prove Theorem 14.

SOLUTION   By the limit definition, we have

$$\frac{d}{dx}\Big(f(x)g(x)\Big) = \lim_{h \to 0} \frac{f(x+h)g(x+h) - f(x)g(x)}{h}.$$

We now do something a bit unexpected; add 0 to the numerator (so that nothing is changed) in the form of $-f(x+h)g(x)+f(x+h)g(x)$, then do some regrouping as shown.

$$\frac{d}{dx}\Big(f(x)g(x)\Big) = \lim_{h \to 0} \frac{f(x+h)g(x+h) - f(x)g(x)}{h} \quad \text{(now add 0 to the numerator)}$$

$$= \lim_{h \to 0} \frac{f(x+h)g(x+h) - f(x+h)g(x) + f(x+h)g(x) - f(x)g(x)}{h} \quad \text{(regroup)}$$

$$= \lim_{h \to 0} \frac{\Big(f(x+h)g(x+h) - f(x+h)g(x)\Big) + \Big(f(x+h)g(x) - f(x)g(x)\Big)}{h}$$

$$= \lim_{h \to 0} \frac{f(x+h)g(x+h) - f(x+h)g(x)}{h} + \lim_{h \to 0} \frac{f(x+h)g(x) - f(x)g(x)}{h} \quad \text{(factor)}$$

$$= \lim_{h \to 0} f(x+h)\frac{g(x+h) - g(x)}{h} + \lim_{h \to 0} \frac{f(x+h) - f(x)}{h}g(x) \quad \text{(apply limits)}$$

$$= f(x)g'(x) + f'(x)g(x)$$

It is often true that we can recognize that a theorem is true through its proof yet somehow doubt its applicability to real problems. In the following example, we compute the derivative of a product of functions in two ways to verify that the Product Rule is indeed "right."

Example 51   Exploring alternate derivative methods
Let $y = (x^2 + 3x + 1)(2x^2 - 3x + 1)$. Find y' two ways: first, by expanding the given product and then taking the derivative, and second, by applying the Product Rule. Verify that both methods give the same answer.

SOLUTION   We first expand the expression for y; a little algebra shows that $y = 2x^4 + 3x^3 - 6x^2 + 1$. It is easy to compute y';

$$y' = 8x^3 + 9x^2 - 12x.$$

Notes:

Now apply the Product Rule.

$$y' = (x^2 + 3x + 1)(4x - 3) + (2x + 3)(2x^2 - 3x + 1)$$
$$= \left(4x^3 + 9x^2 - 5x - 3\right) + \left(4x^3 - 7x + 3\right)$$
$$= 8x^3 + 9x^2 - 12x.$$

The uninformed usually assume that "the derivative of the product is the product of the derivatives." Thus we are tempted to say that $y' = (2x+3)(4x-3) = 8x^2 + 6x - 9$. Obviously this is not correct.

Example 52   **Using the Product Rule with a product of three functions**
Let $y = x^3 \ln x \cos x$. Find y'.

     SOLUTION   We have a product of three functions while the Product Rule only specifies how to handle a product of two functions. Our method of handling this problem is to simply group the latter two functions together, and consider $y = x^3 \left(\ln x \cos x \right)$. Following the Product Rule, we have

$$y' = (x^3)\left(\ln x \cos x \right)' + 3x^2 \left(\ln x \cos x \right)$$

To evaluate $\left(\ln x \cos x \right)'$, we apply the Product Rule again:

$$= (x^3)\left(\ln x(- \sin x) + \frac{1}{x} \cos x \right) + 3x^2 \left(\ln x \cos x \right)$$
$$= x^3 \ln x(- \sin x) + x^3 \frac{1}{x} \cos x + 3x^2 \ln x \cos x$$

Recognize the pattern in our answer above: when applying the Product Rule to a product of three functions, there are three terms added together in the final derivative. Each terms contains only one derivative of one of the original functions, and each function's derivative shows up in only one term. It is straightforward to extend this pattern to finding the derivative of a product of 4 or more functions.

We consider one more example before discussing another derivative rule.

Example 53   **Using the Product Rule**
Find the derivatives of the following functions.

1. $f(x) = x \ln x$

2. $g(x) = x \ln x - x$.

Notes:

SOLUTION   Recalling that the derivative of $\ln x$ is $1/x$, we use the Product Rule to find our answers.

1. $\dfrac{d}{dx}\left(x\ln x\right) = x \cdot 1/x + 1 \cdot \ln x = 1 + \ln x$.

2. Using the result from above, we compute

$$\frac{d}{dx}\left(x\ln x - x\right) = 1 + \ln x - 1 = \ln x.$$

This seems significant; if the natural log function $\ln x$ is an important function (it is), it seems worthwhile to know a function whose derivative is $\ln x$. We have found one. (We leave it to the reader to find another; a correct answer will be *very* similar to this one.)

We have learned how to compute the derivatives of sums, differences, and products of functions. We now learn how to find the derivative of a quotient of functions.

Theorem 15   Quotient Rule

Let f and g be functions defined on an open interval I, where $g(x) \neq 0$ on I. Then f/g is differentiable on I, and

$$\frac{d}{dx}\left(\frac{f(x)}{g(x)}\right) = \frac{g(x)f'(x) - f(x)g'(x)}{g(x)^2}.$$

The Quotient Rule is not hard to use, although it might be a bit tricky to remember. A useful mnemonic works as follows. Consider a fraction's numerator and denominator as "HI" and "LO", respectively. Then

$$\frac{d}{dx}\left(\frac{\text{HI}}{\text{LO}}\right) = \frac{\text{LO}\cdot\text{dHI} - \text{HI}\cdot\text{dLO}}{\text{LOLO}},$$

read "low dee high minus high dee low, over low low." Said fast, that phrase can roll off the tongue, making it easy to memorize. The "dee high" and "dee low" parts refer to the derivatives of the numerator and denominator, respectively.

Let's practice using the Quotient Rule.

Example 54   Using the Quotient Rule

Let $f(x) = \dfrac{5x^2}{\sin x}$. Find $f'(x)$.

Notes:

SOLUTION   Directly applying the Quotient Rule gives:

$$\frac{d}{dx}\left(\frac{5x^2}{\sin x}\right) = \frac{\sin x \cdot 10x - 5x^2 \cdot \cos x}{\sin^2 x}$$

$$= \frac{10x\sin x - 5x^2\cos x}{\sin^2 x}.$$

The Quotient Rule allows us to fill in holes in our understanding of derivatives of the common trigonometric functions. We start with finding the derivative of the tangent function.

Example 55   **Using the Quotient Rule to find $\frac{d}{dx}(\tan x)$.**
Find the derivative of $y = \tan x$.

SOLUTION   At first, one might feel unequipped to answer this question. But recall that $\tan x = \sin x/\cos x$, so we can apply the Quotient Rule.

$$\frac{d}{dx}\left(\tan x\right) = \frac{d}{dx}\left(\frac{\sin x}{\cos x}\right)$$

$$= \frac{\cos x \cos x - \sin x(-\sin x)}{\cos^2 x}$$

$$= \frac{\cos^2 x + \sin^2 x}{\cos^2 x}$$

$$= \frac{1}{\cos^2 x}$$

$$= \sec^2 x.$$

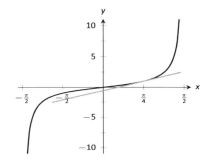

Figure 2.16: A graph of $y = \tan x$ along with its tangent line at $x = \pi/4$.

This is beautiful result. To confirm its truth, we can find the equation of the tangent line to $y = \tan x$ at $x = \pi/4$. The slope is $\sec^2(\pi/4) = 2$; $y = \tan x$, along with its tangent line, is graphed in Figure 2.16.

We include this result in the following theorem about the derivatives of the trigonometric functions. Recall we found the derivative of $y = \sin x$ in Example 38 and stated the derivative of the cosine function in Theorem 12. The derivatives of the cotangent, cosecant and secant functions can all be computed directly using Theorem 12 and the Quotient Rule.

Notes:

Theorem 16   Derivatives of Trigonometric Functions

1. $\dfrac{d}{dx}\left(\sin x\right) = \cos x$ 2. $\dfrac{d}{dx}\left(\cos x\right) = -\sin x$

3. $\dfrac{d}{dx}\left(\tan x\right) = \sec^2 x$ 4. $\dfrac{d}{dx}\left(\cot x\right) = -\csc^2 x$

5. $\dfrac{d}{dx}\left(\sec x\right) = \sec x \tan x$ 6. $\dfrac{d}{dx}\left(\csc x\right) = -\csc x \cot x$

To remember the above, it may be helpful to keep in mind that the derivatives of the trigonometric functions that start with "c" have a minus sign in them.

Example 56   Exploring alternate derivative methods

In Example 54 the derivative of $f(x) = \dfrac{5x^2}{\sin x}$ was found using the Quotient Rule. Rewriting f as $f(x) = 5x^2 \csc x$, find f' using Theorem 16 and verify the two answers are the same.

SOLUTION   We found in Example 54 that the $f'(x) = \dfrac{10x \sin x - 5x^2 \cos x}{\sin^2 x}$.
We now find f' using the Product Rule, considering f as $f(x) = 5x^2 \csc x$.

$$f'(x) = \frac{d}{dx}\left(5x^2 \csc x\right)$$
$$= 5x^2(-\csc x \cot x) + 10x \csc x \qquad \text{(now rewrite trig functions)}$$
$$= 5x^2 \cdot \frac{-1}{\sin x} \cdot \frac{\cos x}{\sin x} + \frac{10x}{\sin x}$$
$$= \frac{-5x^2 \cos x}{\sin^2 x} + \frac{10x}{\sin x} \qquad \text{(get common denominator)}$$
$$= \frac{10x \sin x - 5x^2 \cos x}{\sin^2 x}$$

Finding f' using either method returned the same result. At first, the answers looked different, but some algebra verified they are the same. In general, there is not one final form that we seek; the immediate result from the Product Rule is fine. Work to "simplify" your results into a form that is most readable and useful to you.

The Quotient Rule gives other useful results, as show in the next example.

Notes:

Example 57   **Using the Quotient Rule to expand the Power Rule**

Find the derivatives of the following functions.

1. $f(x) = \dfrac{1}{x}$

2. $f(x) = \dfrac{1}{x^n}$, where $n > 0$ is an integer.

SOLUTION   We employ the Quotient Rule.

1. $f'(x) = \dfrac{x \cdot 0 - 1 \cdot 1}{x^2} = -\dfrac{1}{x^2}.$

2. $f'(x) = \dfrac{x^n \cdot 0 - 1 \cdot nx^{n-1}}{(x^n)^2} = -\dfrac{nx^{n-1}}{x^{2n}} = -\dfrac{n}{x^{n+1}}.$

The derivative of $y = \dfrac{1}{x^n}$ turned out to be rather nice. It gets better. Consider:

$$\frac{d}{dx}\left(\frac{1}{x^n}\right) = \frac{d}{dx}\left(x^{-n}\right) \qquad \text{(apply result from Example 57)}$$

$$= -\frac{n}{x^{n+1}} \qquad \text{(rewrite algebraically)}$$

$$= -nx^{-(n+1)}$$

$$= -nx^{-n-1}.$$

This is reminiscent of the Power Rule: multiply by the power, then subtract 1 from the power. We now add to our previous Power Rule, which had the restriction of $n > 0$.

Theorem 17   **Power Rule with Integer Exponents**

Let $f(x) = x^n$, where $n \neq 0$ is an integer. Then

$$f'(x) = n \cdot x^{n-1}.$$

Taking the derivative of many functions is relatively straightforward. It is clear (with practice) what rules apply and in what order they should be applied. Other functions present multiple paths; different rules may be applied depending on how the function is treated. One of the beautiful things about calculus is that there is not "the" right way; each path, when applied correctly, leads to

---

Notes:

the same result, the derivative. We demonstrate this concept in an example.

Example 58   **Exploring alternate derivative methods**

Let $f(x) = \dfrac{x^2 - 3x + 1}{x}$. Find $f'(x)$ in each of the following ways:

1. By applying the Quotient Rule,

2. by viewing f as $f(x) = (x^2 - 3x + 1) \cdot x^{-1}$ and applying the Product and Power Rules, and

3. by "simplifying" first through division.

Verify that all three methods give the same result.

SOLUTION

1. Applying the Quotient Rule gives:

$$f'(x) = \frac{x \cdot (2x - 3) - (x^2 - 3x + 1) \cdot 1}{x^2} = \frac{x^2 - 1}{x^2} = 1 - \frac{1}{x^2}.$$

2. By rewriting f, we can apply the Product and Power Rules as follows:

$$f'(x) = (x^2 - 3x + 1) \cdot (-1)x^{-2} + (2x - 3) \cdot x^{-1}$$

$$= -\frac{x^2 - 3x + 1}{x^2} + \frac{2x - 3}{x}$$

$$= -\frac{x^2 - 3x + 1}{x^2} + \frac{2x^2 - 3x}{x^2}$$

$$= \frac{x^2 - 1}{x^2} = 1 - \frac{1}{x^2},$$

the same result as above.

3. As $x \neq 0$, we can divide through by x first, giving $f(x) = x - 3 + \dfrac{1}{x}$. Now apply the Power Rule.

$$f'(x) = 1 - \frac{1}{x^2},$$

the same result as before.

Example 58 demonstrates three methods of finding f'. One is hard pressed to argue for a "best method" as all three gave the same result without too much difficulty, although it is clear that using the Product Rule required more steps. Ultimately, the important principle to take away from this is: reduce the answer

Notes:

to a form that seems "simple" and easy to interpret. In that example, we saw different expressions for f', including:

$$1 - \frac{1}{x^2} = \frac{x \cdot (2x - 3) - (x^2 - 3x + 1) \cdot 1}{x^2} = (x^2 - 3x + 1) \cdot (-1)x^{-2} + (2x - 3) \cdot x^{-1}.$$

They are equal; they are all correct; only the first is "clear." Work to make answers clear.

In the next section we continue to learn rules that allow us to more easily compute derivatives than using the limit definition directly. We have to memorize the derivatives of a certain set of functions, such as "the derivative of $\sin x$ is $\cos x$." The Sum/Difference, Constant Multiple, Power, Product and Quotient Rules show us how to find the derivatives of certain combinations of these functions. The next section shows how to find the derivatives when we *compose* these functions together.

Notes:

Exercises 2.4

Terms and Concepts

1. T/F: The Product Rule states that $\dfrac{d}{dx}(x^2 \sin x) = 2x \cos x$.

2. T/F: The Quotient Rule states that $\dfrac{d}{dx}\left(\dfrac{x^2}{\sin x}\right) = \dfrac{\cos x}{2x}$.

3. T/F: The derivatives of the trigonometric functions that start with "c" have minus signs in them.

4. What derivative rule is used to extend the Power Rule to include negative integer exponents?

5. T/F: Regardless of the function, there is always exactly one right way of computing its derivative.

6. In your own words, explain what it means to make your answers "clear."

Problems

In Exercises 7 – 10:

(a) Use the Product Rule to differentiate the function.

(b) Manipulate the function algebraically and differentiate without the Product Rule.

(c) Show that the answers from (a) and (b) are equivalent.

7. $f(x) = x(x^2 + 3x)$

8. $g(x) = 2x^2(5x^3)$

9. $h(s) = (2s - 1)(s + 4)$

10. $f(x) = (x^2 + 5)(3 - x^3)$

In Exercises 11 – 14:

(a) Use the Quotient Rule to differentiate the function.

(b) Manipulate the function algebraically and differentiate without the Quotient Rule.

(c) Show that the answers from (a) and (b) are equivalent.

11. $f(x) = \dfrac{x^2 + 3}{x}$

12. $g(x) = \dfrac{x^3 - 2x^2}{2x^2}$

13. $h(s) = \dfrac{3}{4s^3}$

14. $f(t) = \dfrac{t^2 - 1}{t + 1}$

In Exercises 15 – 29, compute the derivative of the given function.

15. $f(x) = x \sin x$

16. $f(t) = \dfrac{1}{t^2}(\csc t - 4)$

17. $g(x) = \dfrac{x + 7}{x - 5}$

18. $g(t) = \dfrac{t^5}{\cos t - 2t^2}$

19. $h(x) = \cot x - e^x$

20. $h(t) = 7t^2 + 6t - 2$

21. $f(x) = \dfrac{x^4 + 2x^3}{x + 2}$

22. $f(x) = (16x^3 + 24x^2 + 3x)\dfrac{7x - 1}{16x^3 + 24x^2 + 3x}$

23. $f(t) = t^5(\sec t + e^t)$

24. $f(x) = \dfrac{\sin x}{\cos x + 3}$

25. $g(x) = e^2\left(\sin(\pi/4) - 1\right)$

26. $g(t) = 4t^3 e^t - \sin t \cos t$

27. $h(t) = \dfrac{t^2 \sin t + 3}{t^2 \cos t + 2}$

28. $f(x) = x^2 e^x \tan x$

29. $g(x) = 2x \sin x \sec x$

In Exercises 30 – 33, find the equations of the tangent and normal lines to the graph of g at the indicated point.

30. $g(s) = e^s(s^2 + 2)$ at $(0, 2)$.

31. $g(t) = t \sin t$ at $\left(\dfrac{3\pi}{2}, -\dfrac{3\pi}{2}\right)$

32. $g(x) = \dfrac{x^2}{x - 1}$ at $(2, 4)$

33. $g(\theta) = \dfrac{\cos \theta - 8\theta}{\theta + 1}$ at $(0, -5)$

In Exercises 34 – 37, find the x–values where the graph of the function has a horizontal tangent line.

34. $f(x) = 6x^2 - 18x - 24$

35. $f(x) = x \sin x$ on $[-1, 1]$

36. $f(x) = \dfrac{x}{x+1}$

37. $f(x) = \dfrac{x^2}{x+1}$

In Exercises 38 – 41, find the requested derivative.

38. $f(x) = x \sin x$; find $f''(x)$.

39. $f(x) = x \sin x$; find $f^{(4)}(x)$.

40. $f(x) = \csc x$; find $f''(x)$.

41. $f(x) = (x^3 - 5x + 2)(x^2 + x - 7)$; find $f^{(8)}(x)$.

In Exercises 42 – 45, use the graph of $f(x)$ to sketch $f'(x)$.

43.
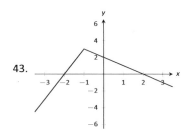

44.
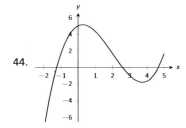

42.
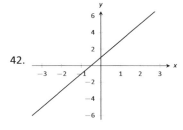

45.
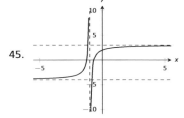

2.5 The Chain Rule

We have covered almost all of the derivative rules that deal with combinations of two (or more) functions. The operations of addition, subtraction, multiplication (including by a constant) and division led to the Sum and Difference rules, the Constant Multiple Rule, the Power Rule, the Product Rule and the Quotient Rule. To complete the list of differentiation rules, we look at the last way two (or more) functions can be combined: the process of composition (i.e. one function "inside" another).

One example of a composition of functions is $f(x) = \cos(x^2)$. We currently do not know how to compute this derivative. If forced to guess, one would likely guess $f'(x) = -\sin(2x)$, where we recognize $-\sin x$ as the derivative of $\cos x$ and $2x$ as the derivative of x^2. However, this is not the case; $f'(x) \neq -\sin(2x)$. In Example 62 we'll see the correct answer, which employs the new rule this section introduces, the **Chain Rule**.

Before we define this new rule, recall the notation for composition of functions. We write $(f \circ g)(x)$ or $f(g(x))$, read as "f of g of x," to denote composing f with g. In shorthand, we simply write $f \circ g$ or $f(g)$ and read it as "f of g." Before giving the corresponding differentiation rule, we note that the rule extends to multiple compositions like $f(g(h(x)))$ or $f(g(h(j(x))))$, etc.

To motivate the rule, let's look at three derivatives we can already compute.

Example 59   **Exploring similar derivatives**
Find the derivatives of $F_1(x) = (1-x)^2$, $F_2(x) = (1-x)^3$, and $F_3(x) = (1-x)^4$. (We'll see later why we are using subscripts for different functions and an uppercase F.)

SOLUTION   In order to use the rules we already have, we must first expand each function as $F_1(x) = 1 - 2x + x^2$, $F_2(x) = 1 - 3x + 3x^2 - x^3$ and $F_3(x) = 1 - 4x + 6x^2 - 4x^3 + x^4$.

It is not hard to see that:

$F'_1(x) = -2 + 2x,$

$F'_2(x) = -3 + 6x - 3x^2$ and

$F'_3(x) = -4 + 12x - 12x^2 + 4x^3.$

An interesting fact is that these can be rewritten as

$$F'_1(x) = -2(1-x), \quad F'_2(x) = -3(1-x)^2 \text{ and } F'_3(x) = -4(1-x)^3.$$

A pattern might jump out at you. Recognize that each of these functions is a

---

Notes:

composition, letting $g(x) = 1 - x$:

$$F_1(x) = f_1(g(x)), \quad \text{where } f_1(x) = x^2,$$
$$F_2(x) = f_2(g(x)), \quad \text{where } f_2(x) = x^3,$$
$$F_3(x) = f_3(g(x)), \quad \text{where } f_3(x) = x^4.$$

We'll come back to this example after giving the formal statements of the Chain Rule; for now, we are just illustrating a pattern.

Theorem 18   The Chain Rule

Let $y = f(u)$ be a differentiable function of u and let $u = g(x)$ be a differentiable function of x. Then $y = f(g(x))$ is a differentiable function of x, and

$$y' = f'(g(x)) \cdot g'(x).$$

To help understand the Chain Rule, we return to Example 59.

Example 60   Using the Chain Rule
Use the Chain Rule to find the derivatives of the following functions, as given in Example 59.

SOLUTION   Example 59 ended with the recognition that each of the given functions was actually a composition of functions. To avoid confusion, we ignore most of the subscripts here.

$F_1(x) = (1 - x)^2$:

We found that

$$y = (1 - x)^2 = f(g(x)), \text{ where } f(x) = x^2 \text{ and } g(x) = 1 - x.$$

To find y', we apply the Chain Rule. We need $f'(x) = 2x$ and $g'(x) = -1$.
Part of the Chain Rule uses $f'(g(x))$. This means substitute $g(x)$ for x in the equation for $f'(x)$. That is, $f'(x) = 2(1 - x)$. Finishing out the Chain Rule we have

$$y' = f'(g(x)) \cdot g'(x) = 2(1 - x) \cdot (-1) = -2(1 - x) = 2x - 2.$$

$F_2(x) = (1 - x)^3$:

Notes:

Let $y = (1 - x)^3 = f(g(x))$, where $f(x) = x^3$ and $g(x) = (1 - x)$. We have $f'(x) = 3x^2$, so $f'(g(x)) = 3(1 - x)^2$. The Chain Rule then states

$$y' = f'(g(x)) \cdot g'(x) = 3(1 - x)^2 \cdot (-1) = -3(1 - x)^2.$$

$F_3(x) = (1 - x)^4$:

Finally, when $y = (1 - x)^4$, we have $f(x) = x^4$ and $g(x) = (1 - x)$. Thus $f'(x) = 4x^3$ and $f'(g(x)) = 4(1 - x)^3$. Thus

$$y' = f'(g(x)) \cdot g'(x) = 4(1 - x)^3 \cdot (-1) = -4(1 - x)^3.$$

Example 60 demonstrated a particular pattern: when $f(x) = x^n$, then $y' = n \cdot (g(x))^{n-1} \cdot g'(x)$. This is called the Generalized Power Rule.

Theorem 19 Generalized Power Rule

Let $g(x)$ be a differentiable function and let $n \neq 0$ be an integer. Then

$$\frac{d}{dx}\left(g(x)^n\right) = n \cdot \left(g(x)\right)^{n-1} \cdot g'(x).$$

This allows us to quickly find the derivative of functions like $y = (3x^2 - 5x + 7 + \sin x)^{20}$. While it may look intimidating, the Generalized Power Rule states that
$$y' = 20(3x^2 - 5x + 7 + \sin x)^{19} \cdot (6x - 5 + \cos x).$$

Treat the derivative–taking process step–by–step. In the example just given, first multiply by 20, the rewrite the inside of the parentheses, raising it all to the 19th power. Then think about the derivative of the expression inside the parentheses, and multiply by that.

We now consider more examples that employ the Chain Rule.

Example 61 Using the Chain Rule

Find the derivatives of the following functions:

1. $y = \sin 2x$
2. $y = \ln(4x^3 - 2x^2)$
3. $y = e^{-x^2}$

SOLUTION

1. Consider $y = \sin 2x$. Recognize that this is a composition of functions, where $f(x) = \sin x$ and $g(x) = 2x$. Thus

$$y' = f'(g(x)) \cdot g'(x) = \cos(2x) \cdot 2 = 2 \cos 2x.$$

Notes:

2. Recognize that $y = \ln(4x^3 - 2x^2)$ is the composition of $f(x) = \ln x$ and $g(x) = 4x^3 - 2x^2$. Also, recall that

$$\frac{d}{dx}\left(\ln x \right) = \frac{1}{x}.$$

This leads us to:

$$y' = \frac{1}{4x^3 - 2x^2} \cdot (12x^2 - 4x) = \frac{12x^2 - 4x}{4x^3 - 2x^2} = \frac{4x(3x - 1)}{2x(2x^2 - x)} = \frac{2(3x - 1)}{2x^2 - x}.$$

3. Recognize that $y = e^{-x^2}$ is the composition of $f(x) = e^x$ and $g(x) = -x^2$. Remembering that $f'(x) = e^x$, we have

$$y' = e^{-x^2} \cdot (-2x) = (-2x)e^{-x^2}.$$

Example 62   Using the Chain Rule to find a tangent line
Let $f(x) = \cos x^2$. Find the equation of the line tangent to the graph of f at $x = 1$.

SOLUTION   The tangent line goes through the point $(1, f(1)) \approx (1, 0.54)$ with slope $f'(1)$. To find f', we need the Chain Rule.
    $f'(x) = -\sin(x^2) \cdot (2x) = -2x \sin x^2$. Evaluated at $x = 1$, we have $f'(1) = -2 \sin 1 \approx -1.68$. Thus the equation of the tangent line is

$$y = -1.68(x - 1) + 0.54.$$

The tangent line is sketched along with f in Figure 2.17.

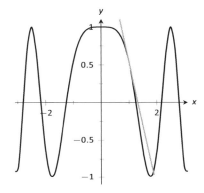

Figure 2.17: $f(x) = \cos x^2$ sketched along with its tangent line at $x = 1$.

The Chain Rule is used often in taking derivatives. Because of this, one can become familiar with the basic process and learn patterns that facilitate finding derivatives quickly. For instance,

$$\frac{d}{dx}\left(\ln(\text{anything}) \right) = \frac{1}{\text{anything}} \cdot (\text{anything})' = \frac{(\text{anything})'}{\text{anything}}.$$

A concrete example of this is

$$\frac{d}{dx}\left(\ln(3x^{15} - \cos x + e^x) \right) = \frac{45x^{14} + \sin x + e^x}{3x^{15} - \cos x + e^x}.$$

While the derivative may look intimidating at first, look for the pattern. The denominator is the same as what was inside the natural log function; the numerator is simply its derivative.

Notes:

This pattern recognition process can be applied to lots of functions. In general, instead of writing "anything", we use u as a generic function of x. We then say

$$\frac{d}{dx}\left(\ln u\right) = \frac{u'}{u}.$$

The following is a short list of how the Chain Rule can be quickly applied to familiar functions.

1. $\dfrac{d}{dx}\left(u^n\right) = n \cdot u^{n-1} \cdot u'.$ 4. $\dfrac{d}{dx}\left(\cos u\right) = -u' \cdot \sin u.$

2. $\dfrac{d}{dx}\left(e^u\right) = u' \cdot e^u.$ 5. $\dfrac{d}{dx}\left(\tan u\right) = u' \cdot \sec^2 u.$

3. $\dfrac{d}{dx}\left(\sin u\right) = u' \cdot \cos u.$

Of course, the Chain Rule can be applied in conjunction with any of the other rules we have already learned. We practice this next.

Example 63   **Using the Product, Quotient and Chain Rules**

Find the derivatives of the following functions.

1. $f(x) = x^5 \sin 2x^3$ 2. $f(x) = \dfrac{5x^3}{e^{-x^2}}.$

SOLUTION

1. We must use the Product and Chain Rules. Do not think that you must be able to "see" the whole answer immediately; rather, just proceed step–by–step.

$$f'(x) = x^5\left(6x^2 \cos 2x^3\right) + 5x^4\left(\sin 2x^3\right) = 6x^7 \cos 2x^3 + 5x^4 \sin 2x^3.$$

2. We must employ the Quotient Rule along with the Chain Rule. Again, proceed step–by–step.

$$f'(x) = \frac{e^{-x^2}\left(15x^2\right) - 5x^3\left((-2x)e^{-x^2}\right)}{\left(e^{-x^2}\right)^2} = \frac{e^{-x^2}\left(10x^4 + 15x^2\right)}{e^{-2x^2}}$$
$$= e^{x^2}\left(10x^4 + 15x^2\right).$$

A key to correctly working these problems is to break the problem down into smaller, more manageable pieces. For instance, when using the Product

---

Notes:

and Chain Rules together, just consider the first part of the Product Rule at first: $f(x)g'(x)$. Just rewrite $f(x)$, then find $g'(x)$. Then move on to the $f'(x)g(x)$ part. Don't attempt to figure out both parts at once.

Likewise, using the Quotient Rule, approach the numerator in two steps and handle the denominator after completing that. Only simplify afterward.

We can also employ the Chain Rule itself several times, as shown in the next example.

Example 64   Using the Chain Rule multiple times
Find the derivative of $y = \tan^5(6x^3 - 7x)$.

SOLUTION   Recognize that we have the $g(x) = \tan(6x^3 - 7x)$ function "inside" the $f(x) = x^5$ function; that is, we have $y = \left(\tan(6x^3 - 7x)\right)^5$. We begin using the Generalized Power Rule; in this first step, we do not fully compute the derivative. Rather, we are approaching this step–by–step.

$$y' = 5\left(\tan(6x^3 - 7x)\right)^4 \cdot g'(x).$$

We now find $g'(x)$. We again need the Chain Rule;

$$g'(x) = \sec^2(6x^3 - 7x) \cdot (18x^2 - 7).$$

Combine this with what we found above to give

$$y' = 5\left(\tan(6x^3 - 7x)\right)^4 \cdot \sec^2(6x^3 - 7x) \cdot (18x^2 - 7)$$
$$= (90x^2 - 35)\sec^2(6x^3 - 7x)\tan^4(6x^3 - 7x).$$

This function is frankly a ridiculous function, possessing no real practical value. It is very difficult to graph, as the tangent function has many vertical asymptotes and $6x^3 - 7x$ grows so very fast. The important thing to learn from this is that the derivative can be found. In fact, it is not "hard;" one must take several simple steps and be careful to keep track of how to apply each of these steps.

It is a traditional mathematical exercise to find the derivatives of arbitrarily complicated functions just to demonstrate that it *can be done*. Just break everything down into smaller pieces.

Example 65   Using the Product, Quotient and Chain Rules
Find the derivative of $f(x) = \dfrac{x\cos(x^{-2}) - \sin^2(e^{4x})}{\ln(x^2 + 5x^4)}$.

SOLUTION   This function likely has no practical use outside of demonstrating derivative skills. The answer is given below without simplification. It

Notes:

employs the Quotient Rule, the Product Rule, and the Chain Rule three times.

$f'(x) =$

$$\frac{\left(\begin{array}{l} \ln(x^2 + 5x^4) \cdot \left[(x \cdot (-\sin(x^{-2})) \cdot (-2x^{-3}) + 1 \cdot \cos(x^{-2})) - 2\sin(e^{4x}) \cdot \cos(e^{4x}) \cdot (4e^{4x}) \right] \\ - \left(x\cos(x^{-2}) - \sin^2(e^{4x}) \right) \cdot \frac{2x + 20x^3}{x^2 + 5x^4} \end{array}\right)}{\left(\ln(x^2 + 5x^4) \right)^2}.$$

The reader is highly encouraged to look at each term and recognize why it is there. (I.e., the Quotient Rule is used; in the numerator, identify the "LOdHI" term, etc.) This example demonstrates that derivatives can be computed systematically, no matter how arbitrarily complicated the function is.

The Chain Rule also has theoretic value. That is, it can be used to find the derivatives of functions that we have not yet learned as we do in the following example.

Example 66   The Chain Rule and exponential functions
Use the Chain Rule to find the derivative of $y = a^x$ where $a > 0$, $a \neq 1$ is constant.

SOLUTION   We only know how to find the derivative of one exponential function: $y = e^x$; this problem is asking us to find the derivative of functions such as $y = 2^x$.

This can be accomplished by rewriting a^x in terms of e. Recalling that e^x and $\ln x$ are inverse functions, we can write

$$a = e^{\ln a} \quad \text{and so} \quad y = a^x = e^{\ln(a^x)}.$$

By the exponent property of logarithms, we can "bring down" the power to get

$$y = a^x = e^{x(\ln a)}.$$

The function is now the composition $y = f(g(x))$, with $f(x) = e^x$ and $g(x) = x(\ln a)$. Since $f'(x) = e^x$ and $g'(x) = \ln a$, the Chain Rule gives

$$y' = e^{x(\ln a)} \cdot \ln a.$$

Recall that the $e^{x(\ln a)}$ term on the right hand side is just a^x, our original function. Thus, the derivative contains the original function itself. We have

$$y' = y \cdot \ln a = a^x \cdot \ln a.$$

Notes:

The Chain Rule, coupled with the derivative rule of e^x, allows us to find the derivatives of all exponential functions.

The previous example produced a result worthy of its own "box."

Theorem 20   **Derivatives of Exponential Functions**

Let $f(x) = a^x$, for $a > 0, a \neq 1$. Then f is differentiable for all real numbers and
$$f'(x) = \ln a \cdot a^x.$$

Alternate Chain Rule Notation

It is instructive to understand what the Chain Rule "looks like" using "$\frac{dy}{dx}$" notation instead of y' notation. Suppose that $y = f(u)$ is a function of u, where $u = g(x)$ is a function of x, as stated in Theorem 18. Then, through the composition $f \circ g$, we can think of y as a function of x, as $y = f(g(x))$. Thus the derivative of y with respect to x makes sense; we can talk about $\frac{dy}{dx}$. This leads to an interesting progression of notation:

$$y' = f'(g(x)) \cdot g'(x)$$
$$\frac{dy}{dx} = y'(u) \cdot u'(x) \qquad \text{(since } y = f(u) \text{ and } u = g(x))$$
$$\frac{dy}{dx} = \frac{dy}{du} \cdot \frac{du}{dx} \qquad \text{(using "fractional" notation for the derivative)}$$

Here the "fractional" aspect of the derivative notation stands out. On the right hand side, it seems as though the "du" terms cancel out, leaving

$$\frac{dy}{dx} = \frac{dy}{dx}.$$

It is important to realize that we *are not* canceling these terms; the derivative notation of $\frac{dy}{dx}$ is one symbol. It is equally important to realize that this notation was chosen precisely because of this behavior. It makes applying the Chain Rule easy with multiple variables. For instance,

$$\frac{dy}{dt} = \frac{dy}{d\bigcirc} \cdot \frac{d\bigcirc}{d\triangle} \cdot \frac{d\triangle}{dt}.$$

where $\bigcirc$ and $\triangle$ are any variables you'd like to use.

Notes:

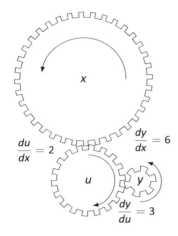

$$\frac{du}{dx} = 2$$

$$\frac{dy}{dx} = 6$$

$$\frac{dy}{du} = 3$$

Figure 2.18: A series of gears to demonstrate the Chain Rule. Note how $\frac{dy}{dx} = \frac{dy}{du} \cdot \frac{du}{dx}$

One of the most common ways of "visualizing" the Chain Rule is to consider a set of gears, as shown in Figure 2.18. The gears have 36, 18, and 6 teeth, respectively. That means for every revolution of the x gear, the u gear revolves twice. That is, the rate at which the u gear makes a revolution is twice as fast as the rate at which the x gear makes a revolution. Using the terminology of calculus, the rate of u-change, with respect to x, is $\frac{du}{dx} = 2$.

Likewise, every revolution of u causes 3 revolutions of y: $\frac{dy}{du} = 3$. How does y change with respect to x? For each revolution of x, y revolves 6 times; that is,

$$\frac{dy}{dx} = \frac{dy}{du} \cdot \frac{du}{dx} = 2 \cdot 3 = 6.$$

We can then extend the Chain Rule with more variables by adding more gears to the picture.

It is difficult to overstate the importance of the Chain Rule. So often the functions that we deal with are compositions of two or more functions, requiring us to use this rule to compute derivatives. It is often used in practice when actual functions are unknown. Rather, through measurement, we can calculate $\frac{dy}{du}$ and $\frac{du}{dx}$. With our knowledge of the Chain Rule, finding $\frac{dy}{dx}$ is straightforward.

In the next section, we use the Chain Rule to justify another differentiation technique. There are many curves that we can draw in the plane that fail the "vertical line test." For instance, consider $x^2 + y^2 = 1$, which describes the unit circle. We may still be interested in finding slopes of tangent lines to the circle at various points. The next section shows how we can find $\frac{dy}{dx}$ without first "solving for y." While we can in this instance, in many other instances solving for y is impossible. In these situations, *implicit differentiation* is indispensable.

Notes:

Exercises 2.5

Terms and Concepts

1. T/F: The Chain Rule describes how to evaluate the derivative of a composition of functions.

2. T/F: The Generalized Power Rule states that $\dfrac{d}{dx}\left(g(x)^n\right) = n\left(g(x)\right)^{n-1}$.

3. T/F: $\dfrac{d}{dx}\left(\ln(x^2)\right) = \dfrac{1}{x^2}$.

4. T/F: $\dfrac{d}{dx}\left(3^x\right) \approx 1.1 \cdot 3^x$.

5. T/F: $\dfrac{dx}{dy} = \dfrac{dx}{dt} \cdot \dfrac{dt}{dy}$

6. T/F: Taking the derivative of $f(x) = x^2 \sin(5x)$ requires the use of both the Product and Chain Rules.

Problems

In Exercises 7 – 28, compute the derivative of the given function.

7. $f(x) = (4x^3 - x)^{10}$

8. $f(t) = (3t - 2)^5$

9. $g(\theta) = (\sin\theta + \cos\theta)^3$

10. $h(t) = e^{3t^2 + t - 1}$

11. $f(x) = \left(x + \frac{1}{x}\right)^4$

12. $f(x) = \cos(3x)$

13. $g(x) = \tan(5x)$

14. $h(t) = \sin^4(2t)$

15. $p(t) = \cos^3(t^2 + 3t + 1)$

16. $f(x) = \ln(\cos x)$

17. $f(x) = \ln(x^2)$

18. $f(x) = 2\ln(x)$

19. $g(r) = 4^r$

20. $g(t) = 5^{\cos t}$

21. $g(t) = 15^2$

22. $m(w) = \dfrac{3^w}{2^w}$

23. $h(t) = \dfrac{2^t + 3}{3^t + 2}$

24. $m(w) = \dfrac{3^w + 1}{2^w}$

25. $f(x) = \dfrac{3^{x^2} + x}{2^{x^2}}$

26. $f(x) = x^2 \sin(5x)$

27. $g(t) = \cos(t^2 + 3t)\sin(5t - 7)$

28. $g(t) = \cos(\frac{1}{t})e^{5t^2}$

In Exercises 29 – 32, find the equations of tangent and normal lines to the graph of the function at the given point. Note: the functions here are the same as in Exercises 7 through 10.

29. $f(x) = (4x^3 - x)^{10}$ at $x = 0$

30. $f(t) = (3t - 2)^5$ at $t = 1$

31. $g(\theta) = (\sin\theta + \cos\theta)^3$ at $\theta = \pi/2$

32. $h(t) = e^{3t^2 + t - 1}$ at $t = -1$

33. Compute $\dfrac{d}{dx}\left(\ln(kx)\right)$ two ways:

   (a) Using the Chain Rule, and

   (b) by first using the logarithm rule $\ln(ab) = \ln a + \ln b$, then taking the derivative.

34. Compute $\dfrac{d}{dx}\left(\ln(x^k)\right)$ two ways:

   (a) Using the Chain Rule, and

   (b) by first using the logarithm rule $\ln(a^p) = p\ln a$, then taking the derivative.

Review

35. The "wind chill factor" is a measurement of how cold it "feels" during cold, windy weather. Let $W(w)$ be the wind chill factor, in degrees Fahrenheit, when it is $25°$F outside with a wind of w mph.

   (a) What are the units of $W'(w)$?

   (b) What would you expect the sign of $W'(10)$ to be?

36. Find the derivatives of the following functions.

   (a) $f(x) = x^2 e^x \cot x$

   (b) $g(x) = 2^x 3^x 4^x$

2.6   Implicit Differentiation

In the previous sections we learned to find the derivative, $\frac{dy}{dx}$, or y', when y is given *explicitly* as a function of x. That is, if we know $y = f(x)$ for some function f, we can find y'. For example, given $y = 3x^2 - 7$, we can easily find $y' = 6x$. (Here we explicitly state how x and y are related. Knowing x, we can directly find y.)

Sometimes the relationship between y and x is not explicit; rather, it is *implicit*. For instance, we might know that $x^2 - y = 4$. This equality defines a relationship between x and y; if we know x, we could figure out y. Can we still find y'? In this case, sure; we solve for y to get $y = x^2 - 4$ (hence we now know y explicitly) and then differentiate to get $y' = 2x$.

Sometimes the *implicit* relationship between x and y is complicated. Suppose we are given $\sin(y) + y^3 = 6 - x^3$. A graph of this implicit function is given in Figure 2.19. In this case there is absolutely no way to solve for y in terms of elementary functions. The surprising thing is, however, that we can still find y' via a process known as **implicit differentiation**.

Implicit differentiation is a technique based on the Chain Rule that is used to find a derivative when the relationship between the variables is given implicitly rather than explicitly (solved for one variable in terms of the other).

We begin by reviewing the Chain Rule. Let f and g be functions of x. Then

$$\frac{d}{dx}\Big(f(g(x))\Big) = f'(g(x)) \cdot g'(x).$$

Suppose now that $y = g(x)$. We can rewrite the above as

$$\frac{d}{dx}\Big(f(y)\Big) = f'(y)) \cdot y', \quad \text{or} \quad \frac{d}{dx}\Big(f(y)\Big) = f'(y) \cdot \frac{dy}{dx}. \tag{2.1}$$

These equations look strange; the key concept to learn here is that we can find y' even if we don't exactly know how y and x relate.

We demonstrate this process in the following example.

Example 67   Using Implicit Differentiation
Find y' given that $\sin(y) + y^3 = 6 - x^3$.

SOLUTION   We start by taking the derivative of both sides (thus maintaining the equality.) We have :

$$\frac{d}{dx}\Big(\sin(y) + y^3\Big) = \frac{d}{dx}\Big(6 - x^3\Big).$$

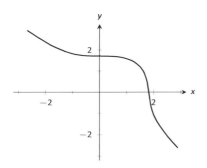

Figure 2.19: A graph of the implicit function $\sin(y) + y^3 = 6 - x^3$.

Notes:

The right hand side is easy; it returns $-3x^2$.

The left hand side requires more consideration. We take the derivative term–by–term. Using the technique derived from Equation 2.1 above, we can see that

$$\frac{d}{dx}\left(\sin y\right) = \cos y \cdot y'.$$

We apply the same process to the y^3 term.

$$\frac{d}{dx}\left(y^3\right) = \frac{d}{dx}\left((y)^3\right) = 3(y)^2 \cdot y'.$$

Putting this together with the right hand side, we have

$$\cos(y)y' + 3y^2 y' = -3x^2.$$

Now solve for y'.

$$\cos(y)y' + 3y^2 y' = -3x^2.$$
$$\left(\cos y + 3y^2\right)y' = -3x^2$$
$$y' = \frac{-3x^2}{\cos y + 3y^2}$$

This equation for y' probably seems unusual for it contains both x and y terms. How is it to be used? We'll address that next.

Implicit functions are generally harder to deal with than explicit functions. With an explicit function, given an x value, we have an explicit formula for computing the corresponding y value. With an implicit function, one often has to find x and y values *at the same time* that satisfy the equation. It is much easier to demonstrate that a given point satisfies the equation than to actually find such a point.

For instance, we can affirm easily that the point $(\sqrt[3]{6}, 0)$ lies on the graph of the implicit function $\sin y + y^3 = 6 - x^3$. Plugging in 0 for y, we see the left hand side is 0. Setting $x = \sqrt[3]{6}$, we see the right hand side is also 0; the equation is satisfied. The following example finds the equation of the tangent line to this function at this point.

Example 68　　Using Implicit Differentiation to find a tangent line
Find the equation of the line tangent to the curve of the implicitly defined function $\sin y + y^3 = 6 - x^3$ at the point $(\sqrt[3]{6}, 0)$.

SOLUTION　　In Example 67 we found that

$$y' = \frac{-3x^2}{\cos y + 3y^2}.$$

Notes:

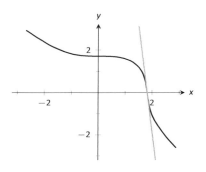

Figure 2.20: The function $\sin y + y^3 = 6 - x^3$ and its tangent line at the point $(\sqrt[3]{6}, 0)$.

We find the slope of the tangent line at the point $(\sqrt[3]{6}, 0)$ by substituting $\sqrt[3]{6}$ for x and 0 for y. Thus at the point $(\sqrt[3]{6}, 0)$, we have the slope as

$$y' = \frac{-3(\sqrt[3]{6})^2}{\cos 0 + 3 \cdot 0^2} = \frac{-3\sqrt[3]{36}}{1} \approx -9.91.$$

Therefore the equation of the tangent line to the implicitly defined function $\sin y + y^3 = 6 - x^3$ at the point $(\sqrt[3]{6}, 0)$ is

$$y = -3\sqrt[3]{36}(x - \sqrt[3]{6}) + 0 \approx -9.91x + 18.$$

The curve and this tangent line are shown in Figure 2.20.

This suggests a general method for implicit differentiation. For the steps below assume y is a function of x.

1. Take the derivative of each term in the equation. Treat the x terms like normal. When taking the derivatives of y terms, the usual rules apply except that, because of the Chain Rule, we need to multiply each term by y'.

2. Get all the y' terms on one side of the equal sign and put the remaining terms on the other side.

3. Factor out y'; solve for y' by dividing.

Practical Note: When working by hand, it may be beneficial to use the symbol $\frac{dy}{dx}$ instead of y', as the latter can be easily confused for y or y^1.

Example 69   Using Implicit Differentiation
Given the implicitly defined function $y^3 + x^2 y^4 = 1 + 2x$, find y'.

SOLUTION   We will take the implicit derivatives term by term. The derivative of y^3 is $3y^2 y'$.

The second term, $x^2 y^4$, is a little tricky. It requires the Product Rule as it is the product of two functions of x: x^2 and y^4. Its derivative is $x^2(4y^3 y') + 2xy^4$. The first part of this expression requires a y' because we are taking the derivative of a y term. The second part does not require it because we are taking the derivative of x^2.

The derivative of the right hand side is easily found to be 2. In all, we get:

$$3y^2 y' + 4x^2 y^3 y' + 2xy^4 = 2.$$

Move terms around so that the left side consists only of the y' terms and the right side consists of all the other terms:

$$3y^2 y' + 4x^2 y^3 y' = 2 - 2xy^4.$$

Notes:

Factor out y' from the left side and solve to get

$$y' = \frac{2 - 2xy^4}{3y^2 + 4x^2y^3}.$$

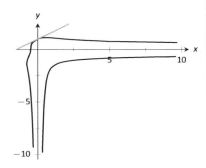

Figure 2.21: A graph of the implicitly defined function $y^3 + x^2y^4 = 1 + 2x$ along with its tangent line at the point $(0, 1)$.

To confirm the validity of our work, let's find the equation of a tangent line to this function at a point. It is easy to confirm that the point $(0, 1)$ lies on the graph of this function. At this point, $y' = 2/3$. So the equation of the tangent line is $y = 2/3(x - 0) + 1$. The function and its tangent line are graphed in Figure 2.21.

Notice how our function looks much different than other functions we have seen. For one, it fails the vertical line test. Such functions are important in many areas of mathematics, so developing tools to deal with them is also important.

Example 70   Using Implicit Differentiation

Given the implicitly defined function $\sin(x^2y^2) + y^3 = x + y$, find y'.

SOLUTION   Differentiating term by term, we find the most difficulty in the first term. It requires both the Chain and Product Rules.

$$\frac{d}{dx}\left(\sin(x^2y^2)\right) = \cos(x^2y^2) \cdot \frac{d}{dx}\left(x^2y^2\right)$$
$$= \cos(x^2y^2) \cdot \left(x^2(2yy') + 2xy^2\right)$$
$$= 2(x^2yy' + xy^2)\cos(x^2y^2).$$

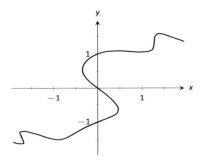

Figure 2.22: A graph of the implicitly defined function $\sin(x^2y^2) + y^3 = x + y$.

We leave the derivatives of the other terms to the reader. After taking the derivatives of both sides, we have

$$2(x^2yy' + xy^2)\cos(x^2y^2) + 3y^2y' = 1 + y'.$$

We now have to be careful to properly solve for y', particularly because of the product on the left. It is best to multiply out the product. Doing this, we get

$$2x^2y\cos(x^2y^2)y' + 2xy^2\cos(x^2y^2) + 3y^2y' = 1 + y'.$$

From here we can safely move around terms to get the following:

$$2x^2y\cos(x^2y^2)y' + 3y^2y' - y' = 1 - 2xy^2\cos(x^2y^2).$$

Then we can solve for y' to get

$$y' = \frac{1 - 2xy^2\cos(x^2y^2)}{2x^2y\cos(x^2y^2) + 3y^2 - 1}.$$

Notes:

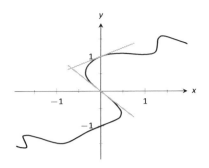

Figure 2.23: A graph of the implicitly defined function $\sin(x^2 y^2) + y^3 = x + y$ and certain tangent lines.

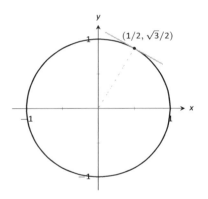

Figure 2.24: The unit circle with its tangent line at $(1/2, \sqrt{3}/2)$.

A graph of this implicit function is given in Figure 2.22. It is easy to verify that the points $(0,0)$, $(0,1)$ and $(0,-1)$ all lie on the graph. We can find the slopes of the tangent lines at each of these points using our formula for y'.

At $(0,0)$, the slope is -1.

At $(0,1)$, the slope is $1/2$.

At $(0,-1)$, the slope is also $1/2$.

The tangent lines have been added to the graph of the function in Figure 2.23.

Quite a few "famous" curves have equations that are given implicitly. We can use implicit differentiation to find the slope at various points on those curves. We investigate two such curves in the next examples.

Example 71 Finding slopes of tangent lines to a circle

Find the slope of the tangent line to the circle $x^2 + y^2 = 1$ at the point $(1/2, \sqrt{3}/2)$.

Solution Taking derivatives, we get $2x + 2yy' = 0$. Solving for y' gives:

$$y' = \frac{-x}{y}.$$

This is a clever formula. Recall that the slope of the line through the origin and the point (x, y) on the circle will be y/x. We have found that the slope of the tangent line to the circle at that point is the opposite reciprocal of y/x, namely, $-x/y$. Hence these two lines are always perpendicular.

At the point $(1/2, \sqrt{3}/2)$, we have the tangent line's slope as

$$y' = \frac{-1/2}{\sqrt{3}/2} = \frac{-1}{\sqrt{3}} \approx -0.577.$$

A graph of the circle and its tangent line at $(1/2, \sqrt{3}/2)$ is given in Figure 2.24, along with a thin dashed line from the origin that is perpendicular to the tangent line. (It turns out that all normal lines to a circle pass through the center of the circle.)

This section has shown how to find the derivatives of implicitly defined functions, whose graphs include a wide variety of interesting and unusual shapes. Implicit differentiation can also be used to further our understanding of "regular" differentiation.

One hole in our current understanding of derivatives is this: what is the derivative of the square root function? That is,

$$\frac{d}{dx}\left(\sqrt{x}\right) = \frac{d}{dx}\left(x^{1/2}\right) = ?$$

Notes:

We allude to a possible solution, as we can write the square root function as a power function with a rational (or, fractional) power. We are then tempted to apply the Power Rule and obtain

$$\frac{d}{dx}\left(x^{1/2}\right) = \frac{1}{2}x^{-1/2} = \frac{1}{2\sqrt{x}}.$$

The trouble with this is that the Power Rule was initially defined only for positive integer powers, $n > 0$. While we did not justify this at the time, generally the Power Rule is proved using something called the Binomial Theorem, which deals only with positive integers. The Quotient Rule allowed us to extend the Power Rule to negative integer powers. Implicit Differentiation allows us to extend the Power Rule to rational powers, as shown below.

Let $y = x^{m/n}$, where m and n are integers with no common factors (so $m = 2$ and $n = 5$ is fine, but $m = 2$ and $n = 4$ is not). We can rewrite this explicit function implicitly as $y^n = x^m$. Now apply implicit differentiation.

$$y = x^{m/n}$$
$$y^n = x^m$$
$$\frac{d}{dx}\left(y^n\right) = \frac{d}{dx}\left(x^m\right)$$
$$n \cdot y^{n-1} \cdot y' = m \cdot x^{m-1}$$
$$y' = \frac{m}{n}\frac{x^{m-1}}{y^{n-1}} \quad \text{(now substitute } x^{m/n} \text{ for } y\text{)}$$
$$= \frac{m}{n}\frac{x^{m-1}}{(x^{m/n})^{n-1}} \quad \text{(apply lots of algebra)}$$
$$= \frac{m}{n}x^{(m-n)/n}$$
$$= \frac{m}{n}x^{m/n-1}.$$

The above derivation is the key to the proof extending the Power Rule to rational powers. Using limits, we can extend this once more to include *all* powers, including irrational (even transcendental!) powers, giving the following theorem.

Theorem 21   Power Rule for Differentiation

Let $f(x) = x^n$, where $n \neq 0$ is a real number. Then f is a differentiable function, and $f'(x) = n \cdot x^{n-1}$.

Notes:

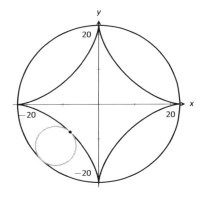

Figure 2.25: An astroid, traced out by a point on the smaller circle as it rolls inside the larger circle.

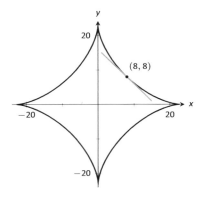

Figure 2.26: An astroid with a tangent line.

This theorem allows us to say the derivative of x^π is $\pi x^{\pi-1}$.

We now apply this final version of the Power Rule in the next example, the second investigation of a "famous" curve.

Example 72   **Using the Power Rule**
Find the slope of $x^{2/3} + y^{2/3} = 8$ at the point $(8, 8)$.

SOLUTION   This is a particularly interesting curve called an *astroid*. It is the shape traced out by a point on the edge of a circle that is rolling around inside of a larger circle, as shown in Figure 2.25.

To find the slope of the astroid at the point $(8, 8)$, we take the derivative implicitly.

$$\frac{2}{3}x^{-1/3} + \frac{2}{3}y^{-1/3}y' = 0$$

$$\frac{2}{3}y^{-1/3}y' = -\frac{2}{3}x^{-1/3}$$

$$y' = -\frac{x^{-1/3}}{y^{-1/3}}$$

$$y' = -\frac{y^{1/3}}{x^{1/3}} = -\sqrt[3]{\frac{y}{x}}.$$

Plugging in $x = 8$ and $y = 8$, we get a slope of -1. The astroid, with its tangent line at $(8, 8)$, is shown in Figure 2.26.

Implicit Differentiation and the Second Derivative

We can use implicit differentiation to find higher order derivatives. In theory, this is simple: first find $\frac{dy}{dx}$, then take its derivative with respect to x. In practice, it is not hard, but it often requires a bit of algebra. We demonstrate this in an example.

Example 73   **Finding the second derivative**
Given $x^2 + y^2 = 1$, find $\dfrac{d^2y}{dx^2} = y''$.

SOLUTION   We found that $y' = \frac{dy}{dx} = -x/y$ in Example 71. To find y'',

———————————————————————

Notes:

we apply implicit differentiation to y'.

$$y'' = \frac{d}{dx}(y')$$
$$= \frac{d}{dx}\left(-\frac{x}{y}\right) \quad \text{(Now use the Quotient Rule.)}$$
$$= -\frac{y(1) - x(y')}{y^2}$$

replace y' with $-x/y$:

$$= -\frac{y - x(-x/y)}{y^2}$$
$$= -\frac{y + x^2/y}{y^2}.$$

While this is not a particularly simple expression, it is usable. We can see that $y'' > 0$ when $y < 0$ and $y'' < 0$ when $y > 0$. In Section 3.4, we will see how this relates to the shape of the graph.

Logarithmic Differentiation

Consider the function $y = x^x$; it is graphed in Figure 2.27. It is well–defined for $x > 0$ and we might be interested in finding equations of lines tangent and normal to its graph. How do we take its derivative?

The function is not a power function: it has a "power" of x, not a constant. It is not an exponential function: it has a "base" of x, not a constant.

A differentiation technique known as *logarithmic differentiation* becomes useful here. The basic principle is this: take the natural log of both sides of an equation $y = f(x)$, then use implicit differentiation to find y'. We demonstrate this in the following example.

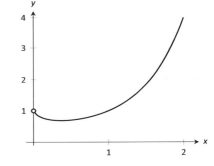

Figure 2.27: A plot of $y = x^x$.

Example 74   **Using Logarithmic Differentiation**
Given $y = x^x$, use logarithmic differentiation to find y'.

    SOLUTION   As suggested above, we start by taking the natural log of

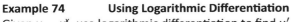

Notes:

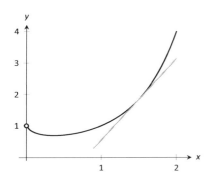

Figure 2.28: A graph of $y = x^x$ and its tangent line at $x = 1.5$.

both sides then applying implicit differentiation.

$$y = x^x$$
$$\ln(y) = \ln(x^x) \qquad \text{(apply logarithm rule)}$$
$$\ln(y) = x \ln x \qquad \text{(now use implicit differentiation)}$$
$$\frac{d}{dx}\Big(\ln(y)\Big) = \frac{d}{dx}\Big(x \ln x\Big)$$
$$\frac{y'}{y} = \ln x + x \cdot \frac{1}{x}$$
$$\frac{y'}{y} = \ln x + 1$$
$$y' = y\big(\ln x + 1\big) \quad \text{(substitute } y = x^x\text{)}$$
$$y' = x^x\big(\ln x + 1\big).$$

To "test" our answer, let's use it to find the equation of the tangent line at $x = 1.5$. The point on the graph our tangent line must pass through is $(1.5, 1.5^{1.5}) \approx (1.5, 1.837)$. Using the equation for y', we find the slope as

$$y' = 1.5^{1.5}\big(\ln 1.5 + 1\big) \approx 1.837(1.405) \approx 2.582.$$

Thus the equation of the tangent line is $y = 1.6833(x - 1.5) + 1.837$. Figure 2.25 graphs $y = x^x$ along with this tangent line.

Implicit differentiation proves to be useful as it allows us to find the instantaneous rates of change of a variety of functions. In particular, it extended the Power Rule to rational exponents, which we then extended to all real numbers. In the next section, implicit differentiation will be used to find the derivatives of *inverse* functions, such as $y = \sin^{-1} x$.

Notes:

Exercises 2.6

Terms and Concepts

1. In your own words, explain the difference between implicit functions and explicit functions.

2. Implicit differentiation is based on what other differentiation rule?

3. T/F: Implicit differentiation can be used to find the derivative of $y = \sqrt{x}$.

4. T/F: Implicit differentiation can be used to find the derivative of $y = x^{3/4}$.

Problems

In Exercises 5 – 12, compute the derivative of the given function.

5. $f(x) = \sqrt{x} + \dfrac{1}{\sqrt{x}}$

6. $f(x) = \sqrt[3]{x} + x^{2/3}$

7. $f(t) = \sqrt{1 - t^2}$

8. $g(t) = \sqrt{t}\sin t$

9. $h(x) = x^{1.5}$

10. $f(x) = x^{\pi} + x^{1.9} + \pi^{1.9}$

11. $g(x) = \dfrac{x + 7}{\sqrt{x}}$

12. $f(t) = \sqrt[5]{t}(\sec t + e^t)$

In Exercises 13 – 25, find $\dfrac{dy}{dx}$ using implicit differentiation.

13. $x^4 + y^2 + y = 7$

14. $x^{2/5} + y^{2/5} = 1$

15. $\cos(x) + \sin(y) = 1$

16. $\dfrac{x}{y} = 10$

17. $\dfrac{y}{x} = 10$

18. $x^2 e^2 + 2^y = 5$

19. $x^2 \tan y = 50$

20. $(3x^2 + 2y^3)^4 = 2$

21. $(y^2 + 2y - x)^2 = 200$

22. $\dfrac{x^2 + y}{x + y^2} = 17$

23. $\dfrac{\sin(x) + y}{\cos(y) + x} = 1$

24. $\ln(x^2 + y^2) = e$

25. $\ln(x^2 + xy + y^2) = 1$

26. Show that $\dfrac{dy}{dx}$ is the same for each of the following implicitly defined functions.

    (a) $xy = 1$

    (b) $x^2 y^2 = 1$

    (c) $\sin(xy) = 1$

    (d) $\ln(xy) = 1$

In Exercises 27 – 31, find the equation of the tangent line to the graph of the implicitly defined function at the indicated points. As a visual aid, each function is graphed.

27. $x^{2/5} + y^{2/5} = 1$

    (a) At $(1, 0)$.

    (b) At $(0.1, 0.281)$ (which does not *exactly* lie on the curve, but is very close).

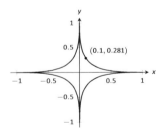

28. $x^4 + y^4 = 1$

    (a) At $(1, 0)$.

    (b) At $(\sqrt{0.6}, \sqrt{0.8})$.

    (c) At $(0, 1)$.

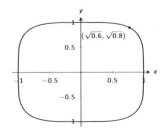

29. $(x^2 + y^2 - 4)^3 = 108y^2$

   (a) At $(0, 4)$.

   (b) At $(2, -\sqrt[4]{108})$.

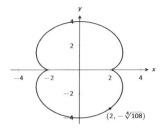

30. $(x^2 + y^2 + x)^2 = x^2 + y^2$

   (a) At $(0, 1)$.

   (b) At $\left(-\dfrac{3}{4}, \dfrac{3\sqrt{3}}{4}\right)$.

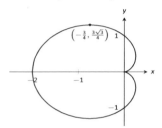

31. $(x - 2)^2 + (y - 3)^2 = 9$

   (a) At $\left(\dfrac{7}{2}, \dfrac{6 + 3\sqrt{3}}{2}\right)$.

   (b) At $\left(\dfrac{4 + 3\sqrt{3}}{2}, \dfrac{3}{2}\right)$.

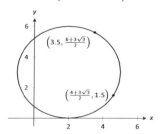

In Exercises 32 – 35, an implicitly defined function is given. Find $\dfrac{d^2y}{dx^2}$. Note: these are the same problems used in Exercises 13 through 16.

32. $x^4 + y^2 + y = 7$

33. $x^{2/5} + y^{2/5} = 1$

34. $\cos x + \sin y = 1$

35. $\dfrac{x}{y} = 10$

In Exercises 36 – 41, use logarithmic differentiation to find $\dfrac{dy}{dx}$, then find the equation of the tangent line at the indicated x–value.

36. $y = (1 + x)^{1/x}, \quad x = 1$

37. $y = (2x)^{x^2}, \quad x = 1$

38. $y = \dfrac{x^x}{x + 1}, \quad x = 1$

39. $y = x^{\sin(x)+2}, \quad x = \pi/2$

40. $y = \dfrac{x + 1}{x + 2}, \quad x = 1$

41. $y = \dfrac{(x + 1)(x + 2)}{(x + 3)(x + 4)}, \quad x = 0$

2.7   Derivatives of Inverse Functions

Recall that a function $y = f(x)$ is said to be *one to one* if it passes the horizontal line test; that is, for two different x values x_1 and x_2, we do *not* have $f(x_1) = f(x_2)$. In some cases the domain of f must be restricted so that it is one to one. For instance, consider $f(x) = x^2$. Clearly, $f(-1) = f(1)$, so f is not one to one on its regular domain, but by restricting f to $(0, \infty)$, f is one to one.

Now recall that one to one functions have *inverses*. That is, if f is one to one, it has an inverse function, denoted by f^{-1}, such that if $f(a) = b$, then $f^{-1}(b) = a$. The domain of f^{-1} is the range of f, and vice-versa. For ease of notation, we set $g = f^{-1}$ and treat g as a function of x.

Since $f(a) = b$ implies $g(b) = a$, when we compose f and g we get a nice result:

$$f\big(g(b)\big) = f(a) = b.$$

In general, $f\big(g(x)\big) = x$ and $g\big(f(x)\big) = x$. This gives us a convenient way to check if two functions are inverses of each other: compose them and if the result is x, then they are inverses (on the appropriate domains.)

When the point (a, b) lies on the graph of f, the point (b, a) lies on the graph of g. This leads us to discover that the graph of g is the reflection of f across the line $y = x$. In Figure 2.29 we see a function graphed along with its inverse. See how the point $(1, 1.5)$ lies on one graph, whereas $(1.5, 1)$ lies on the other. Because of this relationship, whatever we know about f can quickly be transferred into knowledge about g.

For example, consider Figure 2.30 where the tangent line to f at the point (a, b) is drawn. That line has slope $f'(a)$. Through reflection across $y = x$, we can see that the tangent line to g at the point (b, a) should have slope $\dfrac{1}{f'(a)}$.

This then tells us that $g'(b) = \dfrac{1}{f'(a)}$.

Consider:

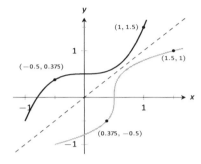

Figure 2.29: A function f along with its inverse f^{-1}. (Note how it does not matter which function we refer to as f; the other is f^{-1}.)

Information about f	Information about $g = f^{-1}$
$(-0.5, 0.375)$ lies on f	$(0.375, -0.5)$ lies on g
Slope of tangent line to f at $x = -0.5$ is $3/4$	Slope of tangent line to g at $x = 0.375$ is $4/3$
$f'(-0.5) = 3/4$	$g'(0.375) = 4/3$

We have discovered a relationship between f' and g' in a mostly graphical way. We can realize this relationship analytically as well. Let $y = g(x)$, where again $g = f^{-1}$. We want to find y'. Since $y = g(x)$, we know that $f(y) = x$. Using the Chain Rule and Implicit Differentiation, take the derivative of both sides of

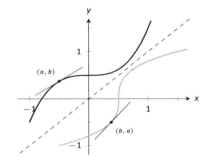

Figure 2.30: Corresponding tangent lines drawn to f and f^{-1}.

Notes:

this last equality.

$$\frac{d}{dx}\Big(f(y)\Big) = \frac{d}{dx}\Big(x\Big)$$
$$f'(y)\cdot y' = 1$$
$$y' = \frac{1}{f'(y)}$$
$$y' = \frac{1}{f'(g(x))}$$

This leads us to the following theorem.

Theorem 22   Derivatives of Inverse Functions

Let f be differentiable and one to one on an open interval I, where $f'(x) \neq 0$ for all x in I, let J be the range of f on I, let g be the inverse function of f, and let $f(a) = b$ for some a in I. Then g is a differentiable function on J, and in particular,

1. $\left(f^{-1}\right)'(b) = g'(b) = \dfrac{1}{f'(a)}$ and   2. $\left(f^{-1}\right)'(x) = g'(x) = \dfrac{1}{f'(g(x))}$

The results of Theorem 22 are not trivial; the notation may seem confusing at first. Careful consideration, along with examples, should earn understanding.

In the next example we apply Theorem 22 to the arcsine function.

Example 75   Finding the derivative of an inverse trigonometric function
Let $y = \arcsin x = \sin^{-1} x$. Find y' using Theorem 22.

SOLUTION   Adopting our previously defined notation, let $g(x) = \arcsin x$ and $f(x) = \sin x$. Thus $f'(x) = \cos x$. Applying the theorem, we have

$$g'(x) = \frac{1}{f'(g(x))}$$
$$= \frac{1}{\cos(\arcsin x)}.$$

This last expression is not immediately illuminating. Drawing a figure will help, as shown in Figure 2.32. Recall that the sine function can be viewed as taking in an angle and returning a ratio of sides of a right triangle, specifically, the ratio "opposite over hypotenuse." This means that the arcsine function takes as input a ratio of sides and returns an angle. The equation $y = \arcsin x$ can be rewritten as $y = \arcsin(x/1)$; that is, consider a right triangle where the

---

Notes:

hypotenuse has length 1 and the side opposite of the angle with measure y has length x. This means the final side has length $\sqrt{1-x^2}$, using the Pythagorean Theorem.

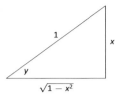

Therefore $\cos(\sin^{-1}x) = \cos y = \sqrt{1-x^2}/1 = \sqrt{1-x^2}$, resulting in

$$\frac{d}{dx}\left(\arcsin x\right) = g'(x) = \frac{1}{\sqrt{1-x^2}}.$$

Figure 2.32: A right triangle defined by $y = \sin^{-1}(x/1)$ with the length of the third leg found using the Pythagorean Theorem.

Remember that the input x of the arcsine function is a ratio of a side of a right triangle to its hypotenuse; the absolute value of this ratio will never be greater than 1. Therefore the inside of the square root will never be negative.

In order to make $y = \sin x$ one to one, we restrict its domain to $[-\pi/2, \pi/2]$; on this domain, the range is $[-1, 1]$. Therefore the domain of $y = \arcsin x$ is $[-1, 1]$ and the range is $[-\pi/2, \pi/2]$. When $x = \pm 1$, note how the derivative of the arcsine function is undefined; this corresponds to the fact that as $x \to \pm 1$, the tangent lines to arcsine approach vertical lines with undefined slopes.

In Figure 2.33 we see $f(x) = \sin x$ and $f^{-1} = \sin^{-1} x$ graphed on their respective domains. The line tangent to $\sin x$ at the point $(\pi/3, \sqrt{3}/2)$ has slope $\cos \pi/3 = 1/2$. The slope of the corresponding point on $\sin^{-1} x$, the point $(\sqrt{3}/2, \pi/3)$, is

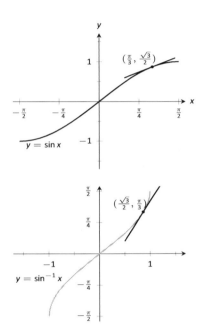

$$\frac{1}{\sqrt{1-(\sqrt{3}/2)^2}} = \frac{1}{\sqrt{1-3/4}} = \frac{1}{\sqrt{1/4}} = \frac{1}{1/2} = 2,$$

verifying yet again that at corresponding points, a function and its inverse have reciprocal slopes.

Figure 2.33: Graphs of $\sin x$ and $\sin^{-1} x$ along with corresponding tangent lines.

Using similar techniques, we can find the derivatives of all the inverse trigonometric functions. In Figure 2.31 we show the restrictions of the domains of the standard trigonometric functions that allow them to be invertible.

Notes:

Function	Domain	Range	Inverse Function	Domain	Range
$\sin x$	$[-\pi/2, \pi/2]$	$[-1, 1]$	$\sin^{-1} x$	$[-1, 1]$	$[-\pi/2, \pi/2]$
$\cos x$	$[0, \pi]$	$[-1, 1]$	$\cos^{-1}(x)$	$[-1, 1]$	$[0, \pi]$
$\tan x$	$(-\pi/2, \pi/2)$	$(-\infty, \infty)$	$\tan^{-1}(x)$	$(-\infty, \infty)$	$(-\pi/2, \pi/2)$
$\csc x$	$[-\pi/2, 0) \cup (0, \pi/2]$	$(-\infty, -1] \cup [1, \infty)$	$\csc^{-1} x$	$(-\infty, -1] \cup [1, \infty)$	$[-\pi/2, 0) \cup (0, \pi/2]$
$\sec x$	$[0, \pi/2) \cup (\pi/2, \pi]$	$(-\infty, -1] \cup [1, \infty)$	$\sec^{-1}(x)$	$(-\infty, -1] \cup [1, \infty)$	$[0, \pi/2) \cup (\pi/2, \pi]$
$\cot x$	$(0, \pi)$	$(-\infty, \infty)$	$\cot^{-1}(x)$	$(-\infty, \infty)$	$(0, \pi)$

Figure 2.31: Domains and ranges of the trigonometric and inverse trigonometric functions.

Theorem 23   Derivatives of Inverse Trigonometric Functions

The inverse trigonometric functions are differentiable on all open sets contained in their domains (as listed in Figure 2.31) and their derivatives are as follows:

1. $\dfrac{d}{dx}\left(\sin^{-1}(x)\right) = \dfrac{1}{\sqrt{1 - x^2}}$

4. $\dfrac{d}{dx}\left(\cos^{-1}(x)\right) = -\dfrac{1}{\sqrt{1 - x^2}}$

2. $\dfrac{d}{dx}\left(\sec^{-1}(x)\right) = \dfrac{1}{|x|\sqrt{x^2 - 1}}$

5. $\dfrac{d}{dx}\left(\csc^{-1}(x)\right) = -\dfrac{1}{|x|\sqrt{x^2 - 1}}$

3. $\dfrac{d}{dx}\left(\tan^{-1}(x)\right) = \dfrac{1}{1 + x^2}$

6. $\dfrac{d}{dx}\left(\cot^{-1}(x)\right) = -\dfrac{1}{1 + x^2}$

Note how the last three derivatives are merely the opposites of the first three, respectively. Because of this, the first three are used almost exclusively throughout this text.

In Section 2.3, we stated without proof or explanation that $\dfrac{d}{dx}(\ln x) = \dfrac{1}{x}$. We can justify that now using Theorem 22, as shown in the example.

Example 76   Finding the derivative of $y = \ln x$

Use Theorem 22 to compute $\dfrac{d}{dx}(\ln x)$.

SOLUTION   View $y = \ln x$ as the inverse of $y = e^x$. Therefore, using our standard notation, let $f(x) = e^x$ and $g(x) = \ln x$. We wish to find $g'(x)$. Theorem

---

Notes:

22 gives:

$$g'(x) = \frac{1}{f'(g(x))}$$

$$= \frac{1}{e^{\ln x}}$$

$$= \frac{1}{x}.$$

In this chapter we have defined the derivative, given rules to facilitate its computation, and given the derivatives of a number of standard functions. We restate the most important of these in the following theorem, intended to be a reference for further work.

Theorem 24   Glossary of Derivatives of Elementary Functions

Let u and v be differentiable functions, and let a, c and n be real numbers, $a > 0, n \neq 0$.

1. $\frac{d}{dx}(cu) = cu'$

2. $\frac{d}{dx}(u \pm v) = u' \pm v'$

3. $\frac{d}{dx}(u \cdot v) = uv' + u'v$

4. $\frac{d}{dx}\left(\frac{u}{v}\right) = \frac{u'v - uv'}{v^2}$

5. $\frac{d}{dx}(u(v)) = u'(v)v'$

6. $\frac{d}{dx}(c) = 0$

7. $\frac{d}{dx}(x) = 1$

8. $\frac{d}{dx}(x^n) = nx^{n-1}$

9. $\frac{d}{dx}(e^x) = e^x$

10. $\frac{d}{dx}(a^x) = \ln a \cdot a^x$

11. $\frac{d}{dx}(\ln x) = \frac{1}{x}$

12. $\frac{d}{dx}(\log_a x) = \frac{1}{\ln a} \cdot \frac{1}{x}$

13. $\frac{d}{dx}(\sin x) = \cos x$

14. $\frac{d}{dx}(\cos x) = -\sin x$

15. $\frac{d}{dx}(\csc x) = -\csc x \cot x$

16. $\frac{d}{dx}(\sec x) = \sec x \tan x$

17. $\frac{d}{dx}(\tan x) = \sec^2 x$

18. $\frac{d}{dx}(\cot x) = -\csc^2 x$

19. $\frac{d}{dx}(\sin^{-1} x) = \frac{1}{\sqrt{1-x^2}}$

20. $\frac{d}{dx}(\cos^{-1} x) = -\frac{1}{\sqrt{1-x^2}}$

21. $\frac{d}{dx}(\csc^{-1} x) = -\frac{1}{|x|\sqrt{x^2-1}}$

22. $\frac{d}{dx}(\sec^{-1} x) = \frac{1}{|x|\sqrt{x^2-1}}$

23. $\frac{d}{dx}(\tan^{-1} x) = \frac{1}{1+x^2}$

24. $\frac{d}{dx}(\cot^{-1} x) = -\frac{1}{1+x^2}$

Notes:

Exercises 2.7

Terms and Concepts

1. T/F: Every function has an inverse.

2. In your own words explain what it means for a function to be "one to one."

3. If $(1, 10)$ lies on the graph of $y = f(x)$, what can be said about the graph of $y = f^{-1}(x)$?

4. If $(1, 10)$ lies on the graph of $y = f(x)$ and $f'(1) = 5$, what can be said about $y = f^{-1}(x)$?

Problems

In Exercises 5 – 8, verify that the given functions are inverses.

5. $f(x) = 2x + 6$ and $g(x) = \frac{1}{2}x - 3$

6. $f(x) = x^2 + 6x + 11$, $x \geq 3$ and
   $g(x) = \sqrt{x - 2} - 3$, $x \geq 2$

7. $f(x) = \dfrac{3}{x - 5}$, $x \neq 5$ and
   $g(x) = \dfrac{3 + 5x}{x}$, $x \neq 0$

8. $f(x) = \dfrac{x + 1}{x - 1}$, $x \neq 1$ and $g(x) = f(x)$

In Exercises 9 – 14, an invertible function $f(x)$ is given along with a point that lies on its graph. Using Theorem 22, evaluate $\left(f^{-1}\right)'(x)$ at the indicated value.

9. $f(x) = 5x + 10$
   Point= $(2, 20)$
   Evaluate $\left(f^{-1}\right)'(20)$

10. $f(x) = x^2 - 2x + 4$, $x \geq 1$
    Point= $(3, 7)$
    Evaluate $\left(f^{-1}\right)'(7)$

11. $f(x) = \sin 2x$, $-\pi/4 \leq x \leq \pi/4$
    Point= $(\pi/6, \sqrt{3}/2)$
    Evaluate $\left(f^{-1}\right)'(\sqrt{3}/2)$

12. $f(x) = x^3 - 6x^2 + 15x - 2$
    Point= $(1, 8)$
    Evaluate $\left(f^{-1}\right)'(8)$

13. $f(x) = \dfrac{1}{1 + x^2}$, $x \geq 0$
    Point= $(1, 1/2)$
    Evaluate $\left(f^{-1}\right)'(1/2)$

14. $f(x) = 6e^{3x}$
    Point= $(0, 6)$
    Evaluate $\left(f^{-1}\right)'(6)$

In Exercises 15 – 24, compute the derivative of the given function.

15. $h(t) = \sin^{-1}(2t)$

16. $f(t) = \sec^{-1}(2t)$

17. $g(x) = \tan^{-1}(2x)$

18. $f(x) = x \sin^{-1} x$

19. $g(t) = \sin t \cos^{-1} t$

20. $f(t) = \ln t e^t$

21. $h(x) = \dfrac{\sin^{-1} x}{\cos^{-1} x}$

22. $g(x) = \tan^{-1}(\sqrt{x})$

23. $f(x) = \sec^{-1}(1/x)$

24. $f(x) = \sin(\sin^{-1} x)$

In Exercises 25 – 27, compute the derivative of the given function in two ways:

   (a) By simplifying first, then taking the derivative, and

   (b) by using the Chain Rule first then simplifying.

Verify that the two answers are the same.

25. $f(x) = \sin(\sin^{-1} x)$

26. $f(x) = \tan^{-1}(\tan x)$

27. $f(x) = \sin(\cos^{-1} x)$

In Exercises 28 – 29, find the equation of the line tangent to the graph of f at the indicated x value.

28. $f(x) = \sin^{-1} x$ at $x = \frac{\sqrt{2}}{2}$

29. $f(x) = \cos^{-1}(2x)$ at $x = \frac{\sqrt{3}}{4}$

Review

30. Find $\frac{dy}{dx}$, where $x^2 y - y^2 x = 1$.

31. Find the equation of the line tangent to the graph of $x^2 + y^2 + xy = 7$ at the point $(1, 2)$.

32. Let $f(x) = x^3 + x$.
    Evaluate $\lim\limits_{s \to 0} \dfrac{f(x + s) - f(x)}{s}$.

3: THE GRAPHICAL BEHAVIOR OF FUNCTIONS

Our study of limits led to continuous functions, which is a certain class of functions that behave in a particularly nice way. Limits then gave us an even nicer class of functions, functions that are differentiable.

This chapter explores many of the ways we can take advantage of the information that continuous and differentiable functions provide.

3.1 Extreme Values

Given any quantity described by a function, we are often interested in the largest and/or smallest values that quantity attains. For instance, if a function describes the speed of an object, it seems reasonable to want to know the fastest/slowest the object traveled. If a function describes the value of a stock, we might want to know how the highest/lowest values the stock attained over the past year. We call such values *extreme values*.

Definition 12   Extreme Values

Let f be defined on an interval I containing c.

1. $f(c)$ is the **minimum** (also, **absolute minimum**) of f on I if $f(c) \leq f(x)$ for all x in I.

2. $f(c)$ is the **maximum** (also, **absolute maximum**) of f on I if $f(c) \geq f(x)$ for all x in I.

The maximum and minimum values are the **extreme values**, or **extrema**, of f on I.

Consider Figure 3.1. The function displayed in (a) has a maximum, but no minimum, as the interval over which the function is defined is open. In (b), the function has a minimum, but no maximum; there is a discontinuity in the "natural" place for the maximum to occur. Finally, the function shown in (c) has both a maximum and a minimum; note that the function is continuous and the interval on which it is defined is closed.

It is possible for discontinuous functions defined on an open interval to have both a maximum and minimum value, but we have just seen examples where they did not. On the other hand, continuous functions on a closed interval *always* have a maximum and minimum value.

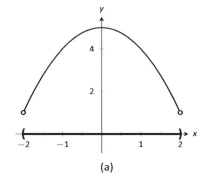

(a)

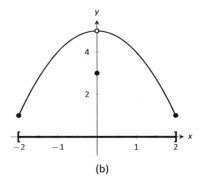

(b)

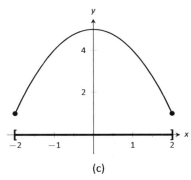

(c)

Figure 3.1: Graphs of functions with and without extreme values.

Note: The extreme values of a function are "y" values, values the function attains, not the input values.

Theorem 25   The Extreme Value Theorem

Let f be a continuous function defined on a closed interval I. Then f has both a maximum and minimum value on I.

This theorem states that f has extreme values, but it does not offer any advice about how/where to find these values. The process can seem to be fairly easy, as the next example illustrates. After the example, we will draw on lessons learned to form a more general and powerful method for finding extreme values.

Example 77   Approximating extreme values
Consider $f(x) = 2x^3 - 9x^2$ on $I = [-1, 5]$, as graphed in Figure 3.2. Approximate the extreme values of f.

SOLUTION   The graph is drawn in such a way to draw attention to certain points. It certainly seems that the smallest y value is -27, found when $x = 3$. It also seems that the largest y value is 25, found at the endpoint of I, $x = 5$. We use the word *seems*, for by the graph alone we cannot be sure the smallest value is not less than -27. Since the problem asks for an approximation, we approximate the extreme values to be 25 and -27.

Notice how the minimum value came at "the bottom of a hill," and the maximum value came at an endpoint. Also note that while 0 is not an extreme value, it would be if we narrowed our interval to $[-1, 4]$. The idea that the point $(0, 0)$ is the location of an extreme value for some interval is important, leading us to a definition.

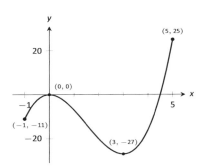

Figure 3.2: A graph of $f(x) = 2x^3 - 9x^2$ as in Example 77.

Definition 13   Relative Minimum and Relative Maximum

Let f be defined on an interval I containing c.

1. If there is an open interval containing c such that $f(c)$ is the minimum value, then $f(c)$ is a **relative minimum** of f. We also say that f has a relative minimum at $(c, f(c))$.

2. If there is an open interval containing c such that $f(c)$ is the maximum value, then $f(c)$ is a **relative maximum** of f. We also say that f has a relative maximum at $(c, f(c))$.

The relative maximum and minimum values comprise the **relative extrema** of f.

Note: The terms *local minimum* and *local maximum* are often used as synonyms for *relative minimum* and *relative maximum*.

Notes:

We briefly practice using these definitions.

Example 78 **Approximating relative extrema**

Consider $f(x) = (3x^4 - 4x^3 - 12x^2 + 5)/5$, as shown in Figure 3.3. Approximate the relative extrema of f. At each of these points, evaluate f'.

SOLUTION We still do not have the tools to exactly find the relative extrema, but the graph does allow us to make reasonable approximations. It seems f has relative minima at $x = -1$ and $x = 2$, with values of $f(-1) = 0$ and $f(2) = -5.4$. It also seems that f has a relative maximum at the point $(0, 1)$.

We approximate the relative minima to be 0 and -5.4; we approximate the relative maximum to be 1.

It is straightforward to evaluate $f'(x) = \frac{1}{5}(12x^3 - 12x^2 - 24x)$ at $x = 0, 1$ and 2. In each case, $f'(x) = 0$.

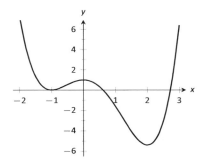

Figure 3.3: A graph of $f(x) = (3x^4 - 4x^3 - 12x^2 + 5)/5$ as in Example 78.

Example 79 **Approximating relative extrema**

Approximate the relative extrema of $f(x) = (x-1)^{2/3} + 2$, shown in Figure 3.4. At each of these points, evaluate f'.

SOLUTION The figure implies that f does not have any relative maxima, but has a relative minimum at $(1, 2)$. In fact, the graph suggests that not only is this point a relative minimum, $y = f(1) = 2$ *the* minimum value of the function.

We compute $f'(x) = \frac{2}{3}(x-1)^{-1/3}$. When $x = 1$, f' is undefined.

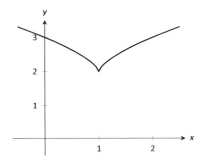

Figure 3.4: A graph of $f(x) = (x-1)^{2/3} + 2$ as in Example 79.

What can we learn from the previous two examples? We were able to visually approximate relative extrema, and at each such point, the derivative was either 0 or it was not defined. This observation holds for all functions, leading to a definition and a theorem.

Definition 14 **Critical Numbers and Critical Points**

Let f be defined at c. The value c is a **critical number** (or **critical value**) of f if $f'(c) = 0$ or $f'(c)$ is not defined.

If c is a critical number of f, then the point $(c, f(c))$ is a **critical point** of f.

Notes:

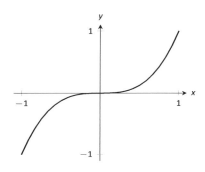

Figure 3.5: A graph of $f(x) = x^3$ which has a critical value of $x = 0$, but no relative extrema.

Be careful to understand that this theorem states "All relative extrema occur at critical points." It does not say "All critical numbers produce relative extrema." For instance, consider $f(x) = x^3$. Since $f'(x) = 3x^2$, it is straightforward to determine that $x = 0$ is a critical number of f. However, f has no relative extrema, as illustrated in Figure 3.5.

Theorem 25 states that a continuous function on a closed interval will have absolute extrema, that is, both an absolute maximum and an absolute minimum. These extrema occur either at the endpoints or at critical values in the interval. We combine these concepts to offer a strategy for finding extrema.

We practice these ideas in the next examples.

Example 80   Finding extreme values
Find the extreme values of $f(x) = 2x^3 + 3x^2 - 12x$ on $[0, 3]$, graphed in Figure 3.6.

SOLUTION   We follow the steps outlined in Key Idea 2. We first evaluate f at the endpoints:
$$f(0) = 0 \quad \text{and} \quad f(3) = 45.$$
Next, we find the critical values of f on $[0, 3]$. $f'(x) = 6x^2 + 6x - 12 = 6(x + 2)(x - 1)$; therefore the critical values of f are $x = -2$ and $x = 1$. Since $x = -2$

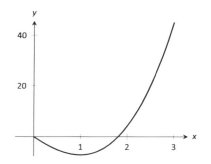

Figure 3.6: A graph of $f(x) = 2x^3 + 3x^2 - 12x$ on $[0, 3]$ as in Example 80.

Notes:

does not lie in the interval $[0, 3]$, we ignore it. Evaluating f at the only critical number in our interval gives: $f(1) = -7$.

The table in Figure 3.7 gives f evaluated at the "important" x values in $[0, 3]$. We can easily see the maximum and minimum values of f: the maximum value is 45 and the minimum value is -7.

x	$f(x)$
0	0
1	-7
3	45

Figure 3.7: Finding the extreme values of f in Example 80.

Note that all this was done without the aid of a graph; this work followed an analytic algorithm and did not depend on any visualization. Figure 3.6 shows f and we can confirm our answer, but it is important to understand that these answers can be found without graphical assistance.

We practice again.

Example 81   Finding extreme values
Find the maximum and minimum values of f on $[-4, 2]$, where

$$f(x) = \begin{cases} (x-1)^2 & x \le 0 \\ x+1 & x > 0 \end{cases}.$$

SOLUTION   Here f is piecewise–defined, but we can still apply Key Idea 2. Evaluating f at the endpoints gives:

$$f(-4) = 25 \quad \text{and} \quad f(2) = 3.$$

x	$f(x)$
-4	25
0	1
2	3

Figure 3.8: Finding the extreme values of f in Example 81.

We now find the critical numbers of f. We have to define f' in a piecewise manner; it is

$$f'(x) = \begin{cases} 2(x-1) & x < 0 \\ 1 & x > 0 \end{cases}.$$

Note that while f is defined for all of $[-4, 2]$, f' is not, as the derivative of f does not exist when $x = 0$. (From the left, the derivative approaches -2; from the right the derivative is 1.) Thus one critical number of f is $x = 0$.

We now set $f'(x) = 0$. When $x > 0$, $f'(x)$ is never 0. When $x < 0$, $f'(x)$ is also never 0. (We may be tempted to say that $f'(x) = 0$ when $x = 1$. However, this is nonsensical, for we only consider $f'(x) = 2(x-1)$ when $x < 0$, so we will ignore a solution that says $x = 1$.)

So we have three important x values to consider: $x = -4, 2$ and 0. Evaluating f at each gives, respectively, 25, 3 and 1, shown in Figure 3.8. Thus the absolute minimum of f is 1; the absolute maximum of f is 25. Our answer is confirmed by the graph of f in Figure 3.9.

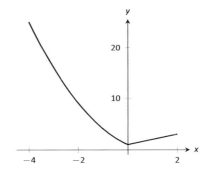

Figure 3.9: A graph of $f(x)$ on $[-4, 2]$ as in Example 81.

Notes:

x	$f(x)$
-2	-0.65
$-\sqrt{\pi}$	-1
0	1
$\sqrt{\pi}$	-1
2	-0.65

Figure 3.10: Finding the extrema of $f(x) = \cos(x^2)$ in Example 82.

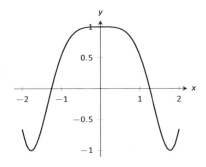

Figure 3.11: A graph of $f(x) = \cos(x^2)$ on $[-2, 2]$ as in Example 82.

x	$f(x)$
-1	0
0	1
1	0

Figure 3.13: Finding the extrema of the half–circle in Example 83.

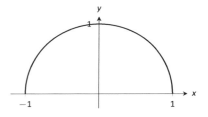

Figure 3.12: A graph of $f(x) = \sqrt{1 - x^2}$ on $[-1, 1]$ as in Example 83.

Note: We implicitly found the derivative of $x^2 + y^2 = 1$, the unit circle, in Example 71 as $\frac{dy}{dx} = -x/y$. In Example 83, half of the unit circle is given as $y = f(x) = \sqrt{1 - x^2}$. We found $f'(x) = \frac{-x}{\sqrt{1-x^2}}$. Recognize that the denominator of this fraction is y; that is, we again found $f'(x) = \frac{dy}{dx} = -x/y$.

Example 82   Finding extreme values
Find the extrema of $f(x) = \cos(x^2)$ on $[-2, 2]$.

SOLUTION   We again use Key Idea 2. Evaluating f at the endpoints of the interval gives: $f(-2) = f(2) = \cos(4) \approx -0.6536$. We now find the critical values of f.

Applying the Chain Rule, we find $f'(x) = -2x\sin(x^2)$. Set $f'(x) = 0$ and solve for x to find the critical values of f.

We have $f'(x) = 0$ when $x = 0$ and when $\sin(x^2) = 0$. In general, $\sin t = 0$ when $t = \ldots - 2\pi, -\pi, 0, \pi, \ldots$ Thus $\sin(x^2) = 0$ when $x^2 = 0, \pi, 2\pi, \ldots$ (x^2 is always positive so we ignore $-\pi$, etc.) So $\sin(x^2) = 0$ when $x = 0, \pm\sqrt{\pi}, \pm\sqrt{2\pi}, \ldots$. The only values to fall in the given interval of $[-2, 2]$ are $-\sqrt{\pi}$ and $\sqrt{\pi}$, approximately ± 1.77.

We again construct a table of important values in Figure 3.10. In this example we have 5 values to consider: $x = 0, \pm 2, \pm\sqrt{\pi}$.

From the table it is clear that the maximum value of f on $[-2, 2]$ is 1; the minimum value is -1. The graph in Figure 3.11 confirms our results.

We consider one more example.

Example 83   Finding extreme values
Find the extreme values of $f(x) = \sqrt{1 - x^2}$.

SOLUTION   A closed interval is not given, so we find the extreme values of f on its domain. f is defined whenever $1 - x^2 \geq 0$; thus the domain of f is $[-1, 1]$. Evaluating f at either endpoint returns 0.

Using the Chain Rule, we find $f'(x) = \frac{-x}{\sqrt{1 - x^2}}$. The critical points of f are found when $f'(x) = 0$ or when f' is undefined. It is straightforward to find that $f'(x) = 0$ when $x = 0$, and f' is undefined when $x = \pm 1$, the endpoints of the interval. The table of important values is given in Figure 3.13. The maximum value is 1, and the minimum value is 0.

We have seen that continuous functions on closed intervals always have a maximum and minimum value, and we have also developed a technique to find these values. In the next section, we further our study of the information we can glean from "nice" functions with the Mean Value Theorem. On a closed interval, we can find the *average rate of change* of a function (as we did at the beginning of Chapter 2). We will see that differentiable functions always have a point at which their *instantaneous* rate of change is same as the *average* rate of change. This is surprisingly useful, as we'll see.

Notes:

Exercises 3.1

Terms and Concepts

1. Describe what an "extreme value" of a function is in your own words.

2. Sketch the graph of a function f on $(-1, 1)$ that has both a maximum and minimum value.

3. Describe the difference between absolute and relative maxima in your own words.

4. Sketch the graph of a function f where f has a relative maximum at $x = 1$ and $f'(1)$ is undefined.

5. T/F: If c is a critical value of a function f, then f has either a relative maximum or relative minimum at $x = c$.

Problems

In Exercises 6 – 7, identify each of the marked points as being an absolute maximum or minimum, a relative maximum or minimum, or none of the above.

6.
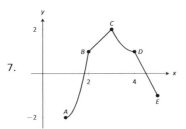

7.
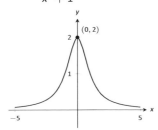

In Exercises 8 – 14, evaluate $f'(x)$ at the points indicated in the graph.

8. $f(x) = \dfrac{2}{x^2 + 1}$

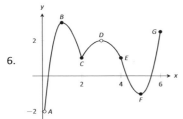

9. $f(x) = x^2\sqrt{6 - x^2}$

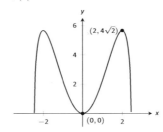

10. $f(x) = \sin x$

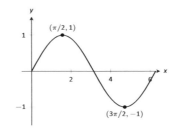

11. $f(x) = x^2\sqrt{4 - x}$

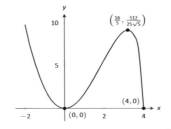

12. $f(x) = \begin{cases} x^2 & x \le 0 \\ x^5 & x > 0 \end{cases}$

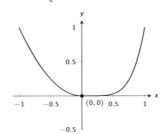

13. $f(x) = \begin{cases} x^2 & x \le 0 \\ x & x > 0 \end{cases}$

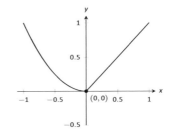

14. $f(x) = \dfrac{(x-2)^{2/3}}{x}$

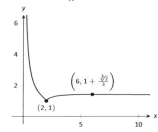

In Exercises 15 – 24, find the extreme values of the function on the given interval.

15. $f(x) = x^2 + x + 4$ on $[-1, 2]$.

16. $f(x) = x^3 - \dfrac{9}{2}x^2 - 30x + 3$ on $[0, 6]$.

17. $f(x) = 3\sin x$ on $[\pi/4, 2\pi/3]$.

18. $f(x) = x^2\sqrt{4 - x^2}$ on $[-2, 2]$.

19. $f(x) = x + \dfrac{3}{x}$ on $[1, 5]$.

20. $f(x) = \dfrac{x^2}{x^2 + 5}$ on $[-3, 5]$.

21. $f(x) = e^x \cos x$ on $[0, \pi]$.

22. $f(x) = e^x \sin x$ on $[0, \pi]$.

23. $f(x) = \dfrac{\ln x}{x}$ on $[1, 4]$.

24. $f(x) = x^{2/3} - x$ on $[0, 2]$.

Review

25. Find $\frac{dy}{dx}$, where $x^2 y - y^2 x = 1$.

26. Find the equation of the line tangent to the graph of $x^2 + y^2 + xy = 7$ at the point $(1, 2)$.

27. Let $f(x) = x^3 + x$.

    Evaluate $\lim\limits_{s \to 0} \dfrac{f(x + s) - f(x)}{s}$.

3.2   The Mean Value Theorem

We motivate this section with the following question: Suppose you leave your house and drive to your friend's house in a city 100 miles away, completing the trip in two hours. At any point during the trip do you necessarily have to be going 50 miles per hour?

In answering this question, it is clear that the *average* speed for the entire trip is 50 mph (i.e. 100 miles in 2 hours), but the question is whether or not your *instantaneous* speed is ever exactly 50 mph. More simply, does your speedometer ever read exactly 50 mph?. The answer, under some very reasonable assumptions, is "yes."

Let's now see why this situation is in a calculus text by translating it into mathematical symbols.

First assume that the function $y = f(t)$ gives the distance (in miles) traveled from your home at time t (in hours) where $0 \leq t \leq 2$. In particular, this gives $f(0) = 0$ and $f(2) = 100$. The slope of the secant line connecting the starting and ending points $(0, f(0))$ and $(2, f(2))$ is therefore

$$\frac{\Delta f}{\Delta t} = \frac{f(2) - f(0)}{2 - 0} = \frac{100 - 0}{2} = 50 \, \text{mph}.$$

The slope at any point on the graph itself is given by the derivative $f'(t)$. So, since the answer to the question above is "yes," this means that at some time during the trip, the derivative takes on the value of 50 mph. Symbolically,

$$f'(c) = \frac{f(2) - f(0)}{2 - 0} = 50$$

for some time $0 \leq c \leq 2$.

How about more generally? Given any function $y = f(x)$ and a range $a \leq x \leq b$ does the value of the derivative at some point between a and b have to match the slope of the secant line connecting the points $(a, f(a))$ and $(b, f(b))$? Or equivalently, does the equation $f'(c) = \frac{f(b) - f(a)}{b - a}$ have to hold for some $a < c < b$?

Let's look at two functions in an example.

Example 84   **Comparing average and instantaneous rates of change**
Consider functions

$$f_1(x) = \frac{1}{x^2} \quad \text{and} \quad f_2(x) = |x|$$

with $a = -1$ and $b = 1$ as shown in Figure 3.14(a) and (b), respectively. Both functions have a value of 1 at a and b. Therefore the slope of the secant line

---

Notes:

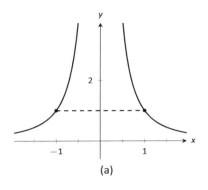

(a)

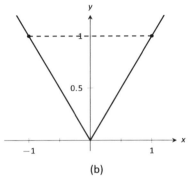

(b)

Figure 3.14: A graph of $f_1(x) = 1/x^2$ and $f_2(x) = |x|$ in Example 84.

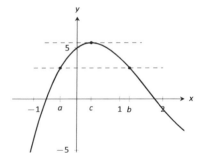

Figure 3.15: A graph of $f(x) = x^3 - 5x^2 + 3x + 5$, where $f(a) = f(b)$. Note the existence of c, where $a < c < b$, where $f'(c) = 0$.

connecting the end points is 0 in each case. But if you look at the plots of each, you can see that there are no points on either graph where the tangent lines have slope zero. Therefore we have found that there is no c in $[-1, 1]$ such that

$$f'(c) = \frac{f(1) - f(-1)}{1 - (-1)} = 0.$$

So what went "wrong"? It may not be surprising to find that the discontinuity of f_1 and the corner of f_2 play a role. If our functions had been continuous and differentiable, would we have been able to find that special value c? This is our motivation for the following theorem.

Theorem 27   The Mean Value Theorem of Differentiation

Let $y = f(x)$ be continuous function on the closed interval $[a, b]$ and differentiable on the open interval (a, b). There exists a value c, $a < c < b$, such that

$$f'(c) = \frac{f(b) - f(a)}{b - a}.$$

That is, there is a value c in (a, b) where the instantaneous rate of change of f at c is equal to the average rate of change of f on $[a, b]$.

Note that the reasons that the functions in Example 84 fail are indeed that f_1 has a discontinuity on the interval $[-1, 1]$ and f_2 is not differentiable at the origin.

We will give a proof of the Mean Value Theorem below. To do so, we use a fact, called Rolle's Theorem, stated here.

Theorem 28   Rolle's Theorem

Let f be continuous on $[a, b]$ and differentiable on (a, b), where $f(a) = f(b)$. There is some c in (a, b) such that $f'(c) = 0$.

Consider Figure 3.15 where the graph of a function f is given, where $f(a) = f(b)$. It should make intuitive sense that if f is differentiable (and hence, continuous) that there would be a value c in (a, b) where $f'(c) = 0$; that is, there would be a relative maximum or minimum of f in (a, b). Rolle's Theorem guarantees at least one; there may be more.

Notes:

Rolle's Theorem is really just a special case of the Mean Value Theorem. If $f(a) = f(b)$, then the *average* rate of change on (a, b) is 0, and the theorem guarantees some c where $f'(c) = 0$. We will prove Rolle's Theorem, then use it to prove the Mean Value Theorem.

Proof of Rolle's Theorem

Let f be differentiable on (a, b) where $f(a) = f(b)$. We consider two cases.
Case 1: Consider the case when f is constant on $[a, b]$; that is, $f(x) = f(a) = f(b)$ for all x in $[a, b]$. Then $f'(x) = 0$ for all x in $[a, b]$, showing there is at least one value c in (a, b) where $f'(c) = 0$.
Case 2: Now assume that f is not constant on $[a, b]$. The Extreme Value Theorem guarantees that f has a maximal and minimal value on $[a, b]$, found either at the endpoints or at a critical value in (a, b). Since $f(a) = f(b)$ and f is not constant, it is clear that the maximum and minimum cannot *both* be found at the endpoints. Assume, without loss of generality, that the maximum of f is not found at the endpoints. Therefore there is a c in (a, b) such that $f(c)$ is the maximum value of f. By Theorem 26, c must be a critical number of f; since f is differentiable, we have that $f'(c) = 0$, completing the proof of the theorem. □

We can now prove the Mean Value Theorem.

Proof of the Mean Value Theorem

Define the function

$$g(x) = f(x) - \frac{f(b) - f(a)}{b - a}x.$$

We know g is differentiable on (a, b) and continuous on $[a, b]$ since f is. We can show $g(a) = g(b)$ (it is actually easier to show $g(b) - g(a) = 0$, which suffices). We can then apply Rolle's theorem to guarantee the existence of $c \in (a, b)$ such that $g'(c) = 0$. But note that

$$0 = g'(c) = f'(c) - \frac{f(b) - f(a)}{b - a} \ ;$$

hence

$$f'(c) = \frac{f(b) - f(a)}{b - a},$$

which is what we sought to prove. □

Going back to the very beginning of the section, we see that the only assumption we would need about our distance function $f(t)$ is that it be continuous and differentiable for t from 0 to 2 hours (both reasonable assumptions). By the Mean Value Theorem, we are guaranteed a time during the trip where our

Notes:

instantaneous speed is 50 mph. This fact is used in practice. Some law enforcement agencies monitor traffic speeds while in aircraft. They do not measure speed with radar, but rather by timing individual cars as they pass over lines painted on the highway whose distances apart are known. The officer is able to measure the *average* speed of a car between the painted lines; if that average speed is greater than the posted speed limit, the officer is assured that the driver exceeded the speed limit at some time.

Note that the Mean Value Theorem is an *existence* theorem. It states that a special value *c exists*, but it does not give any indication about how to find it. It turns out that when we need the Mean Value Theorem, existence is all we need.

Example 85   Using the Mean Value Theorem
Consider $f(x) = x^3 + 5x + 5$ on $[-3, 3]$. Find c in $[-3, 3]$ that satisfies the Mean Value Theorem.

SOLUTION   The average rate of change of f on $[-3, 3]$ is:

$$\frac{f(3) - f(-3)}{3 - (-3)} = \frac{84}{6} = 14.$$

We want to find c such that $f'(c) = 14$. We find $f'(x) = 3x^2 + 5$. We set this equal to 14 and solve for x.

$$f'(x) = 14$$
$$3x^2 + 5 = 14$$
$$x^2 = 3$$
$$x = \pm\sqrt{3} \approx \pm 1.732$$

We have found 2 values c in $[-3, 3]$ where the instantaneous rate of change is equal to the average rate of change; the Mean Value Theorem guaranteed at least one. In Figure 3.16 f is graphed with a dashed line representing the average rate of change; the lines tangent to f at $x = \pm\sqrt{3}$ are also given. Note how these lines are parallel (i.e., have the same slope) as the dashed line.

While the Mean Value Theorem has practical use (for instance, the speed monitoring application mentioned before), it is mostly used to advance other theory. We will use it in the next section to relate the shape of a graph to its derivative.

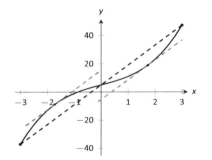

Figure 3.16: Demonstrating the Mean Value Theorem in Example 85.

Notes:

Exercises 3.2

Terms and Concepts

1. Explain in your own words what the Mean Value Theorem states.

2. Explain in your own words what Rolle's Theorem states.

Problems

In Exercises 3 – 10, a function $f(x)$ and interval $[a, b]$ are given. Check if Rolle's Theorem can be applied to f on $[a, b]$; if so, find c in $[a, b]$ such that $f'(c) = 0$.

3. $f(x) = 6$ on $[-1, 1]$.

4. $f(x) = 6x$ on $[-1, 1]$.

5. $f(x) = x^2 + x - 6$ on $[-3, 2]$.

6. $f(x) = x^2 + x - 2$ on $[-3, 2]$.

7. $f(x) = x^2 + x$ on $[-2, 2]$.

8. $f(x) = \sin x$ on $[\pi/6, 5\pi/6]$.

9. $f(x) = \cos x$ on $[0, \pi]$.

10. $f(x) = \dfrac{1}{x^2 - 2x + 1}$ on $[0, 2]$.

In Exercises 11 – 20, a function $f(x)$ and interval $[a, b]$ are given. Check if the Mean Value Theorem can be applied to f on $[a, b]$; if so, find a value c in $[a, b]$ guaranteed by the Mean Value Theorem.

11. $f(x) = x^2 + 3x - 1$ on $[-2, 2]$.

12. $f(x) = 5x^2 - 6x + 8$ on $[0, 5]$.

13. $f(x) = \sqrt{9 - x^2}$ on $[0, 3]$.

14. $f(x) = \sqrt{25 - x}$ on $[0, 9]$.

15. $f(x) = \dfrac{x^2 - 9}{x^2 - 1}$ on $[0, 2]$.

16. $f(x) = \ln x$ on $[1, 5]$.

17. $f(x) = \tan x$ on $[-\pi/4, \pi/4]$.

18. $f(x) = x^3 - 2x^2 + x + 1$ on $[-2, 2]$.

19. $f(x) = 2x^3 - 5x^2 + 6x + 1$ on $[-5, 2]$.

20. $f(x) = \sin^{-1} x$ on $[-1, 1]$.

Review

21. Find the extreme values of $f(x) = x^2 - 3x + 9$ on $[-2, 5]$.

22. Describe the critical points of $f(x) = \cos x$.

23. Describe the critical points of $f(x) = \tan x$.

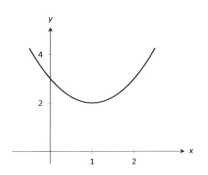

Figure 3.17: A graph of a function f used to illustrate the concepts of *increasing* and *decreasing*.

3.3   Increasing and Decreasing Functions

Our study of "nice" functions f in this chapter has so far focused on individual points: points where f is maximal/minimal, points where $f'(x) = 0$ or f' does not exist, and points c where $f'(c)$ is the average rate of change of f on some interval.

In this section we begin to study how functions behave *between* special points; we begin studying in more detail the shape of their graphs.

We start with an intuitive concept. Given the graph in Figure 3.17, where would you say the function is *increasing*? *Decreasing*? Even though we have not defined these terms mathematically, one likely answered that f is increasing when $x > 1$ and decreasing when $x < 1$. We formally define these terms here.

Definition 15   Increasing and Decreasing Functions

Let f be a function defined on an interval I.

1. f is **increasing** on I if for every $a < b$ in I, $f(a) \leq f(b)$.

2. f is **decreasing** on I if for every $a < b$ in I, $f(a) \geq f(b)$.

A function is **strictly increasing** when $a < b$ in I implies $f(a) < f(b)$, with a similar definition holding for **strictly decreasing**.

Informally, a function is increasing if as x gets larger (i.e., looking left to right) $f(x)$ gets larger.

Our interest lies in finding intervals in the domain of f on which f is either increasing or decreasing. Such information should seem useful. For instance, if f describes the speed of an object, we might want to know when the speed was increasing or decreasing (i.e., when the object was accelerating vs. decelerating). If f describes the population of a city, we should be interested in when the population is growing or declining.

To find such intervals, we again consider secant lines. Let f be an increasing, differentiable function on an open interval I, such as the one shown in Figure 3.18, and let $a < b$ be given in I. The secant line on the graph of f from $x = a$ to $x = b$ is drawn; it has a slope of $(f(b) - f(a))/(b - a)$. But note:

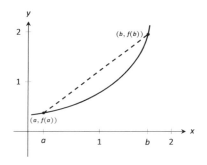

Figure 3.18: Examining the secant line of an increasing function.

$$\frac{f(b) - f(a)}{b - a} \Rightarrow \frac{\text{numerator} > 0}{\text{denominator} > 0} \Rightarrow \begin{array}{c} \text{slope of the} \\ \text{secant line} > 0 \end{array} \Rightarrow \begin{array}{c} \text{Average rate of} \\ \text{change of } f \text{ on} \\ [a, b] \text{ is} > 0. \end{array}$$

We have shown mathematically what may have already been obvious: when f is increasing, its secant lines will have a positive slope. Now recall the Mean

---

Notes:

Value Theorem guarantees that there is a number c, where $a < c < b$, such that

$$f'(c) = \frac{f(b) - f(a)}{b - a} > 0.$$

By considering all such secant lines in I, we strongly imply that $f'(x) \geq 0$ on I. A similar statement can be made for decreasing functions.

Our above logic can be summarized as "If f is increasing, then f' is probably positive." Theorem 29 below turns this around by stating "If f' is postive, then f is increasing." This leads us to a method for finding when functions are increasing and decreasing.

Theorem 29   Test For Increasing/Decreasing Functions

Let f be a continuous function on $[a, b]$ and differentiable on (a, b).

1. If $f'(c) > 0$ for all c in (a, b), then f is increasing on $[a, b]$.

2. If $f'(c) < 0$ for all c in (a, b), then f is decreasing on $[a, b]$.

3. If $f'(c) = 0$ for all c in (a, b), then f is constant on $[a, b]$.

Note: Theorem 29 also holds if $f'(c) = 0$ for a finite number of values of c in I.

Let a and b be in I where $f'(a) > 0$ and $f'(b) < 0$. It follows from the Intermediate Value Theorem that there must be some value c between a and b where $f'(c) = 0$. This leads us to the following method for finding intervals on which a function is increasing or decreasing.

Key Idea 3   Finding Intervals on Which f is Increasing or Decreasing

Let f be a differentiable function on an interval I. To find intervals on which f is increasing and decreasing:

1. Find the critical values of f. That is, find all c in I where $f'(c) = 0$ or f' is not defined.

2. Use the critical values to divide I into subintervals.

3. Pick any point p in each subinterval, and find the sign of $f'(p)$.

   (a) If $f'(p) > 0$, then f is increasing on that subinterval.

   (b) If $f'(p) < 0$, then f is decreasing on that subinterval.

Notes:

We demonstrate using this process in the following example.

Example 86   Finding intervals of increasing/decreasing
Let $f(x) = x^3 + x^2 - x + 1$. Find intervals on which f is increasing or decreasing.

SOLUTION   Using Key Idea 3, we first find the critical values of f. We have $f'(x) = 3x^2 + 2x - 1 = (3x - 1)(x + 1)$, so $f'(x) = 0$ when $x = -1$ and when $x = 1/3$. f' is never undefined.

Since an interval was not specified for us to consider, we consider the entire domain of f which is $(-\infty, \infty)$. We thus break the whole real line into three subintervals based on the two critical values we just found: $(-\infty, -1)$, $(-1, 1/3)$ and $(1/3, \infty)$. This is shown in Figure 3.19.

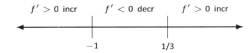

Figure 3.19: Number line for f in Example 86.

We now pick a value p in each subinterval and find the sign of $f'(p)$. All we care about is the sign, so we do not actually have to fully compute $f'(p)$; pick "nice" values that make this simple.

Subinterval 1, $(-\infty, -1)$:   We (arbitrarily) pick $p = -2$. We can compute $f'(-2)$ directly: $f'(-2) = 3(-2)^2 + 2(-2) - 1 = 7 > 0$. We conclude that f is increasing on $(-\infty, -1)$.

Note we can arrive at the same conclusion without computation. For instance, we could choose $p = -100$. The first term in $f'(-100)$, i.e., $3(-100)^2$ is clearly positive and very large. The other terms are small in comparison, so we know $f'(-100) > 0$. All we need is the sign.

Subinterval 2, $(-1, 1/3)$:   We pick $p = 0$ since that value seems easy to deal with. $f'(0) = -1 < 0$. We conclude f is decreasing on $(-1, 1/3)$.

Subinterval 3, $(1/3, \infty)$:   Pick an arbitrarily large value for $p > 1/3$ and note that $f'(p) = 3p^2 + 2p - 1 > 0$. We conclude that f is increasing on $(1/3, \infty)$.

We can verify our calculations by considering Figure 3.20, where f is graphed. The graph also presents f'; note how $f' > 0$ when f is increasing and $f' < 0$ when f is decreasing.

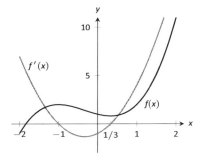

Figure 3.20: A graph of $f(x)$ in Example 86, showing where f is increasing and decreasing.

One is justified in wondering why so much work is done when the graph seems to make the intervals very clear. We give three reasons why the above work is worthwhile.

First, the points at which f switches from increasing to decreasing are not precisely known given a graph. The graph shows us something significant hap-

Notes:

pens near $x = -1$ and $x = 0.3$, but we cannot determine exactly where from the graph.

One could argue that just finding critical values is important; once we know the significant points are $x = -1$ and $x = 1/3$, the graph shows the increasing/decreasing traits just fine. That is true. However, the technique prescribed here helps reinforce the relationship between increasing/decreasing and the sign of f'. Once mastery of this concept (and several others) is obtained, one finds that either (a) just the critical points are computed and the graph shows all else that is desired, or (b) a graph is never produced, because determining increasing/decreasing using f' is straightforward and the graph is unnecessary. So our second reason why the above work is worthwhile is this: once mastery of a subject is gained, one has *options* for finding needed information. We are working to develop mastery.

Finally, our third reason: many problems we face "in the real world" are very complex. Solutions are tractable only through the use of computers to do many calculations for us. Computers do not solve problems "on their own," however; they need to be taught (i.e., *programmed*) to do the right things. It would be beneficial to give a function to a computer and have it return maximum and minimum values, intervals on which the function is increasing and decreasing, the locations of relative maxima, etc. The work that we are doing here is easily programmable. It is hard to teach a computer to "look at the graph and see if it is going up or down." It is easy to teach a computer to "determine if a number is greater than or less than 0."

In Section 3.1 we learned the definition of relative maxima and minima and found that they occur at critical points. We are now learning that functions can switch from increasing to decreasing (and vice–versa) at critical points. This new understanding of increasing and decreasing creates a great method of determining whether a critical point corresponds to a maximum, minimum, or neither. Imagine a function increasing until a critical point at $x = c$, after which it decreases. A quick sketch helps confirm that $f(c)$ must be a relative maximum. A similar statement can be made for relative minimums. We formalize this concept in a theorem.

Theorem 30   **First Derivative Test**

Let f be differentiable on I and let c be a critical number in I.

1. If the sign of f' switches from positive to negative at c, then $f(c)$ is a relative maximum of f.

2. If the sign of f' switches from negative to positive at c, then $f(c)$ is a relative minimum of f.

3. If the sign of f' does not change at c, then $f(c)$ is not a relative extrema of f.

Notes:

Example 87 **Using the First Derivative Test**

Find the intervals on which f is increasing and decreasing, and use the First Derivative Test to determine the relative extrema of f, where

$$f(x) = \frac{x^2 + 3}{x - 1}.$$

SOLUTION We start by noting the domain of f: $(-\infty, 1) \cup (1, \infty)$. Key Idea 3 describes how to find intervals where f is increasing and decreasing *when the domain of f is an interval*. Since the domain of f in this example is the union of two intervals, we apply the techniques of Key Idea 3 to both intervals of the domain of f.

Since f is not defined at $x = 1$, the increasing/decreasing nature of f could switch at this value. We do not formally consider $x = 1$ to be a critical value of f, but we will include it in our list of critical values that we find next.

Using the Quotient Rule, we find

$$f'(x) = \frac{x^2 - 2x - 3}{(x - 1)^2}.$$

We need to find the critical values of f; we want to know when $f'(x) = 0$ and when f' is not defined. That latter is straightforward: when the denominator of $f'(x)$ is 0, f' is undefined. That occurs when $x = 1$, which we've already recognized as an important value.

$f'(x) = 0$ when the numerator of $f'(x)$ is 0. That occurs when $x^2 - 2x - 3 = (x - 3)(x + 1) = 0$; i.e., when $x = -1, 3$.

We have found that f has two critical numbers, $x = -1, 3$, and at $x = 1$ something important might also happen. These three numbers divide the real number line into 4 subintervals:

$$(-\infty, -1), \quad (-1, 1), \quad (1, 3) \quad \text{and} \quad (3, \infty).$$

Pick a number p from each subinterval and test the sign of f' at p to determine whether f is increasing or decreasing on that interval. Again, we do well to avoid complicated computations; notice that the denominator of f' is *always* positive so we can ignore it during our work.

Interval 1, $(-\infty, -1)$: Choosing a very small number (i.e., a negative number with a large magnitude) p returns $p^2 - 2p - 3$ in the numerator of f'; that will be positive. Hence f is increasing on $(-\infty, -1)$.

Interval 2, $(-1, 1)$: Choosing 0 seems simple: $f'(0) = -3 < 0$. We conclude f is decreasing on $(-1, 1)$.

Notes:

Interval 3, $(1, 3)$: Choosing 2 seems simple: $f'(2) = -3 < 0$. Again, f is decreasing.

Interval 4, $(3, \infty)$: Choosing an very large number p from this subinterval will give a positive numerator and (of course) a positive denominator. So f is increasing on $(3, \infty)$.

In summary, f is increasing on the set $(-\infty, -1) \cup (3, \infty)$ and is decreasing on the set $(-1, 1) \cup (1, 3)$. Since at $x = -1$, the sign of f' switched from positive to negative, Theorem 30 states that $f(-1)$ is a relative maximum of f. At $x = 3$, the sign of f' switched from negative to positive, meaning $f(3)$ is a relative minimum. At $x = 1$, f is not defined, so there is no relative extrema at $x = 1$.

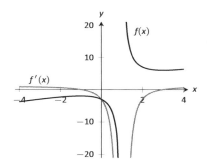

Figure 3.22: A graph of $f(x)$ in Example 87, showing where f is increasing and decreasing.

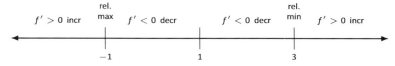

Figure 3.21: Number line for f in Example 87.

This is summarized in the number line shown in Figure 3.21. Also, Figure 3.22 shows a graph of f, confirming our calculations. This figure also shows f', again demonstrating that f is increasing when $f' > 0$ and decreasing when $f' < 0$.

One is often tempted to think that functions always alternate "increasing, decreasing, increasing, decreasing,..." around critical values. Our previous example demonstrated that this is not always the case. While $x = 1$ was not technically a critical value, it was an important value we needed to consider. We found that f was decreasing on "both sides of $x = 1$."

We examine one more example.

Example 88   Using the First Derivative Test
Find the intervals on which $f(x) = x^{8/3} - 4x^{2/3}$ is increasing and decreasing and identify the relative extrema.

SOLUTION   We again start with taking derivatives. Since we know we want to solve $f'(x) = 0$, we will do some algebra after taking derivatives.

$$
\begin{aligned}
f(x) &= x^{\frac{8}{3}} - 4x^{\frac{2}{3}} \\
f'(x) &= \frac{8}{3}x^{\frac{5}{3}} - \frac{8}{3}x^{-\frac{1}{3}} \\
&= \frac{8}{3}x^{-\frac{1}{3}}\left(x^{\frac{6}{3}} - 1\right) \\
&= \frac{8}{3}x^{-\frac{1}{3}}(x^2 - 1) \\
&= \frac{8}{3}x^{-\frac{1}{3}}(x - 1)(x + 1).
\end{aligned}
$$

Notes:

This derivation of f' shows that $f'(x) = 0$ when $x = \pm 1$ and f' is not defined when $x = 0$. Thus we have 3 critical values, breaking the number line into 4 subintervals as shown in Figure 3.23.

Interval 1, $(\infty, -1)$: We choose $p = -2$; we can easily verify that $f'(-2) < 0$. So f is decreasing on $(-\infty, -1)$.

Interval 2, $(-1, 0)$: Choose $p = -1/2$. Once more we practice finding the sign of $f'(p)$ without computing an actual value. We have $f'(p) = (8/3)p^{-1/3}(p - 1)(p + 1)$; find the sign of each of the three terms.

$$f'(p) = \frac{8}{3} \cdot \underbrace{p^{-\frac{1}{3}}}_{<0} \cdot \underbrace{(p - 1)}_{<0} \underbrace{(p + 1)}_{>0}.$$

We have a "negative $\times$ negative $\times$ positive" giving a positive number; f is increasing on $(-1, 0)$.

Interval 3, $(0, 1)$: We do a similar sign analysis as before, using p in $(0, 1)$.

$$f'(p) = \frac{8}{3} \cdot \underbrace{p^{-\frac{1}{3}}}_{>0} \cdot \underbrace{(p - 1)}_{<0} \underbrace{(p + 1)}_{>0}.$$

We have 2 positive factors and one negative factor; $f'(p) < 0$ and so f is decreasing on $(0, 1)$.

Interval 4, $(1, \infty)$: Similar work to that done for the other three intervals shows that $f'(x) > 0$ on $(1, \infty)$, so f is increasing on this interval.

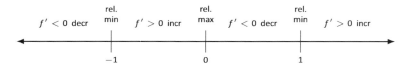

Figure 3.23: Number line for f in Example 88.

We conclude by stating that f is increasing on $(-1, 0) \cup (1, \infty)$ and decreasing on $(-\infty, -1) \cup (0, 1)$. The sign of f' changes from negative to positive around $x = -1$ and $x = 1$, meaning by Theorem 30 that $f(-1)$ and $f(1)$ are relative minima of f. As the sign of f' changes from positive to negative at $x = 0$, we have a relative maximum at $f(0)$. Figure 3.24 shows a graph of f, confirming our result. We also graph f', highlighting once more that f is increasing when $f' > 0$ and is decreasing when $f' < 0$.

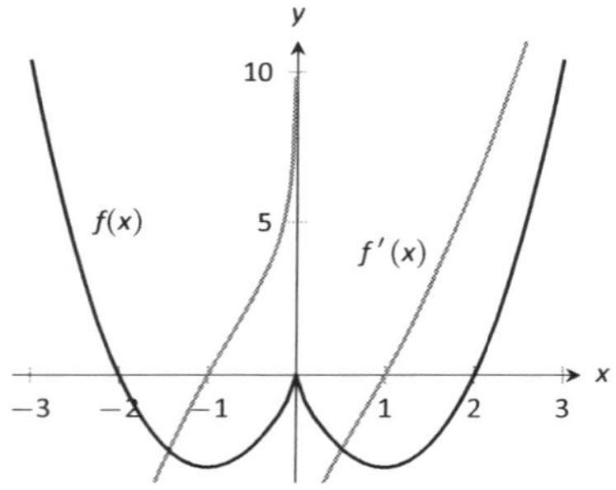

We have seen how the first derivative of a function helps determine when the function is going "up" or "down." In the next section, we will see how the second derivative helps determine how the graph of a function curves.

Figure 3.24: A graph of $f(x)$ in Example 88, showing where f is increasing and decreasing.

Notes:

Exercises 3.3

Terms and Concepts

1. In your own words describe what it means for a function to be increasing.

2. What does a decreasing function "look like"?

3. Sketch a graph of a function on $[0, 2]$ that is increasing but not strictly increasing.

4. Give an example of a function describing a situation where it is "bad" to be increasing and "good" to be decreasing.

5. A function f has derivative $f'(x) = (\sin x + 2)e^{x^2+1}$, where $f'(x) > 1$ for all x. Is f increasing, decreasing, or can we not tell from the given information?

Problems

In Exercises 6 – 13, a function $f(x)$ is given.

(a) Compute $f'(x)$.

(b) Graph f and f' on the same axes (using technology is permitted) and verify Theorem 29.

6. $f(x) = 2x + 3$

7. $f(x) = x^2 - 3x + 5$

8. $f(x) = \cos x$

9. $f(x) = \tan x$

10. $f(x) = x^3 - 5x^2 + 7x - 1$

11. $f(x) = 2x^3 - x^2 + x - 1$

12. $f(x) = x^4 - 5x^2 + 4$

13. $f(x) = \dfrac{1}{x^2 + 1}$

In Exercises 14 – 23, a function $f(x)$ is given.

(a) Give the domain of f.

(b) Find the critical numbers of f.

(c) Create a number line to determine the intervals on which f is increasing and decreasing.

(d) Use the First Derivative Test to determine whether each critical point is a relative maximum, minimum, or neither.

14. $f(x) = x^2 + 2x - 3$

15. $f(x) = x^3 + 3x^2 + 3$

16. $f(x) = 2x^3 + x^2 - x + 3$

17. $f(x) = x^3 - 3x^2 + 3x - 1$

18. $f(x) = \dfrac{1}{x^2 - 2x + 2}$

19. $f(x) = \dfrac{x^2 - 4}{x^2 - 1}$

20. $f(x) = \dfrac{x}{x^2 - 2x - 8}$

21. $f(x) = \dfrac{(x - 2)^{2/3}}{x}$

22. $f(x) = \sin x \cos x$ on $(-\pi, \pi)$.

23. $f(x) = x^5 - 5x$

Review

24. Consider $f(x) = x^2 - 3x + 5$ on $[-1, 2]$; find c guaranteed by the Mean Value Theorem.

25. Consider $f(x) = \sin x$ on $[-\pi/2, \pi/2]$; find c guaranteed by the Mean Value Theorem.

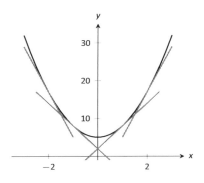

Figure 3.25: A function f with a concave up graph. Notice how the slopes of the tangent lines, when looking from left to right, are increasing.

Note: We often state that "f is concave up" instead of "the graph of f is concave up" for simplicity.

Note: A mnemonic for remembering what concave up/down means is: "Concave up is like a cup; concave down is like a frown." It is admittedly terrible, but it works.

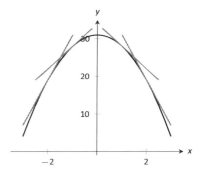

Figure 3.26: A function f with a concave down graph. Notice how the slopes of the tangent lines, when looking from left to right, are decreasing.

3.4   Concavity and the Second Derivative

Our study of "nice" functions continues. The previous section showed how the first derivative of a function, f', can relay important information about f. We now apply the same technique to f' itself, and learn what this tells us about f.

The key to studying f' is to consider its derivative, namely f'', which is the second derivative of f. When $f'' > 0$, f' is increasing. When $f'' < 0$, f' is decreasing. f' has relative maxima and minima where $f'' = 0$ or is undefined.

This section explores how knowing information about f'' gives information about f.

Concavity

We begin with a definition, then explore its meaning.

Definition 16   Concave Up and Concave Down

Let f be differentiable on an interval I. The graph of f is **concave up** on I if f' is increasing. The graph of f is **concave down** on I if f' is decreasing. If f' is constant then the graph of f is said to have **no concavity**.

The graph of a function f is *concave up* when f' is *increasing*. That means as one looks at a concave up graph from left to right, the slopes of the tangent lines will be increasing. Consider Figure 3.25, where a concave up graph is shown along with some tangent lines. Notice how the tangent line on the left is steep, downward, corresponding to a small value of f'. On the right, the tangent line is steep, upward, corresponding to a large value of f'.

If a function is decreasing and concave up, then its rate of decrease is slowing; it is "leveling off." If the function is increasing and concave up, then the *rate* of increase is increasing. The function is increasing at a faster and faster rate.

Now consider a function which is concave down. We essentially repeat the above paragraphs with slight variation.

The graph of a function f is *concave down* when f' is *decreasing*. That means as one looks at a concave down graph from left to right, the slopes of the tangent lines will be decreasing. Consider Figure 3.26, where a concave down graph is shown along with some tangent lines. Notice how the tangent line on the left is steep, upward, corresponding to a large value of f'. On the right, the tangent line is steep, downward, corresponding to a small value of f'.

If a function is increasing and concave down, then its rate of increase is slowing; it is "leveling off." If the function is decreasing and concave down, then the *rate* of decrease is decreasing. The function is decreasing at a faster and faster rate.

Notes:

Our definition of concave up and concave down is given in terms of when the first derivative is increasing or decreasing. We can apply the results of the previous section and to find intervals on which a graph is concave up or down. That is, we recognize that f' is increasing when $f'' > 0$, etc.

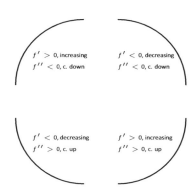

Theorem 31   Test for Concavity

Let f be twice differentiable on an interval I. The graph of f is concave up if $f'' > 0$ on I, and is concave down if $f'' < 0$ on I.

Figure 3.27: Demonstrating the 4 ways that concavity interacts with increasing/decreasing, along with the relationships with the first and second derivatives.

If knowing where a graph is concave up/down is important, it makes sense that the places where the graph changes from one to the other is also important. This leads us to a definition.

Definition 17   Point of Inflection

A **point of inflection** is a point on the graph of f at which the concavity of f changes.

Figure 3.28 shows a graph of a function with inflection points labeled.

If the concavity of f changes at a point $(c, f(c))$, then f' is changing from increasing to decreasing (or, decreasing to increasing) at $x = c$. That means that the sign of f'' is changing from positive to negative (or, negative to positive) at $x = c$. This leads to the following theorem.

Note: Geometrically speaking, a function is concave up if its graph lies above its tangent lines. A function is concave down if its graph lies below its tangent lines.

Theorem 32   Points of Inflection

If $(c, f(c))$ is a point of inflection on the graph of f, then either $f'' = 0$ or f'' is not defined at c.

We have identified the concepts of concavity and points of inflection. It is now time to practice using these concepts; given a function, we should be able to find its points of inflection and identify intervals on which it is concave up or down. We do so in the following examples.

Example 89   Finding intervals of concave up/down, inflection points
Let $f(x) = x^3 - 3x + 1$. Find the inflection points of f and the intervals on which it is concave up/down.

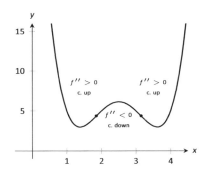

Figure 3.28: A graph of a function with its inflection points marked. The intervals where concave up/down are also indicated.

Notes:

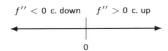

Figure 3.29: A number line determining the concavity of f in Example 89.

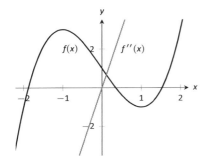

Figure 3.30: A graph of $f(x)$ used in Example 89.

SOLUTION   We start by finding $f'(x) = 3x^2 - 3$ and $f''(x) = 6x$. To find the inflection points, we use Theorem 32 and find where $f''(x) = 0$ or where f'' is undefined. We find f'' is always defined, and is 0 only when $x = 0$. So the point $(0, 1)$ is the only possible point of inflection.

This possible inflection point divides the real line into two intervals, $(-\infty, 0)$ and $(0, \infty)$. We use a process similar to the one used in the previous section to determine increasing/decreasing. Pick any $c < 0$; $f''(c) < 0$ so f is concave down on $(-\infty, 0)$. Pick any $c > 0$; $f''(c) > 0$ so f is concave up on $(0, \infty)$. Since the concavity changes at $x = 0$, the point $(0, 1)$ is an inflection point.

The number line in Figure 3.29 illustrates the process of determining concavity; Figure 3.30 shows a graph of f and f'', confirming our results. Notice how f is concave down precisely when $f''(x) < 0$ and concave up when $f''(x) > 0$.

Example 90   Finding intervals of concave up/down, inflection points
Let $f(x) = x/(x^2 - 1)$. Find the inflection points of f and the intervals on which it is concave up/down.

SOLUTION   We need to find f' and f''. Using the Quotient Rule and simplifying, we find

$$f'(x) = \frac{-(1 + x^2)}{(x^2 - 1)^2} \quad \text{and} \quad f''(x) = \frac{2x(x^2 + 3)}{(x^2 - 1)^3}.$$

To find the possible points of inflection, we seek to find where $f''(x) = 0$ and where f'' is not defined. Solving $f''(x) = 0$ reduces to solving $2x(x^2 + 3) = 0$; we find $x = 0$. We find that f'' is not defined when $x = \pm 1$, for then the denominator of f'' is 0. We also note that f itself is not defined at $x = \pm 1$, having a domain of $(-\infty, -1) \cup (-1, 1) \cup (1, \infty)$. Since the domain of f is the union of three intervals, it makes sense that the concavity of f could switch across intervals. We technically cannot say that f has a point of inflection at $x = \pm 1$ as they are not part of the domain, but we must still consider these x-values to be important and will include them in our number line.

The important x-values at which concavity might switch are $x = -1$, $x = 0$ and $x = 1$, which split the number line into four intervals as shown in Figure 3.31. We determine the concavity on each. Keep in mind that all we are concerned with is the *sign* of f'' on the interval.

Interval 1, $(-\infty, -1)$: Select a number c in this interval with a large magnitude (for instance, $c = -100$). The denominator of $f''(x)$ will be positive. In the numerator, the $(c^2 + 3)$ will be positive and the $2c$ term will be negative. Thus the numerator is negative and $f''(c)$ is negative. We conclude f is concave down on $(-\infty, -1)$.

Notes:

Interval 2, $(-1, 0)$: For any number c in this interval, the term $2c$ in the numerator will be negative, the term $(c^2 + 3)$ in the numerator will be positive, and the term $(c^2 - 1)^3$ in the denominator will be negative. Thus $f''(c) > 0$ and f is concave up on this interval.

Interval 3, $(0, 1)$: Any number c in this interval will be positive and "small." Thus the numerator is positive while the denominator is negative. Thus $f''(c) < 0$ and f is concave down on this interval.

Interval 4, $(1, \infty)$: Choose a large value for c. It is evident that $f''(c) > 0$, so we conclude that f is concave up on $(1, \infty)$.

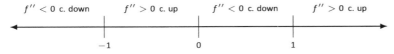

Figure 3.31: Number line for f in Example 90.

We conclude that f is concave up on $(-1, 0) \cup (1, \infty)$ and concave down on $(-\infty, -1) \cup (0, 1)$. There is only one point of inflection, $(0, 0)$, as f is not defined at $x = \pm 1$. Our work is confirmed by the graph of f in Figure 3.32. Notice how f is concave up whenever f'' is positive, and concave down when f'' is negative.

Recall that relative maxima and minima of f are found at critical points of f; that is, they are found when $f'(x) = 0$ or when f' is undefined. Likewise, the relative maxima and minima of f' are found when $f''(x) = 0$ or when f'' is undefined; note that these are the inflection points of f.

What does a "relative maximum of f'" mean? The derivative measures the rate of change of f; maximizing f' means finding the where f is increasing the most – where f has the steepest tangent line. A similar statement can be made for minimizing f'; it corresponds to where f has the steepest negatively–sloped tangent line.

We utilize this concept in the next example.

Example 91   Understanding inflection points

The sales of a certain product over a three-year span are modeled by $S(t) = t^4 - 8t^2 + 20$, where t is the time in years, shown in Figure 3.33. Over the first two years, sales are decreasing. Find the point at which sales are decreasing at their greatest rate.

SOLUTION   We want to maximize the rate of decrease, which is to say, we want to find where S' has a minimum. To do this, we find where S'' is 0. We find $S'(t) = 4t^3 - 16t$ and $S''(t) = 12t^2 - 16$. Setting $S''(t) = 0$ and solving, we get $t = \sqrt{4/3} \approx 1.16$ (we ignore the negative value of t since it does not lie in the domain of our function S).

This is both the inflection point and the point of maximum decrease. This

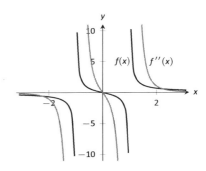

Figure 3.32: A graph of $f(x)$ and $f''(x)$ in Example 90.

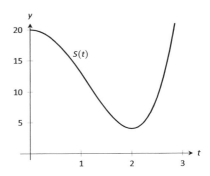

Figure 3.33: A graph of $S(t)$ in Example 91, modeling the sale of a product over time.

Notes:

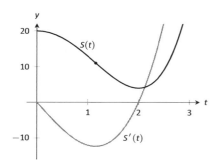

Figure 3.34: A graph of $S(t)$ in Example 91 along with $S'(t)$.

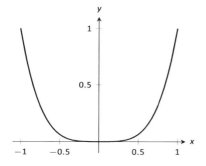

Figure 3.35: A graph of $f(x) = x^4$. Clearly f is always concave up, despite the fact that $f''(x) = 0$ when $x = 0$. It this example, the *possible* point of inflection $(0,0)$ is not a point of inflection.

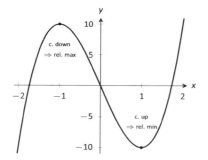

Figure 3.36: Demonstrating the fact that relative maxima occur when the graph is concave down and relative minima occur when the graph is concave up.

is the point at which things first start looking up for the company. After the inflection point, it will still take some time before sales start to increase, but at least sales are not decreasing quite as quickly as they had been.

A graph of $S(t)$ and $S'(t)$ is given in Figure 3.34. When $S'(t) < 0$, sales are decreasing; note how at $t \approx 1.16$, $S'(t)$ is minimized. That is, sales are decreasing at the fastest rate at $t \approx 1.16$. On the interval of $(1.16, 2)$, S is decreasing but concave up, so the decline in sales is "leveling off."

Not every critical point corresponds to a relative extrema; $f(x) = x^3$ has a critical point at $(0,0)$ but no relative maximum or minimum. Likewise, just because $f''(x) = 0$ we cannot conclude concavity changes at that point. We were careful before to use terminology "*possible* point of inflection" since we needed to check to see if the concavity changed. The canonical example of $f''(x) = 0$ *without* concavity changing is $f(x) = x^4$. At $x = 0$, $f''(x) = 0$ but f is always concave up, as shown in Figure 3.35.

The Second Derivative Test

The first derivative of a function gave us a test to find if a critical value corresponded to a relative maximum, minimum, or neither. The second derivative gives us another way to test if a critical point is a local maximum or minimum. The following theorem officially states something that is intuitive: if a critical value occurs in a region where a function f is concave up, then that critical value must correspond to a relative minimum of f, etc. See Figure 3.36 for a visualization of this.

Theorem 33   **The Second Derivative Test**

Let c be a critical value of f where $f''(c)$ is defined.

1. If $f''(c) > 0$, then f has a local minimum at $(c, f(c))$.

2. If $f''(c) < 0$, then f has a local maximum at $(c, f(c))$.

The Second Derivative Test relates to the First Derivative Test in the following way. If $f''(c) > 0$, then the graph is concave up at a critical point c and f' itself is growing. Since $f'(c) = 0$ and f' is growing at c, then it must go from negative to positive at c. This means the function goes from decreasing to increasing, indicating a local minimum at c.

Notes:

Example 92 **Using the Second Derivative Test**

Let $f(x) = 100/x + x$. Find the critical points of f and use the Second Derivative Test to label them as relative maxima or minima.

SOLUTION We find $f'(x) = -100/x^2 + 1$ and $f''(x) = 200/x^3$. We set $f'(x) = 0$ and solve for x to find the critical values (note that f' is not defined at $x = 0$, but neither is f so this is not a critical value.) We find the critical values are $x = \pm 10$. Evaluating f'' at $x = 10$ gives $0.1 > 0$, so there is a local minimum at $x = 10$. Evaluating $f''(-10) = -0.1 < 0$, determining a relative maximum at $x = -10$. These results are confirmed in Figure 3.37.

We have been learning how the first and second derivatives of a function relate information about the graph of that function. We have found intervals of increasing and decreasing, intervals where the graph is concave up and down, along with the locations of relative extrema and inflection points. In Chapter 1 we saw how limits explained asymptotic behavior. In the next section we combine all of this information to produce accurate sketches of functions.

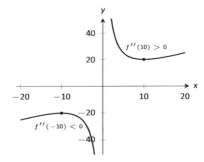

Figure 3.37: A graph of $f(x)$ in Example 92. The second derivative is evaluated at each critical point. When the graph is concave up, the critical point represents a local minimum; when the graph is concave down, the critical point represents a local maximum.

Notes:

Exercises 3.4

Terms and Concepts

1. Sketch a graph of a function $f(x)$ that is concave up on $(0, 1)$ and is concave down on $(1, 2)$.

2. Sketch a graph of a function $f(x)$ that is:

   (a) Increasing, concave up on $(0, 1)$,

   (b) increasing, concave down on $(1, 2)$,

   (c) decreasing, concave down on $(2, 3)$ and

   (d) increasing, concave down on $(3, 4)$.

3. Is is possible for a function to be increasing and concave down on $(0, \infty)$ with a horizontal asymptote of $y = 1$? If so, give a sketch of such a function.

4. Is is possible for a function to be increasing and concave up on $(0, \infty)$ with a horizontal asymptote of $y = 1$? If so, give a sketch of such a function.

Problems

In Exercises 5 – 15, a function $f(x)$ is given.

   (a) **Compute $f''(x)$.**

   (b) **Graph f and f'' on the same axes (using technology is permitted) and verify Theorem 31.**

5. $f(x) = -7x + 3$

6. $f(x) = -4x^2 + 3x - 8$

7. $f(x) = 4x^2 + 3x - 8$

8. $f(x) = x^3 - 3x^2 + x - 1$

9. $f(x) = -x^3 + x^2 - 2x + 5$

10. $f(x) = \cos x$

11. $f(x) = \sin x$

12. $f(x) = \tan x$

13. $f(x) = \dfrac{1}{x^2 + 1}$

14. $f(x) = \dfrac{1}{x}$

15. $f(x) = \dfrac{1}{x^2}$

In Exercises 16 – 28, a function $f(x)$ is given.

   (a) **Find the possible points of inflection of f.**

   (b) **Create a number line to determine the intervals on which f is concave up or concave down.**

16. $f(x) = x^2 - 2x + 1$

17. $f(x) = -x^2 - 5x + 7$

18. $f(x) = x^3 - x + 1$

19. $f(x) = 2x^3 - 3x^2 + 9x + 5$

20. $f(x) = \dfrac{x^4}{4} + \dfrac{x^3}{3} - 2x + 3$

21. $f(x) = -3x^4 + 8x^3 + 6x^2 - 24x + 2$

22. $f(x) = x^4 - 4x^3 + 6x^2 - 4x + 1$

23. $f(x) = \dfrac{1}{x^2 + 1}$

24. $f(x) = \dfrac{x}{x^2 - 1}$

25. $f(x) = \sin x + \cos x$ on $(-\pi, \pi)$

26. $f(x) = x^2 e^x$

27. $f(x) = x^2 \ln x$

28. $f(x) = e^{-x^2}$

In Exercises 29 – 41, a function $f(x)$ is given. Find the critical points of f and use the Second Derivative Test, when possible, to determine the relative extrema. (Note: these are the same functions as in Exercises 16 – 28.)

29. $f(x) = x^2 - 2x + 1$

30. $f(x) = -x^2 - 5x + 7$

31. $f(x) = x^3 - x + 1$

32. $f(x) = 2x^3 - 3x^2 + 9x + 5$

33. $f(x) = \dfrac{x^4}{4} + \dfrac{x^3}{3} - 2x + 3$

34. $f(x) = -3x^4 + 8x^3 + 6x^2 - 24x + 2$

35. $f(x) = x^4 - 4x^3 + 6x^2 - 4x + 1$

36. $f(x) = \dfrac{1}{x^2 + 1}$

37. $f(x) = \dfrac{x}{x^2 - 1}$

38. $f(x) = \sin x + \cos x$ on $(-\pi, \pi)$

39. $f(x) = x^2 e^x$

40. $f(x) = x^2 \ln x$

41. $f(x) = e^{-x^2}$

In Exercises 42 – 54, a function $f(x)$ is given. Find the x values where $f'(x)$ has a relative maximum or minimum. (Note: these are the same functions as in Exercises 16 – 28.)

42. $f(x) = x^2 - 2x + 1$

43. $f(x) = -x^2 - 5x + 7$

44. $f(x) = x^3 - x + 1$

45. $f(x) = 2x^3 - 3x^2 + 9x + 5$

46. $f(x) = \dfrac{x^4}{4} + \dfrac{x^3}{3} - 2x + 3$

47. $f(x) = -3x^4 + 8x^3 + 6x^2 - 24x + 2$

48. $f(x) = x^4 - 4x^3 + 6x^2 - 4x + 1$

49. $f(x) = \dfrac{1}{x^2 + 1}$

50. $f(x) = \dfrac{x}{x^2 - 1}$

51. $f(x) = \sin x + \cos x$ on $(-\pi, \pi)$

52. $f(x) = x^2 e^x$

53. $f(x) = x^2 \ln x$

54. $f(x) = e^{-x^2}$

3.5   Curve Sketching

We have been learning how we can understand the behavior of a function based on its first and second derivatives. While we have been treating the properties of a function separately (increasing and decreasing, concave up and concave down, etc.), we combine them here to produce an accurate graph of the function without plotting lots of extraneous points.

Why bother? Graphing utilities are very accessible, whether on a computer, a hand–held calculator, or a smartphone. These resources are usually very fast and accurate. We will see that our method is not particularly fast – it will require time (but it is not *hard*). So again: why bother?

We are attempting to understand the behavior of a function f based on the information given by its derivatives. While all of a function's derivatives relay information about it, it turns out that "most" of the behavior we care about is explained by f' and f''. Understanding the interactions between the graph of f and f' and f'' is important. To gain this understanding, one might argue that all that is needed is to look at lots of graphs. This is true to a point, but is somewhat similar to stating that one understands how an engine works after looking only at pictures. It is true that the basic ideas will be conveyed, but "hands–on" access increases understanding.

The following Key Idea summarizes what we have learned so far that is applicable to sketching graphs of functions and gives a framework for putting that information together. It is followed by several examples.

Key Idea 4   Curve Sketching

To produce an accurate sketch a given function f, consider the following steps.

1. Find the domain of f. Generally, we assume that the domain is the entire real line then find restrictions, such as where a denominator is 0 or where negatives appear under the radical.

2. Find the critical values of f.

3. Find the possible points of inflection of f.

4. Find the location of any vertical asymptotes of f (usually done in conjunction with item 1 above).

5. Consider the limits $\lim_{x \to -\infty} f(x)$ and $\lim_{x \to \infty} f(x)$ to determine the end behavior of the function.

(continued)

Notes:

Key Idea 4   **Curve Sketching – Continued**

6. Create a number line that includes all critical points, possible points of inflection, and locations of vertical asymptotes. For each interval created, determine whether f is increasing or decreasing, concave up or down.

7. Evaluate f at each critical point and possible point of inflection. Plot these points on a set of axes. Connect these points with curves exhibiting the proper concavity. Sketch asymptotes and x and y intercepts where applicable.

Example 93   **Curve sketching**
Use Key Idea 4 to sketch $f(x) = 3x^3 - 10x^2 + 7x + 5$.

SOLUTION   We follow the steps outlined in the Key Idea.

1. The domain of f is the entire real line; there are no values x for which $f(x)$ is not defined.

2. Find the critical values of f. We compute $f'(x) = 9x^2 - 20x + 7$. Use the Quadratic Formula to find the roots of f':

$$x = \frac{20 \pm \sqrt{(-20)^2 - 4(9)(7)}}{2(9)} = \frac{1}{9}\left(10 \pm \sqrt{37}\right) \Rightarrow x \approx 0.435, 1.787.$$

3. Find the possible points of inflection of f. Compute $f''(x) = 18x - 20$. We have
$$f''(x) = 0 \Rightarrow x = 10/9 \approx 1.111.$$

4. There are no vertical asymptotes.

5. We determine the end behavior using limits as x approaches $\pm$infinity.

$$\lim_{x \to -\infty} f(x) = -\infty \qquad \lim_{x \to \infty} f(x) = \infty.$$

We do not have any horizontal asymptotes.

6. We place the values $x = (10 \pm \sqrt{37})/9$ and $x = 10/9$ on a number line, as shown in Figure 3.38. We mark each subinterval as increasing or

---

Notes:

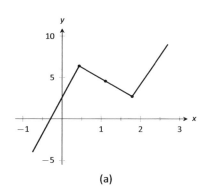

(a)

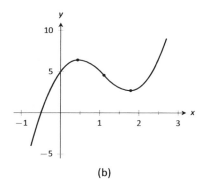

(b)

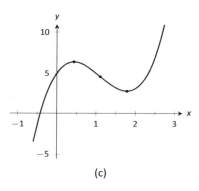

(c)

Figure 3.39: Sketching f in Example 93.

decreasing, concave up or down, using the techniques used in Sections 3.3 and 3.4.

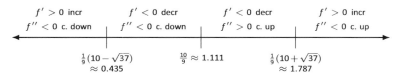

Figure 3.38: Number line for f in Example 93.

7. We plot the appropriate points on axes as shown in Figure 3.39(a) and connect the points with straight lines. In Figure 3.39(b) we adjust these lines to demonstrate the proper concavity. Our curve crosses the y axis at $y = 5$ and crosses the x axis near $x = -0.424$. In Figure 3.39(c) we show a graph of f drawn with a computer program, verifying the accuracy of our sketch.

Example 94   **Curve sketching**
Sketch $f(x) = \dfrac{x^2 - x - 2}{x^2 - x - 6}$.

SOLUTION   We again follow the steps outlined in Key Idea 4.

1. In determining the domain, we assume it is all real numbers and looks for restrictions. We find that at $x = -2$ and $x = 3$, $f(x)$ is not defined. So the domain of f is $D = \{$real numbers $x \mid x \neq -2, 3\}$.

2. To find the critical values of f, we first find $f'(x)$. Using the Quotient Rule, we find

$$f'(x) = \frac{-8x + 4}{(x^2 + x - 6)^2} = \frac{-8x + 4}{(x - 3)^2(x + 2)^2}.$$

$f'(x) = 0$ when $x = 1/2$, and f' is undefined when $x = -2, 3$. Since f' is undefined only when f is, these are not critical values. The only critical value is $x = 1/2$.

3. To find the possible points of inflection, we find $f''(x)$, again employing the Quotient Rule:

$$f''(x) = \frac{24x^2 - 24x + 56}{(x - 3)^3(x + 2)^3}.$$

We find that $f''(x)$ is never 0 (setting the numerator equal to 0 and solving for x, we find the only roots to this quadratic are imaginary) and f'' is

Notes:

undefined when $x = -2, 3$. Thus concavity will possibly only change at $x = -2$ and $x = 3$.

4. The vertical asymptotes of f are at $x = -2$ and $x = 3$, the places where f is undefined.

5. There is a horizontal asymptote of $y = 1$, as $\lim_{x \to -\infty} f(x) = 1$ and $\lim_{x \to \infty} f(x) = 1$.

6. We place the values $x = 1/2$, $x = -2$ and $x = 3$ on a number line as shown in Figure 3.40. We mark in each interval whether f is increasing or decreasing, concave up or down. We see that f has a relative maximum at $x = 1/2$; concavity changes only at the vertical asymptotes.

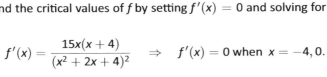

Figure 3.40: Number line for f in Example 94.

7. In Figure 3.41(a), we plot the points from the number line on a set of axes and connect the points with straight lines to get a general idea of what the function looks like (these lines effectively only convey increasing/decreasing information). In Figure 3.41(b), we adjust the graph with the appropriate concavity. We also show f crossing the x axis at $x = -1$ and $x = 2$.

Figure 3.41(c) shows a computer generated graph of f, which verifies the accuracy of our sketch.

Example 95   Curve sketching

Sketch $f(x) = \dfrac{5(x-2)(x+1)}{x^2 + 2x + 4}$.

SOLUTION   We again follow Key Idea 4.

1. We assume that the domain of f is all real numbers and consider restrictions. The only restrictions come when the denominator is 0, but this never occurs. Therefore the domain of f is all real numbers, $\mathbb{R}$.

2. We find the critical values of f by setting $f'(x) = 0$ and solving for x. We find

$$f'(x) = \frac{15x(x+4)}{(x^2+2x+4)^2} \quad \Rightarrow \quad f'(x) = 0 \text{ when } x = -4, 0.$$

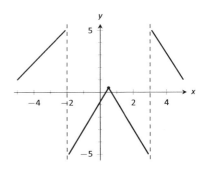

(a)

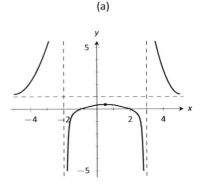

(b)

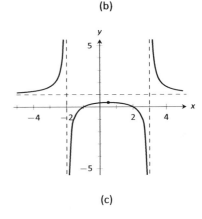

(c)

Figure 3.41: Sketching f in Example 94.

Notes:

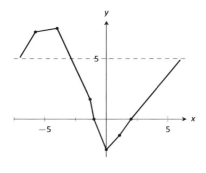

(a)

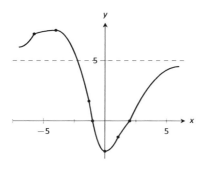

(b)

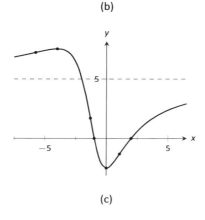

(c)

Figure 3.43: Sketching f in Example 95.

3. We find the possible points of inflection by solving $f''(x) = 0$ for x. We find

$$f''(x) = -\frac{30x^3 + 180x^2 - 240}{(x^2 + 2x + 4)^3}.$$

The cubic in the numerator does not factor very "nicely." We instead approximate the roots at $x = -5.759$, $x = -1.305$ and $x = 1.064$.

4. There are no vertical asymptotes.

5. We have a horizontal asymptote of $y = 5$, as $\lim\limits_{x \to -\infty} f(x) = \lim\limits_{x \to \infty} f(x) = 5$.

6. We place the critical points and possible points on a number line as shown in Figure 3.42 and mark each interval as increasing/decreasing, concave up/down appropriately.

Figure 3.42: Number line for f in Example 95.

7. In Figure 3.43(a) we plot the significant points from the number line as well as the two roots of f, $x = -1$ and $x = 2$, and connect the points with straight lines to get a general impression about the graph. In Figure 3.43(b), we add concavity. Figure 3.43(c) shows a computer generated graph of f, affirming our results.

In each of our examples, we found a few, significant points on the graph of f that corresponded to changes in increasing/decreasing or concavity. We connected these points with straight lines, then adjusted for concavity, and finished by showing a very accurate, computer generated graph.

Why are computer graphics so good? It is not because computers are "smarter" than we are. Rather, it is largely because computers are much faster at computing than we are. In general, computers graph functions much like most students do when first learning to draw graphs: they plot equally spaced points, then connect the dots using lines. By using lots of points, the connecting lines are short and the graph looks smooth.

This does a fine job of graphing in most cases (in fact, this is the method used for many graphs in this text). However, in regions where the graph is very "curvy," this can generate noticeable sharp edges on the graph unless a large number of points are used. High quality computer algebra systems, such as

Notes:

Mathematica, use special algorithms to plot lots of points only where the graph is "curvy."

In Figure 3.44, a graph of $y = \sin x$ is given, generated by *Mathematica*. The small points represent each of the places *Mathematica* sampled the function. Notice how at the "bends" of $\sin x$, lots of points are used; where $\sin x$ is relatively straight, fewer points are used. (Many points are also used at the endpoints to ensure the "end behavior" is accurate.)

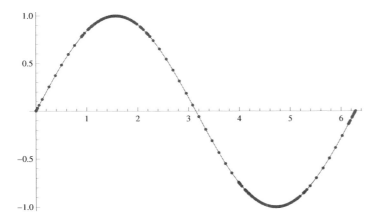

Figure 3.44: A graph of $y = \sin x$ generated by *Mathematica*.

How does *Mathematica* know where the graph is "curvy"? Calculus. When we study *curvature* in a later chapter, we will see how the first and second derivatives of a function work together to provide a measurement of "curviness." *Mathematica* employs algorithms to determine regions of "high curvature" and plots extra points there.

Again, the goal of this section is not "How to graph a function when there is no computer to help." Rather, the goal is "Understand that the shape of the graph of a function is largely determined by understanding the behavior of the function at a few key places." In Example 95, we were able to accurately sketch a complicated graph using only 5 points and knowledge of asymptotes!

There are many applications of our understanding of derivatives beyond curve sketching. The next chapter explores some of these applications, demonstrating just a few kinds of problems that can be solved with a basic knowledge of differentiation.

Notes:

Exercises 3.5

Terms and Concepts

1. Why is sketching curves by hand beneficial even though technology is ubiquitous?

2. What does "ubiquitous" mean?

3. T/F: When sketching graphs of functions, it is useful to find the critical points.

4. T/F: When sketching graphs of functions, it is useful to find the possible points of inflection.

5. T/F: When sketching graphs of functions, it is useful to find the horizontal and vertical asymptotes.

Problems

In Exercises 6 – 11, practice using Key Idea 4 by applying the principles to the given functions with familiar graphs.

6. $f(x) = 2x + 4$

7. $f(x) = -x^2 + 1$

8. $f(x) = \sin x$

9. $f(x) = e^x$

10. $f(x) = \dfrac{1}{x}$

11. $f(x) = \dfrac{1}{x^2}$

In Exercises 12 – 25, sketch a graph of the given function using Key Idea 4. Show all work; check your answer with technology.

12. $f(x) = x^3 - 2x^2 + 4x + 1$

13. $f(x) = -x^3 + 5x^2 - 3x + 2$

14. $f(x) = x^3 + 3x^2 + 3x + 1$

15. $f(x) = x^3 - x^2 - x + 1$

16. $f(x) = (x - 2)\ln(x - 2)$

17. $f(x) = (x - 2)^2 \ln(x - 2)$

18. $f(x) = \dfrac{x^2 - 4}{x^2}$

19. $f(x) = \dfrac{x^2 - 4x + 3}{x^2 - 6x + 8}$

20. $f(x) = \dfrac{x^2 - 2x + 1}{x^2 - 6x + 8}$

21. $f(x) = x\sqrt{x + 1}$

22. $f(x) = x^2 e^x$

23. $f(x) = \sin x \cos x$ on $[-\pi, \pi]$

24. $f(x) = (x - 3)^{2/3} + 2$

25. $f(x) = \dfrac{(x - 1)^{2/3}}{x}$

In Exercises 26 – 28, a function with the parameters a and b are given. Describe the critical points and possible points of inflection of f in terms of a and b.

26. $f(x) = \dfrac{a}{x^2 + b^2}$

27. $f(x) = \sin(ax + b)$

28. $f(x) = (x - a)(x - b)$

29. Given $x^2 + y^2 = 1$, use implicit differentiation to find $\frac{dy}{dx}$ and $\frac{d^2y}{dx^2}$. Use this information to justify the sketch of the unit circle.

4: APPLICATIONS OF THE DERIVATIVE

In Chapter 3, we learned how the first and second derivatives of a function influence its graph. In this chapter we explore other applications of the derivative.

4.1 Newton's Method

Solving equations is one of the most important things we do in mathematics, yet we are surprisingly limited in what we can solve analytically. For instance, equations as simple as $x^5 + x + 1 = 0$ or $\cos x = x$ cannot be solved by algebraic methods in terms of familiar functions. Fortunately, there are methods that can give us *approximate* solutions to equations like these. These methods can usually give an approximation correct to as many decimal places as we like. In Section 1.5 we learned about the Bisection Method. This section focuses on another technique (which generally works faster), called Newton's Method.

Newton's Method is built around tangent lines. The main idea is that if x is sufficiently close to a root of $f(x)$, then the tangent line to the graph at $(x, f(x))$ will cross the x-axis at a point closer to the root than x.

We start Newton's Method with an initial guess about roughly where the root is. Call this x_0. (See Figure 4.1(a).) Draw the tangent line to the graph at $(x_0, f(x_0))$ and see where it meets the x-axis. Call this point x_1. Then repeat the process – draw the tangent line to the graph at $(x_1, f(x_1))$ and see where it meets the x-axis. (See Figure 4.1(b).) Call this point x_2. Repeat the process again to get x_3, x_4, etc. This sequence of points will often converge rather quickly to a root of f.

We can use this *geometric* process to create an *algebraic* process. Let's look at how we found x_1. We started with the tangent line to the graph at $(x_0, f(x_0))$. The slope of this tangent line is $f'(x_0)$ and the equation of the line is

$$y = f'(x_0)(x - x_0) + f(x_0).$$

This line crosses the x-axis when $y = 0$, and the x–value where it crosses is what we called x_1. So let $y = 0$ and replace x with x_1, giving the equation:

$$0 = f'(x_0)(x_1 - x_0) + f(x_0).$$

Now solve for x_1:

$$x_1 = x_0 - \frac{f(x_0)}{f'(x_0)}.$$

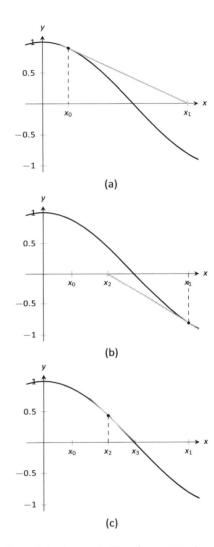

Figure 4.1: Demonstrating the geometric concept behind Newton's Method.

Since we repeat the same geometric process to find x_2 from x_1, we have

$$x_2 = x_1 - \frac{f(x_1)}{f'(x_1)}.$$

In general, given an approximation x_n, we can find the next approximation, x_{n+1} as follows:

$$x_{n+1} = x_n - \frac{f(x_n)}{f'(x_n)}.$$

We summarize this process as follows.

Key Idea 5   Newton's Method

Let f be a differentiable function on an interval I with a root in I. To approximate the value of the root, accurate to d decimal places:

1. Choose a value x_0 as an initial approximation of the root. (This is often done by looking at a graph of f.)

2. Create successive approximations iteratively; given an approximation x_n, compute the next approximation x_{n+1} as

$$x_{n+1} = x_n - \frac{f(x_n)}{f'(x_n)}.$$

3. Stop the iterations when successive approximations do not differ in the first d places after the decimal point.

Note: Newton's Method is not infallible. The sequence of approximate values may not converge, or it may converge so slowly that one is "tricked" into thinking a certain approximation is better than it actually is. These issues will be discussed at the end of the section.

Let's practice Newton's Method with a concrete example.

Example 96   Using Newton's Method
Approximate the real root of $x^3 - x^2 - 1 = 0$, accurate to the first 3 places after the decimal, using Newton's Method and an initial approximation of $x_0 = 1$.

SOLUTION   To begin, we compute $f'(x) = 3x^2 - 2x$. Then we apply the

---

Notes:

Newton's Method algorithm, outlined in Key Idea 5.

$$x_1 = 1 - \frac{f(1)}{f'(1)} = 1 - \frac{1^3 - 1^2 - 1}{3 \cdot 1^2 - 2 \cdot 1} = 2,$$

$$x_2 = 2 - \frac{f(2)}{f'(2)} = 2 - \frac{2^3 - 2^2 - 1}{3 \cdot 2^2 - 2 \cdot 2} = 1.625,$$

$$x_3 = 1.625 - \frac{f(1.625)}{f'(1.625)} = 1.625 - \frac{1.625^3 - 1.625^2 - 1}{3 \cdot 1.625^2 - 2 \cdot 1.625} \approx 1.48579.$$

$$x_4 = 1.48579 - \frac{f(1.48579)}{f'(1.48579)} \approx 1.46596$$

$$x_5 = 1.46596 - \frac{f(1.46596)}{f'(1.46596)} \approx 1.46557$$

We performed 5 iterations of Newton's Method to find a root accurate to the first 3 places after the decimal; our final approximation is 1.465. The exact value of the root, to six decimal places, is 1.465571; It turns out that our x_5 is accurate to more than just 3 decimal places.

A graph of $f(x)$ is given in Figure 4.2. We can see from the graph that our initial approximation of $x_0 = 1$ was not particularly accurate; a closer guess would have been $x_0 = 1.5$. Our choice was based on ease of initial calculation, and shows that Newton's Method can be robust enough that we do not have to make a very accurate initial approximation.

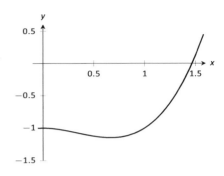

Figure 4.2: A graph of $f(x) = x^3 - x^2 - 1$ in Example 96.

We can automate this process on a calculator that has an Ans key that returns the result of the previous calculation. Start by pressing 1 and then Enter. (We have just entered our initial guess, $x_0 = 1$.) Now compute

$$\text{Ans} - \frac{f(\text{Ans})}{f'(\text{Ans})}$$

by entering the following and repeatedly press the Enter key:

```
Ans-(Ans^3-Ans^2-1)/(3*Ans^2-2*Ans)
```

Each time we press the Enter key, we are finding the successive approximations, x_1, x_2, ..., and each one is getting closer to the root. In fact, once we get past around x_7 or so, the approximations don't appear to be changing. They actually are changing, but the change is far enough to the right of the decimal point that it doesn't show up on the calculator's display. When this happens, we can be pretty confident that we have found an accurate approximation.

Using a calculator in this manner makes the calculations simple; many iterations can be computed very quickly.

Notes:

Example 97   Using Newton's Method to find where functions intersect

Use Newton's Method to approximate a solution to $\cos x = x$, accurate to 5 places after the decimal.

SOLUTION   Newton's Method provides a method of solving $f(x) = 0$; it is not (directly) a method for solving equations like $f(x) = g(x)$. However, this is not a problem; we can rewrite the latter equation as $f(x) - g(x) = 0$ and then use Newton's Method.

So we rewrite $\cos x = x$ as $\cos x - x = 0$. Written this way, we are finding a root of $f(x) = \cos x - x$. We compute $f'(x) = -\sin x - 1$. Next we need a starting value, x_0. Consider Figure 4.3, where $f(x) = \cos x - x$ is graphed. It seems that $x_0 = 0.75$ is pretty close to the root, so we will use that as our x_0. (The figure also shows the graphs of $y = \cos x$ and $y = x$, drawn with dashed lines. Note how they intersect at the same x value as when $f(x) = 0$.)

We now compute x_1, x_2, etc. The formula for x_1 is

$$x_1 = 0.75 - \frac{\cos(0.75) - 0.75}{-\sin(0.75) - 1} \approx 0.7391111388.$$

Apply Newton's Method again to find x_2:

$$x_2 = 0.7391111388 - \frac{\cos(0.7391111388) - 0.7391111388}{-\sin(0.7391111388) - 1} \approx 0.7390851334.$$

We can continue this way, but it is really best to automate this process. On a calculator with an Ans key, we would start by pressing 0.75, then Enter, inputting our initial approximation. We then enter:

$$\text{Ans} - (\cos(\text{Ans})-\text{Ans})/(-\sin(\text{Ans})-1).$$

Repeatedly pressing the Enter key gives successive approximations. We quickly find:

$$x_3 = 0.7390851332$$
$$x_4 = 0.7390851332.$$

Our approximations x_2 and x_3 did not differ for at least the first 5 places after the decimal, so we could have stopped. However, using our calculator in the manner described is easy, so finding x_4 was not hard. It is interesting to see how we found an approximation, accurate to as many decimal places as our calculator displays, in just 4 iterations.

If you know how to program, you can translate the following pseudocode into your favorite language to perform the computation in this problem.

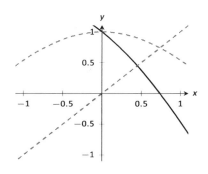

Figure 4.3: A graph of $f(x) = \cos x - x$ used to find an initial approximation of its root.

Notes:

```
x = .75
while true
    oldx = x
    x = x - (cos(x)-x)/(-sin(x)-1)
    print x
    if abs(x-oldx) < .0000000001
        break
```

This code calculates x_1, x_2, etc., storing each result in the variable x. The previous approximation is stored in the variable `oldx`. We continue looping until the difference between two successive approximations, `abs(x-oldx)`, is less than some small tolerance, in this case, `.0000000001`.

Convergence of Newton's Method

What should one use for the initial guess, x_0? Generally, the closer to the actual root the initial guess is, the better. However, some initial guesses should be avoided. For instance, consider Example 96 where we sought the root to $f(x) = x^3 - x^2 - 1$. Choosing $x_0 = 0$ would have been a particularly poor choice. Consider Figure 4.4, where $f(x)$ is graphed along with its tangent line at $x = 0$. Since $f'(0) = 0$, the tangent line is horizontal and does not intersect the x–axis. Graphically, we see that Newton's Method fails.

We can also see analytically that it fails. Since

$$x_1 = 0 - \frac{f(0)}{f'(0)}$$

and $f'(0) = 0$, we see that x_1 is not well defined.

This problem can also occur if, for instance, it turns out that $f'(x_5) = 0$. Adjusting the initial approximation x_0 by a very small amount will likely fix the problem.

It is also possible for Newton's Method to not converge while each successive approximation is well defined. Consider $f(x) = x^{1/3}$, as shown in Figure 4.5. It is clear that the root is $x = 0$, but let's approximate this with $x_0 = 0.1$. Figure 4.5(a) shows graphically the calculation of x_1; notice how it is farther from the root than x_0. Figures 4.5(b) and (c) show the calculation of x_2 and x_3, which are even farther away; our successive approximations are getting worse. (It turns out that in this particular example, each successive approximation is twice as far from the true answer as the previous approximation.)

There is no "fix" to this problem; Newton's Method simply will not work and another method must be used.

While Newton's Method does not always work, it does work "most of the time," and it is generally very fast. Once the approximations get close to the root,

Notes:

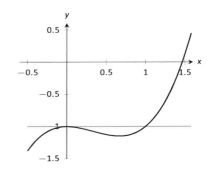

Figure 4.4: A graph of $f(x) = x^3 - x^2 - 1$, showing why an initial approximation of $x_0 = 0$ with Newton's Method fails.

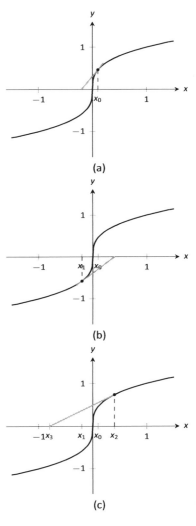

(a)

(b)

(c)

Figure 4.5: Newton's Method fails to find a root of $f(x) = x^{1/3}$, regardless of the choice of x_0.

Newton's Method can as much as double the number of correct decimal places with each successive approximation. A course in Numerical Analysis will introduce the reader to more iterative root finding methods, as well as give greater detail about the strengths and weaknesses of Newton's Method.

Notes:

Exercises 4.1

Terms and Concepts

1. T/F: Given a function $f(x)$, Newton's Method produces an exact solution to $f(x) = 0$.

2. T/F: In order to get a solution to $f(x) = 0$ accurate to d places after the decimal, at least $d + 1$ iterations of Newtons' Method must be used.

Problems

In Exercises 3 – 7, the roots of $f(x)$ are known or are easily found. Use 5 iterations of Newton's Method with the given initial approximation to approximate the root. Compare it to the known value of the root.

3. $f(x) = \cos x$, $x_0 = 1.5$

4. $f(x) = \sin x$, $x_0 = 1$

5. $f(x) = x^2 + x - 2$, $x_0 = 0$

6. $f(x) = x^2 - 2$, $x_0 = 1.5$

7. $f(x) = \ln x$, $x_0 = 2$

In Exercises 8 – 11, use Newton's Method to approximate all roots of the given functions accurate to 3 places after the decimal. If an interval is given, find only the roots that lie in that interval. Use technology to obtain good initial approximations.

8. $f(x) = x^3 + 5x^2 - x - 1$

9. $f(x) = x^4 + 2x^3 - 7x^2 - x + 5$

10. $f(x) = x^{17} - 2x^{13} - 10x^8 + 10$ on $(-2, 2)$

11. $f(x) = x^2 \cos x + (x - 1) \sin x$ on $(-3, 3)$

In Exercises 12 – 15, use Newton's Method to approximate when the given functions are equal, accurate to 3 places after the decimal. Use technology to obtain good initial approximations.

12. $f(x) = x^2$, $g(x) = \cos x$

13. $f(x) = x^2 - 1$, $g(x) = \sin x$

14. $f(x) = e^{x^2}$, $g(x) = \cos x$

15. $f(x) = x$, $g(x) = \tan x$ on $[-6, 6]$

16. Why does Newton's Method fail in finding a root of $f(x) = x^3 - 3x^2 + x + 3$ when $x_0 = 1$?

17. Why does Newton's Method fail in finding a root of $f(x) = -17x^4 + 130x^3 - 301x^2 + 156x + 156$ when $x_0 = 1$?

4.2 Related Rates

When two quantities are related by an equation, knowing the value of one quantity can determine the value of the other. For instance, the circumference and radius of a circle are related by $C = 2\pi r$; knowing that $C = 6\pi$ in determines the radius must be 3in.

The topic of **related rates** takes this one step further: knowing the *rate* at which one quantity is changing can determine the *rate* at which the other changes.

We demonstrate the concepts of related rates through examples.

Example 98   **Understanding related rates**
The radius of a circle is growing at a rate of 5in/hr. At what rate is the circumference growing?

SOLUTION   The circumference and radius of a circle are related by $C = 2\pi r$. We are given information about how the length of r changes with respect to time; that is, we are told $\frac{dr}{dt} = 5$in/hr. We want to know how the length of C changes with respect to time, i.e., we want to know $\frac{dC}{dt}$.

Implicitly differentiate both sides of $C = 2\pi r$ with respect to t:

$$C = 2\pi r$$
$$\frac{d}{dt}(C) = \frac{d}{dt}(2\pi r)$$
$$\frac{dC}{dt} = 2\pi \frac{dr}{dt}.$$

As we know $\frac{dr}{dt} = 5$in/hr, we know

$$\frac{dC}{dt} = 2\pi 5 = 10\pi \approx 31.4\text{in/hr}.$$

Note: This section relies heavily on implicit differentiation, so referring back to Section 2.6 may help.

Consider another, similar example.

Example 99   **Finding related rates**
Water streams out of a faucet at a rate of 2in^3/s onto a flat surface at a constant rate, forming a circular puddle that is 1/8in deep.

1. At what rate is the area of the puddle growing?

2. At what rate is the radius of the circle growing?

Notes:

SOLUTION

1. We can answer this question two ways: using "common sense" or related rates. The common sense method states that the volume of the puddle is growing by 2in^3/s, where

$$\text{volume of puddle} = \text{area of circle} \times \text{depth}.$$

Since the depth is constant at 1/8in, the area must be growing by 16in^2/s.

This approach reveals the underlying related–rates principle. Let V and A represent the Volume and Area of the puddle. We know $V = A \times \frac{1}{8}$. Take the derivative of both sides with respect to t, employing implicit differentiation.

$$V = \frac{1}{8}A$$
$$\frac{d}{dt}(V) = \frac{d}{dt}\left(\frac{1}{8}A\right)$$
$$\frac{dV}{dt} = \frac{1}{8}\frac{dA}{dt}$$

As $\frac{dV}{dt} = 2$, we know $2 = \frac{1}{8}\frac{dA}{dt}$, and hence $\frac{dA}{dt} = 16$. Thus the area is growing by 16in^2/s.

2. To start, we need an equation that relates what we know to the radius. We just learned something about the surface area of the circular puddle, and we know $A = \pi r^2$. We should be able to learn about the rate at which the radius is growing with this information.

Implicitly derive both sides of $A = \pi r^2$ with respect to t:

$$A = \pi r^2$$
$$\frac{d}{dt}(A) = \frac{d}{dt}\left(\pi r^2\right)$$
$$\frac{dA}{dt} = 2\pi r\frac{dr}{dt}$$

Our work above told us that $\frac{dA}{dt} = 16$in^2/s. Solving for $\frac{dr}{dt}$, we have

$$\frac{dr}{dt} = \frac{8}{\pi r}.$$

Note how our answer is not a number, but rather a function of r. In other words, *the rate at which the radius is growing depends on how big the*

Notes:

circle already is. If the circle is very large, adding $2in^3$ of water will not make the circle much bigger at all. If the circle dime–sized, adding the same amount of water will make a radical change in the radius of the circle.

In some ways, our problem was (intentionally) ill–posed. We need to specify a current radius in order to know a rate of change. When the puddle has a radius of 10in, the radius is growing at a rate of

$$\frac{dr}{dt} = \frac{8}{10\pi} = \frac{4}{5\pi} \approx 0.25in/s.$$

Example 100   **Studying related rates**
Radar guns measure the rate of distance change between the gun and the object it is measuring. For instance, a reading of "55mph" means the object is moving away from the gun at a rate of 55 miles per hour, whereas a measurement of "−25mph" would mean that the object is approaching the gun at a rate of 25 miles per hour.

If the radar gun is moving (say, attached to a police car) then radar readouts are only immediately understandable if the gun and the object are moving along the same line. If a police officer is traveling 60mph and gets a readout of 15mph, he knows that the car ahead of him is moving away at a rate of 15 miles an hour, meaning the car is traveling 75mph. (This straight–line principle is one reason officers park on the side of the highway and try to shoot straight back down the road. It gives the most accurate reading.)

Suppose an officer is driving due north at 30 mph and sees a car moving due east, as shown in Figure 4.6. Using his radar gun, he measures a reading of 20mph. By using landmarks, he believes both he and the other car are about 1/2 mile from the intersection of their two roads.

If the speed limit on the other road is 55mph, is the other driver speeding?

    SOLUTION   Using the diagram in Figure 4.6, let's label what we know about the situation. As both the police officer and other driver are 1/2 mile from the intersection, we have $A = 1/2$, $B = 1/2$, and through the Pythagorean Theorem, $C = 1/\sqrt{2} \approx 0.707$.

We know the police officer is traveling at 30mph; that is, $\frac{dA}{dt} = -30$. The reason this rate of change is negative is that A is getting smaller; the distance between the officer and the intersection is shrinking. The radar measurement is $\frac{dC}{dt} = 20$. We want to find $\frac{dB}{dt}$.

We need an equation that relates B to A and/or C. The Pythagorean Theorem

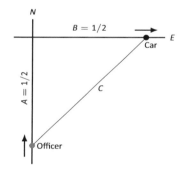

Figure 4.6: A sketch of a police car (at bottom) attempting to measure the speed of a car (at right) in Example 100.

Notes:

is a good choice: $A^2 + B^2 = C^2$. Differentiate both sides with respect to t:

$$A^2 + B^2 = C^2$$

$$\frac{d}{dt}\left(A^2 + B^2\right) = \frac{d}{dt}\left(C^2\right)$$

$$2A\frac{dA}{dt} + 2B\frac{dB}{dt} = 2C\frac{dC}{dt}$$

We have values for everything except $\frac{dB}{dt}$. Solving for this we have

$$\frac{dB}{dt} = \frac{C\frac{dC}{dt} - A\frac{dA}{dt}}{B} \approx 58.28\text{mph}.$$

The other driver appears to be speeding slightly.

Example 101   Studying related rates

A camera is placed on a tripod 10ft from the side of a road. The camera is to turn to track a car that is to drive by at 100mph for a promotional video. The video's planners want to know what kind of motor the tripod should be equipped with in order to properly track the car as it passes by. Figure 4.7 shows the proposed setup.

How fast must the camera be able to turn to track the car?

SOLUTION   We seek information about how fast the camera is to *turn*; therefore, we need an equation that will relate an angle θ to the position of the camera and the speed and position of the car.

Figure 4.7 suggests we use a trigonometric equation. Letting x represent the distance the car is from the point on the road directly in front of the camera, we have

$$\tan\theta = \frac{x}{10}. \tag{4.1}$$

As the car is moving at 100mph, we have $\frac{dx}{dt} = -100$mph (as in the last example, since x is getting smaller as the car travels, $\frac{dx}{dt}$ is negative). We need to convert the measurements so they use the same units; rewrite -100mph in terms of ft/s:

$$\frac{dx}{dt} = -100\frac{m}{hr} = -100\frac{m}{hr} \cdot 5280\frac{ft}{m} \cdot \frac{1}{3600}\frac{hr}{s} = -146.\overline{6}\text{ft/s}.$$

Now take the derivative of both sides of Equation (4.1) using implicit differenti-

Note: Example 100 is both interesting and impractical. It highlights the difficulty in using radar in a non–linear fashion, and explains why "in real life" the police officer would follow the other driver to determine their speed, and not pull out pencil and paper.

The principles here are important, though. Many automated vehicles make judgments about other moving objects based on perceived distances, radar–like measurements and the concepts of related rates.

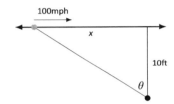

Figure 4.7: Tracking a speeding car (at left) with a rotating camera.

Notes:

ation:

$$\tan\theta = \frac{x}{10}$$

$$\frac{d}{dt}\left(\tan\theta\right) = \frac{d}{dt}\left(\frac{x}{10}\right)$$

$$\sec^2\theta\,\frac{d\theta}{dt} = \frac{1}{10}\frac{dx}{dt}$$

$$\frac{d\theta}{dt} = \frac{\cos^2\theta}{10}\frac{dx}{dt} \qquad (4.2)$$

We want to know the fastest the camera has to turn. Common sense tells us this is when the car is directly in front of the camera (i.e., when $\theta = 0$). Our mathematics bears this out. In Equation (4.2) we see this is when $\cos^2\theta$ is largest; this is when $\cos\theta = 1$, or when $\theta = 0$.

With $\frac{dx}{dt} \approx -146.67$ft/s, we have

$$\frac{d\theta}{dt} = -\frac{1\text{rad}}{10\text{ft}}146.67\text{ft/s} = -14.667\text{radians/s}.$$

We find that $\frac{d\theta}{dt}$ is negative; this matches our diagram in Figure 4.7 for θ is getting smaller as the car approaches the camera.

What is the practical meaning of -14.667radians/s? Recall that 1 circular revolution goes through 2π radians, thus 14.667rad/s means $14.667/(2\pi) \approx 2.33$ revolutions per second. The negative sign indicates the camera is rotating in a clockwise fashion.

We introduced the derivative as a function that gives the slopes of tangent lines of functions. This chapter emphasizes using the derivative in other ways. Newton's Method uses the derivative to approximate roots of functions; this section stresses the "rate of change" aspect of the derivative to find a relationship between the rates of change of two related quantities.

In the next section we use Extreme Value concepts to *optimize* quantities.

Notes:

Exercises 4.2

Terms and Concepts

1. T/F: Implicit differentiation is often used when solving "related rates" type problems.

2. T/F: A study of related rates is part of the standard police officer training.

Problems

3. Water flows onto a flat surface at a rate of 5cm³/s forming a circular puddle 10mm deep. How fast is the radius growing when the radius is:

    (a) 1 cm?

    (b) 10 cm?

    (c) 100 cm?

4. A circular balloon is inflated with air flowing at a rate of 10cm³/s. How fast is the radius of the balloon increasing when the radius is:

    (a) 1 cm?

    (b) 10 cm?

    (c) 100 cm?

5. Consider the traffic situation introduced in Example 100. How fast is the "other car" traveling if the officer and the other car are each 1/2 mile from the intersection, the other car is traveling *due west*, the officer is traveling north at 50mph, and the radar reading is −80mph?

6. Consider the traffic situation introduced in Example 100. Calculate how fast the "other car" is traveling in each of the following situations.

    (a) The officer is traveling due north at 50mph and is 1/2 mile from the intersection, while the other car is 1 mile from the intersection traveling west and the radar reading is −80mph?

    (b) The officer is traveling due north at 50mph and is 1 mile from the intersection, while the other car is 1/2 mile from the intersection traveling west and the radar reading is −80mph?

7. An F-22 aircraft is flying at 500mph with an elevation of 10,000ft on a straight–line path that will take it directly over an anti–aircraft gun.

How fast must the gun be able to turn to accurately track the aircraft when the plane is:

    (a) 1 mile away?

    (b) 1/5 mile away?

    (c) Directly overhead?

8. An F-22 aircraft is flying at 500mph with an elevation of 100ft on a straight–line path that will take it directly over an anti–aircraft gun as in Exercise 7 (note the lower elevation here).

    How fast must the gun be able to turn to accurately track the aircraft when the plane is:

    (a) 1000 feet away?

    (b) 100 feet away?

    (c) Directly overhead?

9. A 24ft. ladder is leaning against a house while the base is pulled away at a constant rate of 1ft/s.

At what rate is the top of the ladder sliding down the side of the house when the base is:

    (a) 1 foot from the house?

    (b) 10 feet from the house?

    (c) 23 feet from the house?

    (d) 24 feet from the house?

10. A boat is being pulled into a dock at a constant rate of 30ft/min by a winch located 10ft above the deck of the boat.

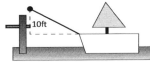

At what rate is the boat approaching the dock when the boat is:

    (a) 50 feet out?

    (b) 15 feet out?

    (c) 1 foot from the dock?

    (d) What happens when the length of rope pulling in the boat is less than 10 feet long?

11. An inverted cylindrical cone, 20ft deep and 10ft across at the top, is being filled with water at a rate of 10ft³/min. At what rate is the water rising in the tank when the depth of the water is:

    (a) 1 foot?

    (b) 10 feet?

    (c) 19 feet?

    How long will the tank take to fill when starting at empty?

12. A rope, attached to a weight, goes up through a pulley at the ceiling and back down to a worker. The man holds the rope at the same height as the connection point between rope and weight.

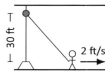

Suppose the man stands directly next to the weight (i.e., a total rope length of 60 ft) and begins to walk away at a rate of 2ft/s. How fast is the weight rising when the man has walked:

   (a) 10 feet?

   (b) 40 feet?

How far must the man walk to raise the weight all the way to the pulley?

13. Consider the situation described in Exercise 12. Suppose the man starts 40ft from the weight and begins to walk away at a rate of 2ft/s.

   (a) How long is the rope?

   (b) How fast is the weight rising after the man has walked 10 feet?

   (c) How fast is the weight rising after the man has walked 40 feet?

   (d) How far must the man walk to raise the weight all the way to the pulley?

14. A hot air balloon lifts off from ground rising vertically. From 100 feet away, a 5' woman tracks the path of the balloon. When her sightline with the balloon makes a 45° angle with the horizontal, she notes the angle is increasing at about 5°/min.

   (a) What is the elevation of the balloon?

   (b) How fast is it rising?

15. A company that produces landscaping materials is dumping sand into a conical pile. The sand is being poured at a rate of 5ft^3/sec; the physical properties of the sand, in conjunction with gravity, ensure that the cone's height is roughly 2/3 the length of the diameter of the circular base.

   How fast is the cone rising when it has a height of 30 feet?

4.3 Optimization

In Section 3.1 we learned about extreme values – the largest and smallest values a function attains on an interval. We motivated our interest in such values by discussing how it made sense to want to know the highest/lowest values of a stock, or the fastest/slowest an object was moving. In this section we apply the concepts of extreme values to solve "word problems," i.e., problems stated in terms of situations that require us to create the appropriate mathematical framework in which to solve the problem.

We start with a classic example which is followed by a discussion of the topic of optimization.

Example 102   Optimization: perimeter and area
A man has 100 feet of fencing, a large yard, and a small dog. He wants to create a rectangular enclosure for his dog with the fencing that provides the maximal area. What dimensions provide the maximal area?

SOLUTION   One can likely guess the correct answer – that is great. We will proceed to show how calculus can provide this answer in a context that proves this answer is correct.

It helps to make a sketch of the situation. Our enclosure is sketched twice in Figure 4.8, either with green grass and nice fence boards or as a simple rectangle. Either way, drawing a rectangle forces us to realize that we need to know the dimensions of this rectangle so we can create an area function – after all, we are trying to maximize the area.

We let x and y denote the lengths of the sides of the rectangle. Clearly,

$$\text{Area} = xy.$$

We do not yet know how to handle functions with 2 variables; we need to reduce this down to a single variable. We know more about the situation: the man has 100 feet of fencing. By knowing the perimeter of the rectangle must be 100, we can create another equation:

$$\text{Perimeter} = 100 = 2x + 2y.$$

We now have 2 equations and 2 unknowns. In the latter equation, we solve for y:

$$y = 50 - x.$$

Now substitute this expression for y in the area equation:

$$\text{Area} = A(x) = x(50 - x).$$

Note we now have an equation of one variable; we can truly call the Area a function of x.

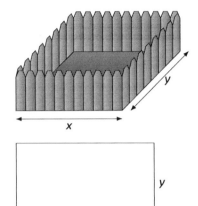

Figure 4.8: A sketch of the enclosure in Example 102.

Notes:

This function only makes sense when $0 \le x \le 50$, otherwise we get negative values of area. So we find the extreme values of $A(x)$ on the interval $[0, 50]$.

To find the critical points, we take the derivative of $A(x)$ and set it equal to 0, then solve for x.

$$A(x) = x(50 - x)$$
$$= 50x - x^2$$
$$A'(x) = 50 - 2x$$

We solve $50 - 2x = 0$ to find $x = 25$; this is the only critical point. We evaluate $A(x)$ at the endpoints of our interval and at this critical point to find the extreme values; in this case, all we care about is the maximum.

Clearly $A(0) = 0$ and $A(50) = 0$, whereas $A(25) = 625\text{ft}^2$. This is the maximum. Since we earlier found $y = 50 - x$, we find that y is also 25. Thus the dimensions of the rectangular enclosure with perimeter of 100 ft. with maximum area is a square, with sides of length 25 ft.

This example is very simplistic and a bit contrived. (After all, most people create a design then buy fencing to meet their needs, and not buy fencing and plan later.) But it models well the necessary process: create equations that describe a situation, reduce an equation to a single variable, then find the needed extreme value.

"In real life," problems are much more complex. The equations are often *not* reducible to a single variable (hence multi–variable calculus is needed) and the equations themselves may be difficult to form. Understanding the principles here will provide a good foundation for the mathematics you will likely encounter later.

We outline here the basic process of solving these optimization problems.

Key Idea 6   Solving Optimization Problems

1. Understand the problem. Clearly identify what quantity is to be maximized or minimized. Make a sketch if helpful.

2. Create equations relevant to the context of the problem, using the information given. (One of these should describe the quantity to be optimized. We'll call this the *fundamental equation*.)

3. If the fundamental equation defines the quantity to be optimized as a function of more than one variable, reduce it to a single variable function using substitutions derived from the other equations.

(continued). . .

Notes:

Key Idea 6   Solving Optimization Problems – Continued

4. Identify the domain of this function, keeping in mind the context of the problem.

5. Find the extreme values of this function on the determined domain.

6. Identify the values of all relevant quantities of the problem.

We will use Key Idea 6 in a variety of examples.

Example 103   Optimization: perimeter and area

Here is another classic calculus problem: A woman has a 100 feet of fencing, a small dog, and a large yard that contains a stream (that is mostly straight). She wants to create a rectangular enclosure with maximal area that uses the stream as one side. (Apparently her dog won't swim away.) What dimensions provide the maximal area?

SOLUTION   We will follow the steps outlined by Key Idea 6.

1. We are maximizing *area*. A sketch of the region will help; Figure 4.9 gives two sketches of the proposed enclosed area. A key feature of the sketches is to acknowledge that one side is not fenced.

2. We want to maximize the area; as in the example before,

$$\text{Area} = xy.$$

This is our fundamental equation. This defines area as a function of two variables, so we need another equation to reduce it to one variable.

We again appeal to the perimeter; here the perimeter is

$$\text{Perimeter} = 100 = x + 2y.$$

Note how this is different than in our previous example.

3. We now reduce the fundamental equation to a single variable. In the perimeter equation, solve for y: $y = 50 - x/2$. We can now write Area as

$$\text{Area} = A(x) = x(50 - x/2) = 50x - \frac{1}{2}x^2.$$

Area is now defined as a function of one variable.

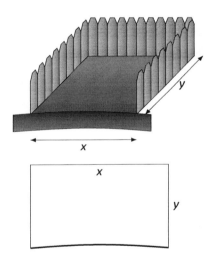

Figure 4.9: A sketch of the enclosure in Example 103.

Notes:

4. We want the area to be nonnegative. Since $A(x) = x(50 - x/2)$, we want $x \geq 0$ and $50 - x/2 \geq 0$. The latter inequality implies that $x \leq 100$, so $0 \leq x \leq 100$.

5. We now find the extreme values. At the endpoints, the minimum is found, giving an area of 0.

   Find the critical points. We have $A'(x) = 50 - x$; setting this equal to 0 and solving for x returns $x = 50$. This gives an area of

$$A(50) = 50(25) = 1250.$$

6. We earlier set $y = 50 - x/2$; thus $y = 25$. Thus our rectangle will have two sides of length 25 and one side of length 50, with a total area of 1250 ft^2.

Keep in mind as we do these problems that we are practicing a *process*; that is, we are learning to turn a situation into a system of equations. These equations allow us to write a certain quantity as a function of one variable, which we then optimize.

Example 104   Optimization: minimizing cost
A power line needs to be run from an power station located on the beach to an offshore facility. Figure 4.10 shows the distances between the power station to the facility.

It costs $50/ft. to run a power line along the land, and $130/ft. to run a power line under water. How much of the power line should be run along the land to minimize the overall cost? What is the minimal cost?

SOLUTION   We will follow the strategy of Key Idea 6 implicitly, without specifically numbering steps.

There are two immediate solutions that we could consider, each of which we will reject through "common sense." First, we could minimize the distance by directly connecting the two locations with a straight line. However, this requires that all the wire be laid underwater, the most costly option. Second, we could minimize the underwater length by running a wire all 5000 ft. along the beach, directly across from the offshore facility. This has the undesired effect of having the longest distance of all, probably ensuring a non–minimal cost.

The optimal solution likely has the line being run along the ground for a while, then underwater, as the figure implies. We need to label our unknown distances – the distance run along the ground and the distance run underwater. Recognizing that the underwater distance can be measured as the hypotenuse of a right triangle, we choose to label the distances as shown in Figure 4.11.

1000 ft

5000 ft

Figure 4.10: Running a power line from the power station to an offshore facility with minimal cost in Example 104.

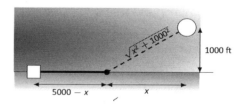

$\sqrt{x^2 + 1000^2}$

1000 ft

5000 − x   x

Figure 4.11: Labeling unknown distances in Example 104.

Notes:

By choosing x as we did, we make the expression under the square root simple. We now create the cost function.

$$\begin{array}{rlcl} \text{Cost} & = & \text{land cost} & + & \text{water cost} \\ & & \$50 \times \text{land distance} & + & \$130 \times \text{water distance} \\ & & 50(5000 - x) & + & 130\sqrt{x^2 + 1000^2}. \end{array}$$

So we have $c(x) = 50(5000 - x) + 130\sqrt{x^2 + 1000^2}$. This function only makes sense on the interval $[0, 5000]$. While we are fairly certain the endpoints will not give a minimal cost, we still evaluate $c(x)$ at each to verify.

$$c(0) = 380,000 \qquad c(5000) \approx 662,873.$$

We now find the critical values of $c(x)$. We compute $c'(x)$ as

$$c'(x) = -50 + \frac{130x}{\sqrt{x^2 + 1000^2}}.$$

Recognize that this is never undefined. Setting $c'(x) = 0$ and solving for x, we have:

$$-50 + \frac{130x}{\sqrt{x^2 + 1000^2}} = 0$$

$$\frac{130x}{\sqrt{x^2 + 1000^2}} = 50$$

$$\frac{130^2 x^2}{x^2 + 1000^2} = 50^2$$

$$130^2 x^2 = 50^2(x^2 + 1000^2)$$

$$130^2 x^2 - 50^2 x^2 = 50^2 \cdot 1000^2$$

$$(130^2 - 50^2)x^2 = 50,000^2$$

$$x^2 = \frac{50,000^2}{130^2 - 50^2}$$

$$x = \frac{50,000}{\sqrt{130^2 - 50^2}}$$

$$x = \frac{50,000}{120} = 416\frac{2}{3} \approx 416.67.$$

Evaluating $c(x)$ at $x = 416.67$ gives a cost of about \$370,000. The distance the power line is laid along land is $5000 - 416.67 = 4583.33$ ft., and the underwater distance is $\sqrt{416.67^2 + 1000^2} \approx 1083$ ft.

Notes:

In the exercises you will see a variety of situations that require you to combine problem–solving skills with calculus. Focus on the *process*; learn how to form equations from situations that can be manipulated into what you need. Eschew memorizing how to do "this kind of problem" as opposed to "that kind of problem." Learning a process will benefit one far longer than memorizing a specific technique.

The next section introduces our final application of the derivative: *differentials*. Given $y = f(x)$, they offer a method of approximating the change in y after x changes by a small amount.

Notes:

Exercises 4.3

Terms and Concepts

1. T/F: An "optimization problem" is essentially an "extreme values" problem in a "story problem" setting.

2. T/F: This section teaches one to find the extreme values of function that have more than one variable.

Problems

3. Find the maximum product of two numbers (not necessarily integers) that have a sum of 100.

4. Find the minimum sum of two numbers whose product is 500.

5. Find the maximum sum of two numbers whose product is 500.

6. Find the maximum sum of two numbers, each of which is in $[0, 300]$ whose product is 500.

7. Find the maximal area of a right triangle with hypotenuse of length 1.

8. A rancher has 1000 feet of fencing in which to construct adjacent, equally sized rectangular pens. What dimensions should these pens have to maximize the enclosed area?

9. A standard soda can is roughly cylindrical and holds 355cm³ of liquid. What dimensions should the cylinder be to minimize the material needed to produce the can? Based on your dimensions, determine whether or not the standard can is produced to minimize the material costs.

10. Find the dimensions of a cylindrical can with a volume of 206in³ that minimizes the surface area.

    The "#10 can" is a standard sized can used by the restaurant industry that holds about 206in³ with a diameter of 6 2/16in and height of 7in. Does it seem these dimensions where chosen with minimization in mind?

11. The United States Postal Service charges more for boxes whose combined length and girth exceeds 108" (the "length" of a package is the length of its longest side; the girth is the perimeter of the cross section, i.e., $2w + 2h$).

What is the maximum volume of a package with a square cross section ($w = h$) that does not exceed the 108" standard?

12. The strength S of a wooden beam is directly proportional to its cross sectional width w and the square of its height h; that is, $S = kwh^2$ for some constant k.

Given a circular log with diameter of 12 inches, what sized beam can be cut from the log with maximum strength?

13. A power line is to be run to an offshore facility in the manner described in Example 104. The offshore facility is 2 miles at sea and 5 miles along the shoreline from the power plant. It costs $50,000 per mile to lay a power line underground and $80,000 to run the line underwater.

How much of the power line should be run underground to minimize the overall costs?

14. A power line is to be run to an offshore facility in the manner described in Example 104. The offshore facility is 5 miles at sea and 2 miles along the shoreline from the power plant. It costs $50,000 per mile to lay a power line underground and $80,000 to run the line underwater.

How much of the power line should be run underground to minimize the overall costs?

15. A woman throws a stick into a lake for her dog to fetch; the stick is 20 feet down the shore line and 15 feet into the water from there. The dog may jump directly into the water and swim, or run along the shore line to get closer to the stick before swimming. The dog runs about 22ft/s and swims about 1.5ft/s.

How far along the shore should the dog run to minimize the time it takes to get to the stick? (Hint: the figure from Example 104 can be useful.)

16. A woman throws a stick into a lake for her dog to fetch; the stick is 15 feet down the shore line and 30 feet into the water from there. The dog may jump directly into the water and swim, or run along the shore line to get closer to the stick before swimming. The dog runs about 22ft/s and swims about 1.5ft/s.

How far along the shore should the dog run to minimize the time it takes to get to the stick? *(Google "calculus dog" to learn more about a dog's ability to minimize times.)*

17. What are the dimensions of the rectangle with largest area that can be drawn inside the unit circle?

4.4   Differentials

In Section 2.2 we explored the meaning and use of the derivative. This section starts by revisiting some of those ideas.

Recall that the derivative of a function f can be used to find the slopes of lines tangent to the graph of f. At $x = c$, the tangent line to the graph of f has equation

$$y = f'(c)(x - c) + f(c).$$

The tangent line can be used to find good approximations of $f(x)$ for values of x near c.

For instance, we can approximate $\sin 1.1$ using the tangent line to the graph of $f(x) = \sin x$ at $x = \pi/3 \approx 1.05$. Recall that $\sin(\pi/3) = \sqrt{3}/2 \approx 0.866$, and $\cos(\pi/3) = 1/2$. Thus the tangent line to $f(x) = \sin x$ at $x = \pi/3$ is:

$$\ell(x) = \frac{1}{2}(x - \pi/3) + 0.866.$$

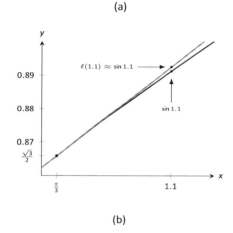

(a)

(b)

Figure 4.12: Graphing $f(x) = \sin x$ and its tangent line at $x = \pi/3$ in order to estimate $\sin 1.1$.

In Figure 4.12(a), we see a graph of $f(x) = \sin x$ graphed along with its tangent line at $x = \pi/3$. The small rectangle shows the region that is displayed in Figure 4.12(b). In this figure, we see how we are approximating $\sin 1.1$ with the tangent line, evaluated at 1.1. Together, the two figures show how close these values are.

Using this line to approximate $\sin 1.1$, we have:

$$\ell(1.1) = \frac{1}{2}(1.1 - \pi/3) + 0.866$$

$$= \frac{1}{2}(0.053) + 0.866 = 0.8925.$$

(We leave it to the reader to see how good of an approximation this is.)

We now generalize this concept. Given $f(x)$ and an x–value c, the tangent line is $\ell(x) = f'(c)(x - c) + f(c)$. Clearly, $f(c) = \ell(c)$. Let Δx be a small number, representing a small change in x value. We assert that:

$$f(c + \Delta x) \approx \ell(c + \Delta x),$$

since the tangent line to a function approximates well the values of that function near $x = c$.

As the x value changes from c to $c + \Delta x$, the y value of f changes from $f(c)$ to $f(c + \Delta x)$. We call this change of y value Δy. That is:

$$\Delta y = f(c + \Delta x) - f(c).$$

Notes:

Replacing $f(c + \Delta x)$ with its tangent line approximation, we have

$$
\begin{aligned}
\Delta y &\approx \ell(c + \Delta x) - f(c) \\
&= f'(c)\big((c + \Delta x) - c\big) + f(c) - f(c) \\
&= f'(c)\Delta x
\end{aligned}
\tag{4.3}
$$

This final equation is important; we'll come back to it in Key Idea 7.

We introduce two new variables, dx and dy in the context of a formal definition.

Definition 18 Differentials of x and y.

Let $y = f(x)$ be differentiable. The **differential of** x, denoted dx, is any nonzero real number (usually taken to be a small number). The **differential of** y, denoted dy, is

$$dy = f'(x)dx.$$

We can solve for $f'(x)$ in the above equation: $f'(x) = dy/dx$. This states that the derivative of f with respect to x is the differential of y divided by the differential of x; this is **not** the alternate notation for the derivative, $\frac{dy}{dx}$. This latter notation was chosen because of the fraction–like qualities of the derivative, but again, it is one symbol and not a fraction.

It is helpful to organize our new concepts and notations in one place.

Key Idea 7 Differential Notation

Let $y = f(x)$ be a differentiable function.

1. Δx represents a small, nonzero change in x value.

2. dx represents a small, nonzero change in x value (i.e., $\Delta x = dx$).

3. Δy is the change in y value as x changes by Δx; hence

$$\Delta y = f(x + \Delta x) - f(x).$$

4. $dy = f'(x)dx$ which, by Equation (4.3), is an *approximation* of the change in y value as x changes by Δx; $dy \approx \Delta y$.

Notes:

What is the value of differentials? Like many mathematical concepts, differentials provide both practical and theoretical benefits. We explore both here.

Example 105   Finding and using differentials
Consider $f(x) = x^2$. Knowing $f(3) = 9$, approximate $f(3.1)$.

SOLUTION   The x value is changing from $x = 3$ to $x = 3.1$; therefore, we see that $dx = 0.1$. If we know how much the y value changes from $f(3)$ to $f(3.1)$ (i.e., if we know Δy), we will know exactly what $f(3.1)$ is (since we already know $f(3)$). We can approximate Δy with dy.

$$\Delta y \approx dy$$
$$= f'(3)dx$$
$$= 2 \cdot 3 \cdot 0.1 = 0.6.$$

We expect the y value to change by about 0.6, so we approximate $f(3.1) \approx 9.6$.

We leave it to the reader to verify this, but the preceding discussion links the differential to the tangent line of $f(x)$ at $x = 3$. One can verify that the tangent line, evaluated at $x = 3.1$, also gives $y = 9.6$.

Of course, it is easy to compute the actual answer (by hand or with a calculator): $3.1^2 = 9.61$. (Before we get too cynical and say "Then why bother?", note our approximation is *really* good!)

So why bother?

In "most" real life situations, we do not know the function that describes a particular behavior. Instead, we can only take measurements of how things change – measurements of the derivative.

Imagine water flowing down a winding channel. It is easy to measure the speed and direction (i.e., the *velocity*) of water at any location. It is very hard to create a function that describes the overall flow, hence it is hard to predict where a floating object placed at the beginning of the channel will end up. However, we can *approximate* the path of an object using differentials. Over small intervals, the path taken by a floating object is essentially linear. Differentials allow us to approximate the true path by piecing together lots of short, linear paths. This technique is called Euler's Method, studied in introductory Differential Equations courses.

We use differentials once more to approximate the value of a function. Even though calculators are very accessible, it is neat to see how these techniques can sometimes be used to easily compute something that looks rather hard.

Notes:

Example 106   **Using differentials to approximate a function value**
Approximate $\sqrt{4.5}$.

SOLUTION   We expect $\sqrt{4.5} \approx 2$, yet we can do better. Let $f(x) = \sqrt{x}$, and let $c = 4$. Thus $f(4) = 2$. We can compute $f'(x) = 1/(2\sqrt{x})$, so $f'(4) = 1/4$.

We approximate the difference between $f(4.5)$ and $f(4)$ using differentials, with $dx = 0.5$:

$$f(4.5) - f(4) = \Delta y \approx dy = f'(4) \cdot dx = 1/4 \cdot 1/2 = 1/8 = 0.125.$$

The approximate change in f from $x = 4$ to $x = 4.5$ is 0.125, so we approximate $\sqrt{4.5} \approx 2.125$.

Differentials are important when we discuss *integration*. When we study that topic, we will use notation such as

$$\int f(x)\, dx$$

quite often. While we don't discuss here what all of that notation means, note the existence of the differential dx. Proper handling of *integrals* comes with proper handling of differentials.

In light of that, we practice finding differentials in general.

Example 107   **Finding differentials**
In each of the following, find the differential dy.

   1. $y = \sin x$ 2. $y = e^x(x^2 + 2)$ 3. $y = \sqrt{x^2 + 3x - 1}$

SOLUTION

1. $y = \sin x$:   As $f(x) = \sin x$, $f'(x) = \cos x$. Thus

$$dy = \cos(x)dx.$$

2. $y = e^x(x^2 + 2)$:   Let $f(x) = e^x(x^2 + 2)$. We need $f'(x)$, requiring the Product Rule.

We have $f'(x) = e^x(x^2 + 2) + 2xe^x$, so

$$dy = \left(e^x(x^2 + 2) + 2xe^x\right)dx.$$

Notes:

3. $y = \sqrt{x^2 + 3x - 1}$:   Let $f(x) = \sqrt{x^2 + 3x - 1}$; we need $f'(x)$, requiring the Chain Rule.

We have $f'(x) = \dfrac{1}{2}(x^2 + 3x - 1)^{-\frac{1}{2}}(2x + 3) = \dfrac{2x + 3}{2\sqrt{x^2 + 3x - 1}}$. Thus

$$dy = \frac{(2x + 3)dx}{2\sqrt{x^2 + 3x - 1}}.$$

Finding the differential dy of $y = f(x)$ is really no harder than finding the derivative of f; we just *multiply* $f'(x)$ by dx. It is important to remember that we are not simply adding the symbol "dx" at the end.

We have seen a practical use of differentials as they offer a good method of making certain approximations. Another use is *error propagation*. Suppose a length is measured to be x, although the actual value is $x + \Delta x$ (where we hope Δx is small). This measurement of x may be used to compute some other value; we can think of this as $f(x)$ for some function f. As the true length is $x + \Delta x$, one really should have computed $f(x + \Delta x)$. The difference between $f(x)$ and $f(x + \Delta x)$ is the propagated error.

How close are $f(x)$ and $f(x + \Delta x)$? This is a difference in "y" values;

$$f(x + \Delta x) - f(x) = \Delta y \approx dy.$$

We can approximate the propagated error using differentials.

Example 108   **Using differentials to approximate propagated error**
A steel ball bearing is to be manufactured with a diameter of 2cm. The manufacturing process has a tolerance of ± 0.1mm in the diameter. Given that the density of steel is about 7.85g/cm^3, estimate the propagated error in the mass of the ball bearing.

SOLUTION   The mass of a ball bearing is found using the equation "mass = volume $\times$ density." In this situation the mass function is a product of the radius of the ball bearing, hence it is $m = 7.85\frac{4}{3}\pi r^3$. The differential of the mass is

$$dm = 31.4\pi r^2 dr.$$

The radius is to be 1cm; the manufacturing tolerance in the radius is ± 0.05mm, or ± 0.005cm. The propagated error is approximately:

$$\Delta m \approx dm$$
$$= 31.4\pi(1)^2(\pm 0.005)$$
$$= \pm 0.493\text{g}$$

Notes:

Is this error significant? It certainly depends on the application, but we can get an idea by computing the *relative error*. The ratio between amount of error to the total mass is

$$
\begin{aligned}
\frac{dm}{m} &= \pm \frac{0.493}{7.85 \frac{4}{3}\pi} \\
&= \pm \frac{0.493}{32.88} \\
&= \pm 0.015,
\end{aligned}
$$

or $\pm 1.5\%$.

We leave it to the reader to confirm this, but if the diameter of the ball was supposed to be 10cm, the same manufacturing tolerance would give a propagated error in mass of ± 12.33g, which corresponds to a *percent error* of $\pm 0.188\%$. While the amount of error is much greater ($12.33 > 0.493$), the percent error is much lower.

We first learned of the derivative in the context of instantaneous rates of change and slopes of tangent lines. We furthered our understanding of the power of the derivative by studying how it relates to the graph of a function (leading to ideas of increasing/decreasing and concavity). This chapter has put the derivative to yet more uses:

- Equation solving (Newton's Method)

- Related Rates (furthering our use of the derivative to find instantaneous rates of change)

- Optimization (applied extreme values), and

- Differentials (useful for various approximations and for something called integration).

In the next chapters, we will consider the "reverse" problem to computing the derivative: given a function f, can we find a function whose derivative is f? Be able to do so opens up an incredible world of mathematics and applications.

Notes:

Exercises 4.4

Terms and Concepts

1. T/F: Given a differentiable function $y = f(x)$, we are generally free to choose a value for dx, which then determines the value of dy.

2. T/F: The symbols "dx" and "Δx" represent the same concept.

3. T/F: The symbols "dy" and "Δy" represent the same concept.

4. T/F: Differentials are important in the study of integration.

5. How are differentials and tangent lines related?

Problems

In Exercises 6 – 17, use differentials to approximate the given value by hand.

6. 2.05^2

7. 5.93^2

8. 5.1^3

9. 6.8^3

10. $\sqrt{16.5}$

11. $\sqrt{24}$

12. $\sqrt[3]{63}$

13. $\sqrt[3]{8.5}$

14. $\sin 3$

15. $\cos 1.5$

16. $e^{0.1}$

In Exercises 17 – 29, compute the differential dy.

17. $y = x^2 + 3x - 5$

18. $y = x^7 - x^5$

19. $y = \dfrac{1}{4x^2}$

20. $y = (2x + \sin x)^2$

21. $y = x^2 e^{3x}$

22. $y = \dfrac{4}{x^4}$

23. $y = \dfrac{2x}{\tan x + 1}$

24. $y = \ln(5x)$

25. $y = e^x \sin x$

26. $y = \cos(\sin x)$

27. $y = \dfrac{x+1}{x+2}$

28. $y = 3^x \ln x$

29. $y = x \ln x - x$

30. A set of plastic spheres are to be made with a diameter of 1cm. If the manufacturing process is accurate to 1mm, what is the propagated error in volume of the spheres?

31. The distance, in feet, a stone drops in t seconds is given by $d(t) = 16t^2$. The depth of a hole is to be approximated by dropping a rock and listening for it to hit the bottom. What is the propagated error if the time measurement is accurate to $2/10^{\text{ths}}$ of a second and the measured time is:

    (a) 2 seconds?

    (b) 5 seconds?

32. What is the propagated error in the measurement of the cross sectional area of a circular log if the diameter is measured at 15″, accurate to $1/4$″?

33. A wall is to be painted that is 8′ high and is measured to be 10′, 7″ long. Find the propagated error in the measurement of the wall's surface area if the measurement is accurate to $1/2$″.

Exercises 34 – 38 explore some issues related to surveying in which distances are approximated using other measured distances and measured angles. (Hint: Convert all angles to radians before computing.)

34. The length l of a long wall is to be approximated. The angle θ, as shown in the diagram (not to scale), is measured to be 85.2°, accurate to 1°. Assume that the triangle formed is a right triangle.

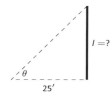

    (a) What is the measured length l of the wall?

    (b) What is the propagated error?

    (c) What is the percent error?

35. Answer the questions of Exercise 34, but with a measured angle of 71.5°, accurate to 1°, measured from a point 100' from the wall.

36. The length l of a long wall is to be calculated by measuring the angle θ shown in the diagram (not to scale). Assume the formed triangle is an isosceles triangle. The measured angle is 143°, accurate to 1°.

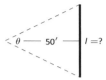

(a) What is the measured length of the wall?

(b) What is the propagated error?

(c) What is the percent error?

37. The length of the walls in Exercises 34 − 36 are essentially the same. Which setup gives the most accurate result?

38. Consider the setup in Exercises 36. This time, assume the angle measurement of 143° is exact but the measured 50' from the wall is accurate to 6″. What is the approximate percent error?

5: INTEGRATION

We have spent considerable time considering the derivatives of a function and their applications. In the following chapters, we are going to starting thinking in "the other direction." That is, given a function $f(x)$, we are going to consider functions $F(x)$ such that $F'(x) = f(x)$. There are numerous reasons this will prove to be useful: these functions will help us compute areas, volumes, mass, force, pressure, work, and much more.

5.1 Antiderivatives and Indefinite Integration

Given a function $y = f(x)$, a *differential equation* is one that incorporates y, x, and the derivatives of y. For instance, a simple differential equation is:

$$y' = 2x.$$

Solving a differential equation amounts to finding a function y that satisfies the given equation. Take a moment and consider that equation; can you find a function y such that $y' = 2x$?

Can you find another?

And yet another?

Hopefully one was able to come up with at least one solution: $y = x^2$. "Finding another" may have seemed impossible until one realizes that a function like $y = x^2 + 1$ also has a derivative of $2x$. Once that discovery is made, finding "yet another" is not difficult; the function $y = x^2 + 123,456,789$ also has a derivative of $2x$. The differential equation $y' = 2x$ has many solutions. This leads us to some definitions.

Definition 19   Antiderivatives and Indefinite Integrals

Let a function $f(x)$ be given. An **antiderivative** of $f(x)$ is a function $F(x)$ such that $F'(x) = f(x)$.

The set of all antiderivatives of $f(x)$ is the **indefinite integral of** f, denoted by

$$\int f(x)\, dx.$$

Make a note about our definition: we refer to *an* antiderivative of f, as opposed to *the* antiderivative of f, since there is *always* an infinite number of them.

We often use upper-case letters to denote antiderivatives.

Knowing one antiderivative of f allows us to find infinitely more, simply by adding a constant. Not only does this give us *more* antiderivatives, it gives us *all* of them.

Theorem 34   Antiderivative Forms

Let $F(x)$ and $G(x)$ be antiderivatives of $f(x)$. Then there exists a constant C such that

$$G(x) = F(x) + C.$$

Given a function f and one of its antiderivatives F, we know *all* antiderivatives of f have the form $F(x) + C$ for some constant C. Using Definition 19, we can say that

$$\int f(x)\, dx = F(x) + C.$$

Let's analyze this indefinite integral notation.

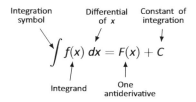

Figure 5.1: Understanding the indefinite integral notation.

Figure 5.1 shows the typical notation of the indefinite integral. The integration symbol, $\int$, is in reality an "elongated S," representing "take the sum." We will later see how *sums* and *antiderivatives* are related.

The function we want to find an antiderivative of is called the *integrand*. It contains the differential of the variable we are integrating with respect to. The $\int$ symbol and the differential dx are not "bookends" with a function sandwiched in between; rather, the symbol $\int$ means "find all antiderivatives of what follows," and the function $f(x)$ and dx are multiplied together; the dx does not "just sit there."

Let's practice using this notation.

Example 109   Evaluating indefinite integrals
Evaluate $\int \sin x\, dx$.

Notes:

SOLUTION   We are asked to find all functions $F(x)$ such that $F'(x) = \sin x$. Some thought will lead us to one solution: $F(x) = -\cos x$, because $\frac{d}{dx}(-\cos x) = \sin x$.

The indefinite integral of $\sin x$ is thus $-\cos x$, plus a constant of integration. So:

$$\int \sin x \, dx = -\cos x + C.$$

A commonly asked question is "What happened to the dx?" The unenlightened response is "Don't worry about it. It just goes away." A full understanding includes the following.

This process of *antidifferentiation* is really solving a *differential* question. The integral

$$\int \sin x \, dx$$

presents us with a differential, $dy = \sin x \, dx$. It is asking: "What is y?" We found lots of solutions, all of the form $y = -\cos x + C$.

Letting $dy = \sin x \, dx$, rewrite

$$\int \sin x \, dx \quad \text{as} \quad \int dy.$$

This is asking: "What functions have a differential of the form dy?" The answer is "Functions of the form $y + C$, where C is a constant." What is y? We have lots of choices, all differing by a constant; the simplest choice is $y = -\cos x$.

Understanding all of this is more important later as we try to find antiderivatives of more complicated functions. In this section, we will simply explore the rules of indefinite integration, and one can succeed for now with answering "What happened to the dx?" with "It went away."

Let's practice once more before stating integration rules.

Example 110   Evaluating indefinite integrals

Evaluate $\int (3x^2 + 4x + 5) \, dx$.

SOLUTION   We seek a function $F(x)$ whose derivative is $3x^2 + 4x + 5$. When taking derivatives, we can consider functions term–by–term, so we can likely do that here.

What functions have a derivative of $3x^2$? Some thought will lead us to a cubic, specifically $x^3 + C_1$, where C_1 is a constant.

What functions have a derivative of $4x$? Here the x term is raised to the first power, so we likely seek a quadratic. Some thought should lead us to $2x^2 + C_2$, where C_2 is a constant.

Notes:

Finally, what functions have a derivative of 5? Functions of the form $5x + C_3$, where C_3 is a constant.

Our answer appears to be

$$\int (3x^2 + 4x + 5) \, dx = x^3 + C_1 + 2x^2 + C_2 + 5x + C_3.$$

We do not need three separate constants of integration; combine them as one constant, giving the final answer of

$$\int (3x^2 + 4x + 5) \, dx = x^3 + 2x^2 + 5x + C.$$

It is easy to verify our answer; take the derivative of $x^3 + 2x^3 + 5x + C$ and see we indeed get $3x^2 + 4x + 5$.

This final step of "verifying our answer" is important both practically and theoretically. In general, taking derivatives is easier than finding antiderivatives so checking our work is easy and vital as we learn.

We also see that taking the derivative of our answer returns the function in the integrand. Thus we can say that:

$$\frac{d}{dx} \left(\int f(x) \, dx \right) = f(x).$$

Differentiation "undoes" the work done by antidifferentiation.

Theorem 24 gave a list of the derivatives of common functions we had learned at that point. We restate part of that list here to stress the relationship between derivatives and antiderivatives. This list will also be useful as a glossary of common antiderivatives as we learn.

Notes:

Theorem 35   Derivatives and Antiderivatives

Common Differentiation Rules   Common Indefinite Integral Rules

1. $\frac{d}{dx}\left(cf(x)\right) = c \cdot f'(x)$

1. $\int c \cdot f(x)\, dx = c \cdot \int f(x)\, dx$

2. $\frac{d}{dx}\left(f(x) \pm g(x)\right) =$ $f'(x) \pm g'(x)$

2. $\int \left(f(x) \pm g(x)\right) dx =$ $\int f(x)\, dx \pm \int g(x)\, dx$

3. $\frac{d}{dx}(C) = 0$

3. $\int 0\, dx = C$

4. $\frac{d}{dx}(x) = 1$

4. $\int 1\, dx = \int dx = x + C$

5. $\frac{d}{dx}(x^n) = n \cdot x^{n-1}$

5. $\int x^n\, dx = \frac{1}{n+1}x^{n+1} + C \quad (n \neq -1)$

6. $\frac{d}{dx}\left(\sin x\right) = \cos x$

6. $\int \cos x\, dx = \sin x + C$

7. $\frac{d}{dx}\left(\cos x\right) = -\sin x$

7. $\int \sin x\, dx = -\cos x + C$

8. $\frac{d}{dx}\left(\tan x\right) = \sec^2 x$

8. $\int \sec^2 x\, dx = \tan x + C$

9. $\frac{d}{dx}\left(\csc x\right) = -\csc x \cot x$

9. $\int \csc x \cot x\, dx = -\csc x + C$

10. $\frac{d}{dx}\left(\sec x\right) = \sec x \tan x$

10. $\int \sec x \tan x\, dx = \sec x + C$

11. $\frac{d}{dx}\left(\cot x\right) = -\csc^2 x$

11. $\int \csc^2 x\, dx = -\cot x + C$

12. $\frac{d}{dx}(e^x) = e^x$

12. $\int e^x\, dx = e^x + C$

13. $\frac{d}{dx}(a^x) = \ln a \cdot a^x$

13. $\int a^x\, dx = \frac{1}{\ln a} \cdot a^x + C$

14. $\frac{d}{dx}\left(\ln x\right) = \frac{1}{x}$

14. $\int \frac{1}{x}\, dx = \ln |x| + C$

We highlight a few important points from Theorem 35:

- Rule #1 states $\int c \cdot f(x)\, dx = c \cdot \int f(x)\, dx$. This is the Constant Multiple Rule: we can temporarily ignore constants when finding antiderivatives, just as we did when computing derivatives (i.e., $\frac{d}{dx}\left(3x^2\right)$ is just as easy to compute as $\frac{d}{dx}\left(x^2\right)$). An example:

$$\int 5 \cos x\, dx = 5 \cdot \int \cos x\, dx = 5 \cdot (\sin x + C) = 5 \sin x + C.$$

In the last step we can consider the constant as also being multiplied by

Notes:

5, but "5 times a constant" is still a constant, so we just write "C".

- Rule #2 is the Sum/Difference Rule: we can split integrals apart when the integrand contains terms that are added/subtracted, as we did in Example 110. So:

$$\int (3x^2 + 4x + 5)\, dx = \int 3x^2\, dx + \int 4x\, dx + \int 5\, dx$$
$$= 3\int x^2\, dx + 4\int x\, dx + \int 5\, dx$$
$$= 3 \cdot \frac{1}{3}x^3 + 4 \cdot \frac{1}{2}x^2 + 5x + C$$
$$= x^3 + 2x^2 + 5x + C$$

In practice we generally do not write out all these steps, but we demonstrate them here for completeness.

- Rule #5 is the Power Rule of indefinite integration. There are two important things to keep in mind:

    1. Notice the restriction that $n \neq -1$. This is important: $\int \frac{1}{x}\, dx \neq$ "$\frac{1}{0}x^0 + C$"; rather, see Rule #14.

    2. We are presenting antidifferentiation as the "inverse operation" of differentiation. Here is a useful quote to remember:

        "Inverse operations do the opposite things in the opposite order."

        When taking a derivative using the Power Rule, we **first** *multiply* by the power, then **second** *subtract* 1 from the power. To find the antiderivative, do the opposite things in the opposite order: **first** *add* one to the power, then **second** *divide* by the power.

- Note that Rule #14 incorporates the absolute value of x. The exercises will work the reader through why this is the case; for now, know the absolute value is important and cannot be ignored.

Initial Value Problems

In Section 2.3 we saw that the derivative of a position function gave a velocity function, and the derivative of a velocity function describes acceleration. We can now go "the other way:" the antiderivative of an acceleration function gives a velocity function, etc. While there is just one derivative of a given function, there are infinite antiderivatives. Therefore we cannot ask "What is *the* velocity of an object whose acceleration is -32ft/s^2?", since there is more than one answer.

Notes:

We can find *the* answer if we provide more information with the question, as done in the following example. Often the additional information comes in the form of an *initial value*, a value of the function that one knows beforehand.

Example 111   **Solving initial value problems**
The acceleration due to gravity of a falling object is -32 ft/s^2. At time $t = 3$, a falling object had a velocity of -10 ft/s. Find the equation of the object's velocity.

    SOLUTION   We want to know a velocity function, $v(t)$. We know two things:

- The acceleration, i.e., $v'(t) = -32$, and

- the velocity at a specific time, i.e., $v(3) = -10$.

Using the first piece of information, we know that $v(t)$ is an antiderivative of $v'(t) = -32$. So we begin by finding the indefinite integral of -32:

$$\int (-32)\, dt = -32t + C = v(t).$$

Now we use the fact that $v(3) = -10$ to find C:

$$v(t) = -32t + C$$
$$v(3) = -10$$
$$-32(3) + C = -10$$
$$C = 86$$

Thus $v(t) = -32t + 86$. We can use this equation to understand the motion of the object: when $t = 0$, the object had a velocity of $v(0) = 86$ ft/s. Since the velocity is positive, the object was moving upward.

When did the object begin moving down? Immediately after $v(t) = 0$:

$$-32t + 86 = 0 \quad \Rightarrow \quad t = \frac{43}{16} \approx 2.69\text{s}.$$

Recognize that we are able to determine quite a bit about the path of the object knowing just its acceleration and its velocity at a single point in time.

Example 112   **Solving initial value problems**
Find $f(t)$, given that $f''(t) = \cos t$, $f'(0) = 3$ and $f(0) = 5$.

    SOLUTION   We start by finding $f'(t)$, which is an antiderivative of $f''(t)$:

$$\int f''(t)\, dt = \int \cos t\, dt = \sin t + C = f'(t).$$

Notes:

So $f'(t) = \sin t + C$ for the correct value of C. We are given that $f'(0) = 3$, so:

$$f'(0) = 3 \quad \Rightarrow \quad \sin 0 + C = 3 \quad \Rightarrow \quad C = 3.$$

Using the initial value, we have found $f'(t) = \sin t + 3$.

We now find $f(t)$ by integrating again.

$$f(t) = \int f'(t)\, dt = \int (\sin t + 3)\, dt = -\cos t + 3t + C.$$

We are given that $f(0) = 5$, so

$$-\cos 0 + 3(0) + C = 5$$
$$-1 + C = 5$$
$$C = 6$$

Thus $f(t) = -\cos t + 3t + 6$.

This section introduced antiderivatives and the indefinite integral. We found they are needed when finding a function given information about its derivative(s). For instance, we found a position function given a velocity function.

In the next section, we will see how position and velocity are unexpectedly related by the areas of certain regions on a graph of the velocity function. Then, in Section 5.4, we will see how areas and antiderivatives are closely tied together.

Notes:

Exercises 5.1

Terms and Concepts

1. Define the term "antiderivative" in your own words.

2. Is it more accurate to refer to "the" antiderivative of $f(x)$ or "an" antiderivative of $f(x)$?

3. Use your own words to define the indefinite integral of $f(x)$.

4. Fill in the blanks: "Inverse operations do the _____ things in the _____ order."

5. What is an "initial value problem"?

6. The derivative of a position function is a _____ function.

7. The antiderivative of an acceleration function is a _____ function.

Problems

In Exercises 8 – 26, evaluate the given indefinite integral.

8. $\int 3x^3 \, dx$

9. $\int x^8 \, dx$

10. $\int (10x^2 - 2) \, dx$

11. $\int dt$

12. $\int 1 \, ds$

13. $\int \frac{1}{3t^2} \, dt$

14. $\int \frac{3}{t^2} \, dt$

15. $\int \frac{1}{\sqrt{x}} \, dx$

16. $\int \sec^2 \theta \, d\theta$

17. $\int \sin \theta \, d\theta$

18. $\int (\sec x \tan x + \csc x \cot x) \, dx$

19. $\int 5e^\theta \, d\theta$

20. $\int 3^t \, dt$

21. $\int \frac{5^t}{2} \, dt$

22. $\int (2t + 3)^2 \, dt$

23. $\int (t^2 + 3)(t^3 - 2t) \, dt$

24. $\int x^2 x^3 \, dx$

25. $\int e^\pi \, dx$

26. $\int a \, dx$

27. This problem investigates why Theorem 35 states that $\int \frac{1}{x} \, dx = \ln|x| + C$.

   (a) What is the domain of $y = \ln x$?
   (b) Find $\frac{d}{dx} \left(\ln x \right)$.
   (c) What is the domain of $y = \ln(-x)$?
   (d) Find $\frac{d}{dx} \left(\ln(-x) \right)$.
   (e) You should find that $1/x$ has two types of antiderivatives, depending on whether $x > 0$ or $x < 0$. In one expression, give a formula for $\int \frac{1}{x} \, dx$ that takes these different domains into account, and explain your answer.

In Exercises 28 – 38, find $f(x)$ described by the given initial value problem.

28. $f'(x) = \sin x$ and $f(0) = 2$

29. $f'(x) = 5e^x$ and $f(0) = 10$

30. $f'(x) = 4x^3 - 3x^2$ and $f(-1) = 9$

31. $f'(x) = \sec^2 x$ and $f(\pi/4) = 5$

32. $f'(x) = 7^x$ and $f(2) = 1$

33. $f''(x) = 5$ and $f'(0) = 7, f(0) = 3$

34. $f''(x) = 7x$ and $f'(1) = -1, f(1) = 10$

35. $f''(x) = 5e^x$ and $f'(0) = 3, f(0) = 5$

36. $f''(\theta) = \sin \theta$ and $f'(\pi) = 2, f(\pi) = 4$

37. $f''(x) = 24x^2 + 2^x - \cos x$ and $f'(0) = 5, f(0) = 0$

38. $f''(x) = 0$ and $f'(1) = 3, f(1) = 1$

Review

39. Use information gained from the first and second derivatives to sketch $f(x) = \dfrac{1}{e^x + 1}$.

40. Given $y = x^2 e^x \cos x$, find dy.

5.2 The Definite Integral

We start with an easy problem. An object travels in a straight line at a constant velocity of 5 ft/s for 10 seconds. How far away from its starting point is the object?

We approach this problem with the familiar "Distance = Rate × Time" equation. In this case, Distance = 5ft/s × 10s = 50 feet.

It is interesting to note that this solution of 50 feet can be represented graphically. Consider Figure 5.2, where the constant velocity of 5ft/s is graphed on the axes. Shading the area under the line from $t = 0$ to $t = 10$ gives a rectangle with an area of 50 square units; when one considers the units of the axes, we can say this area represents 50 ft.

Now consider a slightly harder situation (and not particularly realistic): an object travels in a straight line with a constant velocity of 5ft/s for 10 seconds, then instantly reverses course at a rate of 2ft/s for 4 seconds. (Since the object is traveling in the opposite direction when reversing course, we say the velocity is a constant -2ft/s.) How far away from the starting point is the object – what is its *displacement*?

Here we use "Distance = Rate$_1$ × Time$_1$ + Rate$_2$ × Time$_2$," which is

$$\text{Distance} = 5 \cdot 10 + (-2) \cdot 4 = 42 \text{ ft.}$$

Hence the object is 42 feet from its starting location.

We can again depict this situation graphically. In Figure 5.3 we have the velocities graphed as straight lines on $[0, 10]$ and $[10, 14]$, respectively. The displacement of the object is

"Area above the t–axis — Area below the t–axis,"

which is easy to calculate as $50 - 8 = 42$ feet.

Now consider a more difficult problem.

Example 113   Finding position using velocity
The velocity of an object moving straight up/down under the acceleration of gravity is given as $v(t) = -32t + 48$, where time t is given in seconds and velocity is in ft/s. When $t = 0$, the object had a height of 0 ft.

1. What was the initial velocity of the object?

2. What was the maximum height of the object?

3. What was the height of the object at time $t = 2$?

SOLUTION   It is straightforward to find the initial velocity; at time $t = 0$, $v(0) = -32 \cdot 0 + 48 = 48$ ft/s.

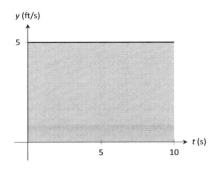

Figure 5.2: The area under a constant velocity function corresponds to distance traveled.

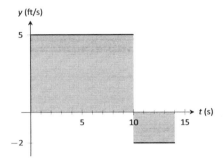

Figure 5.3: The total displacement is the area above the t–axis minus the area below the t–axis.

Notes:

To answer questions about the height of the object, we need to find the object's position function $s(t)$. This is an initial value problem, which we studied in the previous section. We are told the initial height is 0, i.e., $s(0) = 0$. We know $s'(t) = v(t) = -32t + 48$. To find s, we find the indefinite integral of $v(t)$:

$$\int v(t)\, dt = \int (-32t + 48)\, dt = -16t^2 + 48t + C = s(t).$$

Since $s(0) = 0$, we conclude that $C = 0$ and $s(t) = -16t^2 + 48t$.

To find the maximum height of the object, we need to find the maximum of s. Recalling our work finding extreme values, we find the critical points of s by setting its derivative equal to 0 and solving for t:

$$s'(t) = -32t + 48 = 0 \quad \Rightarrow \quad t = 48/32 = 1.5\text{s}.$$

(Notice how we ended up just finding when the velocity was 0ft/s!) The first derivative test shows this is a maximum, so the maximum height of the object is found at

$$s(1.5) = -16(1.5)^2 + 48(1.5) = 36\text{ft}.$$

The height at time $t = 2$ is now straightforward to compute: it is $s(2) = 32$ft.

While we have answered all three questions, let's look at them again graphically, using the concepts of area that we explored earlier.

Figure 5.4 shows a graph of $v(t)$ on axes from $t = 0$ to $t = 3$. It is again straightforward to find $v(0)$. How can we use the graph to find the maximum height of the object?

Recall how in our previous work that the displacement of the object (in this case, its height) was found as the area under the velocity curve, as shaded in the figure. Moreover, the area between the curve and the t–axis that is below the t–axis counted as "negative" area. That is, it represents the object coming back toward its starting position. So to find the maximum distance from the starting point – the maximum height – we find the area under the velocity line that is above the t–axis, i.e., from $t = 0$ to $t = 1.5$. This region is a triangle; its area is

$$\text{Area} = \frac{1}{2}\text{Base} \times \text{Height} = \frac{1}{2} \times 1.5\text{s} \times 48\text{ft/s} = 36\text{ft},$$

which matches our previous calculation of the maximum height.

Finally, we find the total *signed* area under the velocity function from $t = 0$ to $t = 2$ to find the $s(2)$, the height at $t = 2$, which is a displacement, the distance from the current position to the starting position. That is,

Displacement = Area above the t–axis − Area below t–axis.

Figure 5.4: A graph of $v(t) = -32t + 48$; the shaded areas help determine displacement.

The regions are triangles, and we find

$$\text{Displacement} = \frac{1}{2}(1.5\text{s})(48\text{ft/s}) - \frac{1}{2}(.5\text{s})(16\text{ft/s}) = 32\text{ft}.$$

This also matches our previous calculation of the height at $t = 2$.

Notice how we answered each question in this example in two ways. Our first method was to manipulate equations using our understanding of antiderivatives and derivatives. Our second method was geometric: we answered questions looking at a graph and finding the areas of certain regions of this graph.

The above example does not *prove* a relationship between area under a velocity function and displacement, but it does imply a relationship exists. Section 5.4 will fully establish fact that the area under a velocity function is displacement.

Given a graph of a function $y = f(x)$, we will find that there is great use in computing the area between the curve $y = f(x)$ and the x-axis. Because of this, we need to define some terms.

Definition 20   The Definite Integral, Total Signed Area

Let $y = f(x)$ be defined on a closed interval $[a, b]$. The **total signed area from $x = a$ to $x = b$ under f** is:

(area under f and above the x–axis on $[a, b]$) − (area above f and under the x–axis on $[a, b]$).

The **definite integral of f on $[a, b]$** is the total signed area of f on $[a, b]$, denoted

$$\int_a^b f(x)\, dx,$$

where a and b are the **bounds of integration.**

By our definition, the definite integral gives the "signed area under f." We usually drop the word "signed" when talking about the definite integral, and simply say the definite integral gives "the area under f" or, more commonly, "the area under the curve."

The previous section introduced the indefinite integral, which related to antiderivatives. We have now defined the definite integral, which relates to areas under a function. The two are very much related, as we'll see when we learn the Fundamental Theorem of Calculus in Section 5.4. Recall that earlier we said that the "$\int$" symbol was an "elongated S" that represented finding a "sum." In the context of the definite integral, this notation makes a bit more sense, as we are adding up areas under the function f.

Notes:

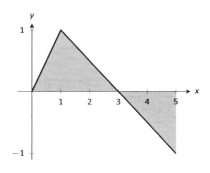

y

1

1 2 3 4 5 x

−1

Figure 5.5: A graph of $f(x)$ in Example 114.

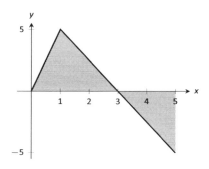

y

5

1 2 3 4 5 x

−5

Figure 5.6: A graph of $5f$ in Example 114. (Yes, it looks just like the graph of f in Figure 5.5, just with a different y-scale.)

We practice using this notation.

Example 114   **Evaluating definite integrals**
Consider the function f given in Figure 5.5.

Find:

1. $\displaystyle\int_0^3 f(x)\,dx$

2. $\displaystyle\int_3^5 f(x)\,dx$

3. $\displaystyle\int_0^5 f(x)\,dx$

4. $\displaystyle\int_0^3 5f(x)\,dx$

5. $\displaystyle\int_1^1 f(x)\,dx$

Solution

1. $\int_0^3 f(x)\,dx$ is the area under f on the interval $[0,3]$. This region is a triangle, so the area is $\int_0^3 f(x)\,dx = \frac{1}{2}(3)(1) = 1.5$.

2. $\int_3^5 f(x)\,dx$ represents the area of the triangle found under the x–axis on $[3,5]$. The area is $\frac{1}{2}(2)(1) = 1$; since it is found *under* the x–axis, this is "negative area." Therefore $\int_3^5 f(x)\,dx = -1$.

3. $\int_0^5 f(x)\,dx$ is the total signed area under f on $[0,5]$. This is $1.5 + (-1) = 0.5$.

4. $\int_0^3 5f(x)\,dx$ is the area under $5f$ on $[0,3]$. This is sketched in Figure 5.6. Again, the region is a triangle, with height 5 times that of the height of the original triangle. Thus the area is $\int_0^3 5f(x)\,dx = 15/2 = 7.5$.

5. $\int_1^1 f(x)\,dx$ is the area under f on the "interval" $[1,1]$. This describes a line segment, not a region; it has no width. Therefore the area is 0.

This example illustrates some of the properties of the definite integral, given here.

Notes:

Theorem 36   Properties of the Definite Integral

Let f and g be defined on a closed interval I that contains the values a, b and c, and let k be a constant. The following hold:

1. $\displaystyle\int_a^a f(x)\,dx = 0$

2. $\displaystyle\int_a^b f(x)\,dx + \int_b^c f(x)\,dx = \int_a^c f(x)\,dx$

3. $\displaystyle\int_a^b f(x)\,dx = -\int_b^a f(x)\,dx$

4. $\displaystyle\int_a^b \big(f(x) \pm g(x)\big)\,dx = \int_a^b f(x)\,dx \pm \int_a^b g(x)\,dx$

5. $\displaystyle\int_a^b k \cdot f(x)\,dx = k \cdot \int_a^b f(x)\,dx$

We give a brief justification of Theorem 36 here.

1. As demonstrated in Example 114, there is no "area under the curve" when the region has no width; hence this definite integral is 0.

2. This states that total area is the sum of the areas of subregions. It is easily considered when we let $a < b < c$. We can break the interval $[a, c]$ into two subintervals, $[a, b]$ and $[b, c]$. The total area over $[a, c]$ is the area over $[a, b]$ plus the area over $[b, c]$.

   It is important to note that this still holds true even if $a < b < c$ is not true. We discuss this in the next point.

3. This property can be viewed a merely a convention to make other properties work well. (Later we will see how this property has a justification all its own, not necessarily in support of other properties.) Suppose $b < a < c$. The discussion from the previous point clearly justifies

$$\int_b^a f(x)\,dx + \int_a^c f(x)\,dx = \int_b^c f(x)\,dx. \tag{5.1}$$

   However, we still claim that, as originally stated,

$$\int_a^b f(x)\,dx + \int_b^c f(x)\,dx = \int_a^c f(x)\,dx. \tag{5.2}$$

Notes:

How do Equations (5.1) and (5.2) relate? Start with Equation (5.1):

$$\int_b^a f(x)\, dx + \int_a^c f(x)\, dx = \int_b^c f(x)\, dx$$

$$\int_a^c f(x)\, dx = -\int_b^a f(x)\, dx + \int_b^c f(x)\, dx$$

Property (3) justifies changing the sign and switching the bounds of integration on the $-\int_b^a f(x)\, dx$ term; when this is done, Equations (5.1) and (5.2) are equivalent.

The conclusion is this: by adopting the convention of Property (3), Property (2) holds no matter the order of a, b and c. Again, in the next section we will see another justification for this property.

4,5. Each of these may be non–intuitive. Property (5) states that when one scales a function by, for instance, 7, the area of the enclosed region also is scaled by a factor of 7. Both Properties (4) and (5) can be proved using geometry. The details are not complicated but are not discussed here.

Example 115   **Evaluating definite integrals using Theorem 36.**
Consider the graph of a function $f(x)$ shown in Figure 5.7. Answer the following:

1. Which value is greater: $\displaystyle\int_a^b f(x)\, dx$ or $\displaystyle\int_b^c f(x)\, dx$?

2. Is $\displaystyle\int_a^c f(x)\, dx$ greater or less than 0?

3. Which value is greater: $\displaystyle\int_a^b f(x)\, dx$ or $\displaystyle\int_c^b f(x)\, dx$?

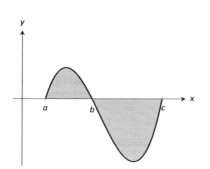

Figure 5.7: A graph of a function in Example 115.

SOLUTION

1. $\int_a^b f(x)\, dx$ has a positive value (since the area is above the x–axis) whereas $\int_b^c f(x)\, dx$ has a negative value. Hence $\int_a^b f(x)\, dx$ is bigger.

2. $\int_a^c f(x)\, dx$ is the total signed area under f between $x = a$ and $x = c$. Since the region below the x–axis looks to be larger than the region above, we conclude that the definite integral has a value less than 0.

3. Note how the second integral has the bounds "reversed." Therefore $\int_c^b f(x)\, dx$ represents a positive number, greater than the area described by the first definite integral. Hence $\int_c^b f(x)\, dx$ is greater.

Notes:

The area definition of the definite integral allows us to use geometry compute the definite integral of some simple functions.

Example 116   Evaluating definite integrals using geometry

Evaluate the following definite integrals:

1. $\int_{-2}^{5} (2x - 4)\, dx$ 2. $\int_{-3}^{3} \sqrt{9 - x^2}\, dx$.

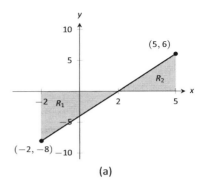

SOLUTION

1. It is useful to sketch the function in the integrand, as shown in Figure 5.8(a). We see we need to compute the areas of two regions, which we have labeled R_1 and R_2. Both are triangles, so the area computation is straightforward:

$$R_1 : \frac{1}{2}(4)(8) = 16 \qquad R_2 : \frac{1}{2}(3)6 = 9.$$

Region R_1 lies under the x–axis, hence it is counted as negative area (we can think of the triangle's height as being "-8"), so

$$\int_{-2}^{5} (2x - 4)\, dx = -16 + 9 = -7.$$

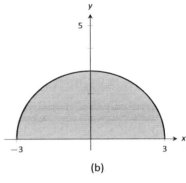

2. Recognize that the integrand of this definite integral describes a half circle, as sketched in Figure 5.8(b), with radius 3. Thus the area is:

$$\int_{-3}^{3} \sqrt{9 - x^2}\, dx = \frac{1}{2}\pi r^2 = \frac{9}{2}\pi.$$

Figure 5.8: A graph of $f(x) = 2x - 4$ in (a) and $f(x) = \sqrt{9 - x^2}$ in (b), from Example 116.

Example 117   Understanding motion given velocity

Consider the graph of a velocity function of an object moving in a straight line, given in Figure 5.9, where the numbers in the given regions gives the area of that region. Assume that the definite integral of a velocity function gives displacement. Find the maximum speed of the object and its maximum displacement from its starting position.

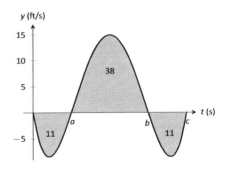

Figure 5.9: A graph of a velocity in Example 117.

SOLUTION   Since the graph gives velocity, finding the maximum speed is simple: it looks to be 15ft/s.

At time $t = 0$, the displacement is 0; the object is at its starting position. At time $t = a$, the object has moved backward 11 feet. Between times $t = a$ and

Notes:

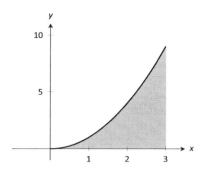

Figure 5.10: What is the area below $y = x^2$ on $[0, 3]$? The region is not a usual geometric shape.

$t = b$, the object moves forward 38 feet, bringing it into a position 27 feet forward of its starting position. From $t = b$ to $t = c$ the object is moving backwards again, hence its maximum displacement is 27 feet from its starting position.

In our examples, we have either found the areas of regions that have nice geometric shapes (such as rectangles, triangles and circles) or the areas were given to us. Consider Figure 5.10, where a region below $y = x^2$ is shaded. What is its area? The function $y = x^2$ is relatively simple, yet the shape it defines has an area that is not simple to find geometrically.

In the next section we will explore how to find the areas of such regions.

Notes:

Exercises 5.2

Terms and Concepts

1. What is "total signed area"?

2. What is "displacement"?

3. What is $\int_3^3 \sin x \, dx$?

4. Give a single definite integral that has the same value as

$$\int_0^1 (2x + 3) \, dx + \int_1^2 (2x + 3) \, dx.$$

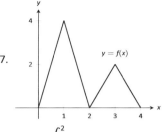

7.

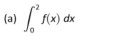

(a) $\int_0^2 f(x) \, dx$

(d) $\int_0^1 4x \, dx$

(b) $\int_2^4 f(x) \, dx$

(e) $\int_2^3 (2x - 4) \, dx$

(c) $\int_2^4 2f(x) \, dx$

(f) $\int_2^3 (4x - 8) \, dx$

Problems

In Exercises 5 – 9, a graph of a function $f(x)$ is given. Using the geometry of the graph, evaluate the definite integrals.

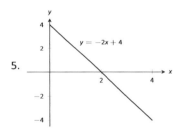

5.

(a) $\int_0^1 (-2x + 4) \, dx$

(d) $\int_1^3 (-2x + 4) \, dx$

(b) $\int_0^2 (-2x + 4) \, dx$

(e) $\int_2^4 (-2x + 4) \, dx$

(c) $\int_0^3 (-2x + 4) \, dx$

(f) $\int_0^1 (-6x + 12) \, dx$

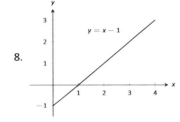

8.

(a) $\int_0^1 (x - 1) \, dx$

(d) $\int_2^3 (x - 1) \, dx$

(b) $\int_0^2 (x - 1) \, dx$

(e) $\int_1^4 (x - 1) \, dx$

(c) $\int_0^3 (x - 1) \, dx$

(f) $\int_1^4 ((x - 1) + 1) \, dx$

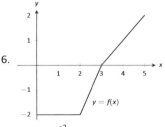

6.

(a) $\int_0^2 f(x) \, dx$

(d) $\int_2^5 f(x) \, dx$

(b) $\int_0^3 f(x) \, dx$

(e) $\int_5^3 f(x) \, dx$

(c) $\int_0^5 f(x) \, dx$

(f) $\int_0^3 -2f(x) \, dx$

9.

(a) $\int_0^2 f(x) \, dx$

(c) $\int_0^4 f(x) \, dx$

(b) $\int_2^4 f(x) \, dx$

(d) $\int_0^4 5f(x) \, dx$

In Exercises 10 – 13, a graph of a function $f(x)$ is given; the numbers inside the shaded regions give the area of that region. Evaluate the definite integrals using this area information.

10.

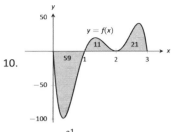

(a) $\displaystyle\int_0^1 f(x)\, dx$

(b) $\displaystyle\int_0^2 f(x)\, dx$

(c) $\displaystyle\int_0^3 f(x)\, dx$

(d) $\displaystyle\int_1^2 -3f(x)\, dx$

11.

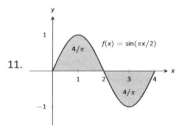

(a) $\displaystyle\int_0^2 f(x)\, dx$

(b) $\displaystyle\int_2^4 f(x)\, dx$

(c) $\displaystyle\int_0^4 f(x)\, dx$

(d) $\displaystyle\int_0^1 f(x)\, dx$

12.

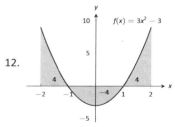

(a) $\displaystyle\int_{-2}^{-1} f(x)\, dx$

(b) $\displaystyle\int_1^2 f(x)\, dx$

(c) $\displaystyle\int_{-1}^1 f(x)\, dx$

(d) $\displaystyle\int_0^1 f(x)\, dx$

13.

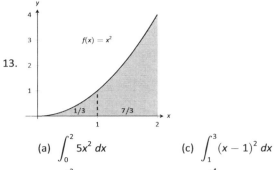

(a) $\displaystyle\int_0^2 5x^2\, dx$

(b) $\displaystyle\int_0^2 (x^2 + 3)\, dx$

(c) $\displaystyle\int_1^3 (x - 1)^2\, dx$

(d) $\displaystyle\int_2^4 \left((x - 2)^2 + 5\right) dx$

In Exercises 14 – 15, a graph of the velocity function of an object moving in a straight line is given. Answer the questions based on that graph.

14.

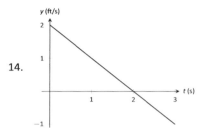

(a) What is the object's maximum velocity?

(b) What is the object's maximum displacement?

(c) What is the object's total displacement on $[0, 3]$?

15.

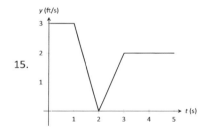

(a) What is the object's maximum velocity?

(b) What is the object's maximum displacement?

(c) What is the object's total displacement on $[0, 5]$?

16. An object is thrown straight up with a velocity, in ft/s, given by $v(t) = -32t + 64$, where t is in seconds, from a height of 48 feet.

(a) What is the object's maximum velocity?

(b) What is the object's maximum displacement?

(c) When does the maximum displacement occur?

(d) When will the object reach a height of 0? (Hint: find when the displacement is -48ft.)

17. An object is thrown straight up with a velocity, in ft/s, given by $v(t) = -32t + 96$, where t is in seconds, from a height of 64 feet.

   (a) What is the object's initial velocity?

   (b) When is the object's displacement 0?

   (c) How long does it take for the object to return to its initial height?

   (d) When will the object reach a height of 210 feet?

In Exercises 18 – 21, let

- $\int_0^2 f(x)\, dx = 5,$

- $\int_0^3 f(x)\, dx = 7,$

- $\int_0^2 g(x)\, dx = -3,$ **and**

- $\int_2^3 g(x)\, dx = 5.$

Use these values to evaluate the given definite integrals.

18. $\int_0^2 \big(f(x) + g(x)\big)\, dx$

19. $\int_0^3 \big(f(x) - g(x)\big)\, dx$

20. $\int_2^3 \big(3f(x) + 2g(x)\big)\, dx$

21. Find values for a and b such that
$$\int_0^3 \big(af(x) + bg(x)\big)\, dx = 0$$

In Exercises 22 – 25, let

- $\int_0^3 s(t)\, dt = 10,$

- $\int_3^5 s(t)\, dt = 8,$

- $\int_3^5 r(t)\, dt = -1,$ **and**

- $\int_0^5 r(t)\, dt = 11.$

Use these values to evaluate the given definite integrals.

22. $\int_0^3 \big(s(t) + r(t)\big)\, dt$

23. $\int_5^0 \big(s(t) - r(t)\big)\, dt$

24. $\int_3^3 \big(\pi s(t) - 7r(t)\big)\, dt$

25. Find values for a and b such that
$$\int_0^5 \big(ar(t) + bs(t)\big)\, dt = 0$$

Review

In Exercises 26 – 29, evaluate the given indefinite integral.

26. $\int \big(x^3 - 2x^2 + 7x - 9\big)\, dx$

27. $\int \big(\sin x - \cos x + \sec^2 x\big)\, dx$

28. $\int \Big(\sqrt[3]{t} + \dfrac{1}{t^2} + 2^t\Big)\, dt$

29. $\int \Big(\dfrac{1}{x} - \csc x \cot x\Big)\, dx$

5.3 Riemann Sums

In the previous section we defined the definite integral of a function on $[a, b]$ to be the signed area between the curve and the x–axis. Some areas were simple to compute; we ended the section with a region whose area was not simple to compute. In this section we develop a technique to find such areas.

A fundamental calculus technique is to first answer a given problem with an approximation, then refine that approximation to make it better, then use limits in the refining process to find the exact answer. That is exactly what we will do here.

Consider the region given in Figure 5.11, which is the area under $y = 4x - x^2$ on $[0, 4]$. What is the signed area of this region – i.e., what is $\int_0^4 (4x - x^2)\, dx$?

We start by approximating. We can surround the region with a rectangle with height and width of 4 and find the area is approximately 16 square units. This is obviously an *over–approximation*; we are including area in the rectangle that is not under the parabola.

We have an approximation of the area, using one rectangle. How can we refine our approximation to make it better? The key to this section is this answer: *use more rectangles*.

Let's use 4 rectangles of equal width of 1. This *partitions* the interval $[0, 4]$ into 4 *subintervals*, $[0, 1]$, $[1, 2]$, $[2, 3]$ and $[3, 4]$. On each subinterval we will draw a rectangle.

There are three common ways to determine the height of these rectangles: the **Left Hand Rule**, the **Right Hand Rule**, and the **Midpoint Rule**. The **Left Hand Rule** says to evaluate the function at the left–hand endpoint of the subinterval and make the rectangle that height. In Figure 5.12, the rectangle drawn on the interval $[2, 3]$ has height determined by the Left Hand Rule; it has a height of $f(2)$. (The rectangle is labeled "LHR.")

The **Right Hand Rule** says the opposite: on each subinterval, evaluate the function at the right endpoint and make the rectangle that height. In the figure, the rectangle drawn on $[0, 1]$ is drawn using $f(1)$ as its height; this rectangle is labeled "RHR.".

The **Midpoint Rule** says that on each subinterval, evaluate the function at the midpoint and make the rectangle that height. The rectangle drawn on $[1, 2]$ was made using the Midpoint Rule, with a height of $f(1.5)$. That rectangle is labeled "MPR."

These are the three most common rules for determining the heights of approximating rectangles, but one is not forced to use one of these three methods. The rectangle on $[3, 4]$ has a height of approximately $f(3.53)$, very close to the Midpoint Rule. It was chosen so that the area of the rectangle is *exactly* the area of the region under f on $[3, 4]$. (Later you'll be able to figure how to do this, too.)

The following example will approximate the value of $\int_0^4 (4x - x^2)\, dx$ using

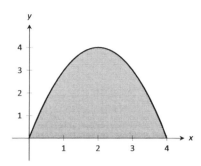

Figure 5.11: A graph of $f(x) = 4x - x^2$. What is the area of the shaded region?

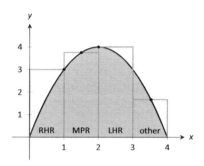

Figure 5.12: Approximating $\int_0^4 (4x - x^2)\, dx$ using rectangles. The heights of the rectangles are determined using different rules.

Notes:

these rules.

Example 118   Using the Left Hand, Right Hand and Midpoint Rules
Approximate the value of $\int_0^4 (4x - x^2)\, dx$ using the Left Hand Rule, the Right Hand Rule, and the Midpoint Rule, using 4 equally spaced subintervals.

SOLUTION   We break the interval $[0, 4]$ into four subintervals as before. In Figure 5.13 we see 4 rectangles drawn on $f(x) = 4x - x^2$ using the Left Hand Rule. (The areas of the rectangles are given in each figure.)
Note how in the first subinterval, $[0, 1]$, the rectangle has height $f(0) = 0$. We add up the areas of each rectangle (height× width) for our Left Hand Rule approximation:

$$f(0) \cdot 1 + f(1) \cdot 1 + f(2) \cdot 1 + f(3) \cdot 1 =$$
$$0 + 3 + 4 + 3 = 10.$$

Figure 5.14 shows 4 rectangles drawn under f using the Right Hand Rule; note how the $[3, 4]$ subinterval has a rectangle of height 0.
In this example, these rectangle seem to be the mirror image of those found in Figure 5.13. (This is because of the symmetry of our shaded region.) Our approximation gives the same answer as before, though calculated a different way:

$$f(1) \cdot 1 + f(2) \cdot 1 + f(3) \cdot 1 + f(4) \cdot 1 =$$
$$3 + 4 + 3 + 0 = 10.$$

Figure 5.15 shows 4 rectangles drawn under f using the Midpoint Rule. This gives an approximation of $\int_0^4 (4x - x^2)\, dx$ as:

$$f(0.5) \cdot 1 + f(1.5) \cdot 1 + f(2.5) \cdot 1 + f(3.5) \cdot 1 =$$
$$1.75 + 3.75 + 3.75 + 1.75 = 11.$$

Our three methods provide two approximations of $\int_0^4 (4x - x^2)\, dx$: 10 and 11.

Summation Notation

It is hard to tell at this moment which is a better approximation: 10 or 11? We can continue to refine our approximation by using more rectangles. The notation can become unwieldy, though, as we add up longer and longer lists of numbers. We introduce **summation notation** to ameliorate this problem.

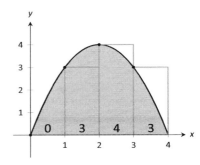

Figure 5.13: Approximating $\int_0^4 (4x - x^2)\, dx$ using the Left Hand Rule in Example 118.

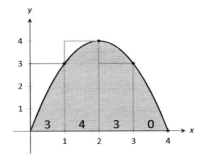

Figure 5.14: Approximating $\int_0^4 (4x - x^2)\, dx$ using the Right Hand Rule in Example 118.

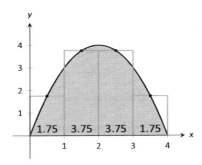

Figure 5.15: Approximating $\int_0^4 (4x - x^2)\, dx$ using the Midpoint Rule in Example 118.

Notes:

Suppose we wish to add up a list of numbers a_1, a_2, a_3, ..., a_9. Instead of writing

$$a_1 + a_2 + a_3 + a_4 + a_5 + a_6 + a_7 + a_8 + a_9,$$

we use summation notation and write

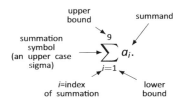

Figure 5.16: Understanding summation notation.

The upper case sigma represents the term "sum." The index of summation in this example is i; any symbol can be used. By convention, the index takes on only the integer values between (and including) the lower and upper bounds.

Let's practice using this notation.

Example 119   Using summation notation
Let the numbers $\{a_i\}$ be defined as $a_i = 2i - 1$ for integers i, where $i \geq 1$. So $a_1 = 1$, $a_2 = 3$, $a_3 = 5$, etc. (The output is the positive odd integers). Evaluate the following summations:

1. $\displaystyle\sum_{i=1}^{6} a_i$ 2. $\displaystyle\sum_{i=3}^{7}(3a_i - 4)$ 3. $\displaystyle\sum_{i=1}^{4}(a_i)^2$

SOLUTION

1.
$$\sum_{i=1}^{6} a_i = a_1 + a_2 + a_3 + a_4 + a_5 + a_6$$
$$= 1 + 3 + 5 + 7 + 9 + 11$$
$$= 36.$$

2. Note the starting value is different than 1:

$$\sum_{i=3}^{7} a_i = (3a_3 - 4) + (3a_4 - 4) + (3a_5 - 4) + (3a_6 - 4) + (3a_7 - 4)$$
$$= 11 + 17 + 23 + 29 + 35$$
$$= 115.$$

Notes:

3.

$$\sum_{i=1}^{4} (a_i)^2 = (a_1)^2 + (a_2)^2 + (a_3)^2 + (a_4)^2$$

$$= 1^2 + 3^2 + 5^2 + 7^2$$

$$= 84$$

It might seem odd to stress a new, concise way of writing summations only to write each term out as we add them up. It is. The following theorem gives some of the properties of summations that allow us to work with them without writing individual terms. Examples will follow.

Theorem 37   Properties of Summations

1. $\displaystyle\sum_{i=1}^{n} c = c \cdot n$, where c is a constant.

2. $\displaystyle\sum_{i=m}^{n} (a_i \pm b_i) = \sum_{i=m}^{n} a_i \pm \sum_{i=m}^{n} b_i$

3. $\displaystyle\sum_{i=m}^{n} c \cdot a_i = c \cdot \sum_{i=m}^{n} a_i$

4. $\displaystyle\sum_{i=m}^{j} a_i + \sum_{i=j+1}^{n} a_i = \sum_{i=m}^{n} a_i$

5. $\displaystyle\sum_{i=1}^{n} i = \frac{n(n+1)}{2}$

6. $\displaystyle\sum_{i=1}^{n} i^2 = \frac{n(n+1)(2n+1)}{6}$

7. $\displaystyle\sum_{i=1}^{n} i^3 = \left(\frac{n(n+1)}{2}\right)^2$

Example 120   Evaluating summations using Theorem 37

Revisit Example 119 and, using Theorem 37, evaluate

$$\sum_{i=1}^{6} a_i = \sum_{i=1}^{6} (2i - 1).$$

Notes:

SOLUTION

$$\sum_{i=1}^{6}(2i-1) = \sum_{i=1}^{6}2i - \sum_{i=1}^{6}(1)$$

$$= \left(2\sum_{i=1}^{6}i\right) - 6$$

$$= 2\frac{6(6+1)}{2} - 6$$

$$= 42 - 6 = 36$$

We obtained the same answer without writing out all six terms. When dealing with small sizes of n, it may be faster to write the terms out by hand. However, Theorem 37 is incredibly important when dealing with large sums as we'll soon see.

Riemann Sums

Consider again $\int_0^4 (4x - x^2)\, dx$. We will approximate this definite integral using 16 equally spaced subintervals and the Right Hand Rule in Example 121. Before doing so, it will pay to do some careful preparation.

Figure 5.17 shows a number line of $[0, 4]$ divided into 16 equally spaced subintervals. We denote 0 as x_1; we have marked the values of x_5, x_9, x_{13} and x_{17}. We could mark them all, but the figure would get crowded. While it is easy to figure that $x_{10} = 2.25$, in general, we want a method of determining the value of x_i without consulting the figure. Consider:

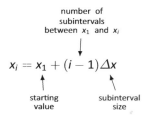

$$x_i = x_1 + (i-1)\Delta x$$

So $x_{10} = x_1 + 9(4/16) = 2.25$.

If we had partitioned $[0, 4]$ into 100 equally spaced subintervals, each subinterval would have length $\Delta x = 4/100 = 0.04$. We could compute x_{32} as

$$x_{32} = x_1 + 31(4/100) = 1.24.$$

(That was far faster than creating a sketch first.)

Figure 5.17: Dividing $[0, 4]$ into 16 equally spaced subintervals.

Notes:

Given any subdivision of $[0, 4]$, the first subinterval is $[x_1, x_2]$; the second is $[x_2, x_3]$; the ith subinterval is $[x_i, x_{i+1}]$.

When using the Left Hand Rule, the height of the ith rectangle will be $f(x_i)$.

When using the Right Hand Rule, the height of the ith rectangle will be $f(x_{i+1})$.

When using the Midpoint Rule, the height of the ith rectangle will be $f\left(\dfrac{x_i + x_{i+1}}{2}\right)$.

Thus approximating $\int_0^4 (4x - x^2)\,dx$ with 16 equally spaced subintervals can be expressed as follows, where $\Delta x = 4/16 = 1/4$:

Left Hand Rule: $\displaystyle\sum_{i=1}^{16} f(x_i)\Delta x$

Right Hand Rule: $\displaystyle\sum_{i=1}^{16} f(x_{i+1})\Delta x$

Midpoint Rule: $\displaystyle\sum_{i=1}^{16} f\left(\dfrac{x_i + x_{i+1}}{2}\right)\Delta x$

We use these formulas in the next two examples. The following example lets us practice using the Right Hand Rule and the summation formulas introduced in Theorem 37.

Example 121   Approximating definite integrals using sums
Approximate $\int_0^4 (4x - x^2)\,dx$ using the Right Hand Rule and summation formulas with 16 and 1000 equally spaced intervals.

SOLUTION   Using the formula derived before, using 16 equally spaced intervals and the Right Hand Rule, we can approximate the definite integral as

$$\sum_{i=1}^{16} f(x_{i+1})\Delta x.$$

We have $\Delta x = 4/16 = 0.25$. Since $x_i = 0 + (i - 1)\Delta x$, we have

$$x_{i+1} = 0 + \big((i+1) - 1\big)\Delta x$$
$$= i\Delta x$$

Notes:

Using the summation formulas, consider:

$$\int_0^4 (4x - x^2)\, dx \approx \sum_{i=1}^{16} f(x_{i+1})\Delta x$$

$$= \sum_{i=1}^{16} f(i\Delta x)\Delta x$$

$$= \sum_{i=1}^{16} \left(4i\Delta x - (i\Delta x)^2\right)\Delta x$$

$$= \sum_{i=1}^{16} (4i\Delta x^2 - i^2\Delta x^3)$$

$$= (4\Delta x^2)\sum_{i=1}^{16} i - \Delta x^3 \sum_{i=1}^{16} i^2 \qquad (5.3)$$

$$= (4\Delta x^2)\frac{16 \cdot 17}{2} - \Delta x^3 \frac{16(17)(33)}{6}$$

$$= 4 \cdot 0.25^2 \cdot 136 - 0.25^3 \cdot 1496$$

$$= 10.625$$

We were able to sum up the areas of 16 rectangles with very little computation. In Figure 5.18 the function and the 16 rectangles are graphed. While some rectangles over–approximate the area, other under–approximate the area (by about the same amount). Thus our approximate area of 10.625 is likely a fairly good approximation.

Notice Equation (5.3); by changing the 16's to 1,000's (and appropriately changing the value of Δx), we can use that equation to sum up 1000 rectangles! We do so here, skipping from the original summand to the equivalent of Equation (5.3) to save space. Note that $\Delta x = 4/1000 = 0.004$.

$$\int_0^4 (4x - x^2)\, dx \approx \sum_{i=1}^{1000} f(x_{i+1})\Delta x$$

$$= (4\Delta x^2)\sum_{i=1}^{1000} i - \Delta x^3 \sum_{i=1}^{1000} i^2$$

$$= (4\Delta x^2)\frac{1000 \cdot 1001}{2} - \Delta x^3 \frac{1000(1001)(2001)}{6}$$

$$= 4 \cdot 0.004^2 \cdot 500500 - 0.004^3 \cdot 333,833,500$$

$$= 10.666656$$

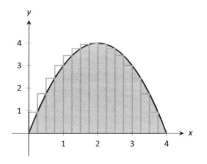

Figure 5.18: Approximating $\int_0^4 (4x-x^2)\, dx$ with the Right Hand Rule and 16 evenly spaced subintervals.

Notes:

Using many, many rectangles, we have a likely good approximation of $\int_0^4 (4x - x^2) \Delta x$. That is,

$$\int_0^4 (4x - x^2)\, dx \approx 10.666656.$$

Before the above example, we stated what the summations for the Left Hand, Right Hand and Midpoint Rules looked like. Each had the same basic structure, which was:

1. each rectangle has the same width, which we referred to as Δx, and

2. each rectangle's height is determined by evaluating f at a particular point in each subinterval. For instance, the Left Hand Rule states that each rectangle's height is determined by evaluating f at the left hand endpoint of the subinterval the rectangle lives on.

One could partition an interval $[a, b]$ with subintervals that did not have the same size. We refer to the length of the first subinterval as Δx_1, the length of the second subinterval as Δx_2, and so on, giving the length of the ith subinterval as Δx_i. Also, one could determine each rectangle's height by evaluating f at any point in the ith subinterval. We refer to the point picked in the first subinterval as c_1, the point picked in the second subinterval as c_2, and so on, with c_i representing the point picked in the ith subinterval. Thus the height of the ith subinterval would be $f(c_i)$, and the area of the ith rectangle would be $f(c_i)\Delta x_i$.

Summations of rectangles with area $f(c_i)\Delta x_i$ are named after mathematician Georg Friedrich Bernhard Riemann, as given in the following definition.

Definition 21   **Riemann Sum**

Let f be defined on the closed interval $[a, b]$ and let Δx be a partition of $[a, b]$, with

$$a = x_1 < x_2 < \ldots < x_n < x_{n+1} = b.$$

Let Δx_i denote the length of the ith subinterval $[x_i, x_{i+1}]$ and let c_i denote any value in the ith subinterval.
The sum

$$\sum_{i=1}^{n} f(c_i)\Delta x_i$$

is a **Riemann sum** of f on $[a, b]$.

Figure 5.19 shows the approximating rectangles of a Riemann sum of $\int_0^4 (4x - x^2)\, dx$. While the rectangles in this example do not approximate well the shaded area, they demonstrate that the subinterval widths may vary and the heights of the rectangles can be determined without following a particular rule.

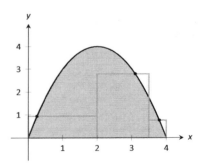

Figure 5.19: An example of a general Riemann sum to approximate $\int_0^4 (4x - x^2)\, dx$.

Notes:

"Usually" Riemann sums are calculated using one of the three methods we have introduced. The uniformity of construction makes computations easier. Before working another example, let's summarize some of what we have learned in a convenient way.

Key Idea 8   Riemann Sum Concepts

Consider $\displaystyle\int_a^b f(x)\,dx \approx \sum_{i=1}^{n} f(c_i)\Delta x_i.$

1. When the n subintervals have equal length, $\Delta x_i = \Delta x = \dfrac{b-a}{n}$.

2. The ith term of the partition is $x_i = a + (i-1)\Delta x$. (This makes $x_{n+1} = b$.)

3. The Left Hand Rule summation is: $\displaystyle\sum_{i=1}^{n} f(x_i)\Delta x.$

4. The Right Hand Rule summation is: $\displaystyle\sum_{i=1}^{n} f(x_{i+1})\Delta x.$

5. The Midpoint Rule summation is: $\displaystyle\sum_{i=1}^{n} f\left(\dfrac{x_i + x_{x+1}}{2}\right)\Delta x.$

Let's do another example.

Example 122   Approximating definite integrals with sums
Approximate $\int_{-2}^{3}(5x + 2)\,dx$ using the Midpoint Rule and 10 equally spaced intervals.

SOLUTION   Following Key Idea 8, we have

$$\Delta x = \frac{3-(-2)}{10} = 1/2 \quad \text{and} \quad x_i = (-2) + (1/2)(i-1) = i/2 - 5/2.$$

As we are using the Midpoint Rule, we will also need x_{i+1} and $\dfrac{x_i + x_{i+1}}{2}$. Since $x_i = i/2 - 5/2$, $x_{i+1} = (i+1)/2 - 5/2 = i/2 - 2$. This gives

$$\frac{x_i + x_{i+1}}{2} = \frac{(i/2 - 5/2) + (i/2 - 2)}{2} = \frac{i - 9/2}{2} = i/2 - 9/4.$$

Notes:

We now construct the Riemann sum and compute its value using summation formulas.

$$\int_{-2}^{3} (5x + 2)\, dx \approx \sum_{i=1}^{10} f\left(\frac{x_i + x_{i+1}}{2}\right) \Delta x$$

$$= \sum_{i=1}^{10} f(i/2 - 9/4)\,\Delta x$$

$$= \sum_{i=1}^{10} \left(5(i/2 - 9/4) + 2\right)\Delta x$$

$$= \Delta x \sum_{i=1}^{10} \left[\left(\frac{5}{2}\right)i - \frac{37}{4}\right]$$

$$= \Delta x \left(\frac{5}{2}\sum_{i=1}^{10}(i) - \sum_{i=1}^{10}\left(\frac{37}{4}\right)\right)$$

$$= \frac{1}{2}\left(\frac{5}{2}\cdot\frac{10(11)}{2} - 10\cdot\frac{37}{4}\right)$$

$$= \frac{45}{2} = 22.5$$

Note the graph of $f(x) = 5x + 2$ in Figure 5.20. The regions whose area is computed by the definite integral are triangles, meaning we can find the exact answer without summation techniques. We find that the exact answer is indeed 22.5. One of the strengths of the Midpoint Rule is that often each rectangle includes area that should not be counted, but misses other area that should. When the partition size is small, these two amounts are about equal and these errors almost "cancel each other out." In this example, since our function is a line, these errors are exactly equal and they do cancel each other out, giving us the exact answer.

Note too that when the function is negative, the rectangles have a "negative" height. When we compute the area of the rectangle, we use $f(c_i)\Delta x$; when f is negative, the area is counted as negative.

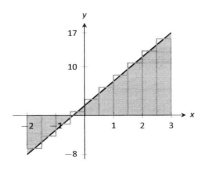

Figure 5.20: Approximating $\int_{-2}^{3}(5x + 2)\, dx$ using the Midpoint Rule and 10 evenly spaced subintervals in Example 122.

Notice in the previous example that while we used 10 equally spaced intervals, the number "10" didn't play a big role in the calculations until the very end. Mathematicians love to abstract ideas; let's approximate the area of another region using n subintervals, where we do not specify a value of n until the very end.

Example 123   Approximating definite integrals with a formula, using sums
Revisit $\int_{0}^{4}(4x - x^2)\, dx$ yet again. Approximate this definite integral using the Right

Notes:

Hand Rule with n equally spaced subintervals.

SOLUTION   Using Key Idea 8, we know $\Delta x = \frac{4-0}{n} = 4/n$. We also find $x_i = 0 + \Delta x(i-1) = 4(i-1)/n$. The Right Hand Rule uses x_{i+1}, which is $x_{i+1} = 4i/n$.

We construct the Right Hand Rule Riemann sum as follows. Be sure to follow each step carefully. If you get stuck, and do not understand how one line proceeds to the next, you may skip to the result and consider how this result is used. You should come back, though, and work through each step for full understanding.

$$\int_0^4 (4x - x^2)\, dx \approx \sum_{i=1}^n f(x_{i+1})\Delta x$$

$$= \sum_{i=1}^n f\left(\frac{4i}{n}\right)\Delta x$$

$$= \sum_{i=1}^n \left[4\frac{4i}{n} - \left(\frac{4i}{n}\right)^2\right]\Delta x$$

$$= \sum_{i=1}^n \left(\frac{16\Delta x}{n}\right)i - \sum_{i=1}^n \left(\frac{16\Delta x}{n^2}\right)i^2$$

$$= \left(\frac{16\Delta x}{n}\right)\sum_{i=1}^n i - \left(\frac{16\Delta x}{n^2}\right)\sum_{i=1}^n i^2$$

$$= \left(\frac{16\Delta x}{n}\right)\cdot\frac{n(n+1)}{2} - \left(\frac{16\Delta x}{n^2}\right)\frac{n(n+1)(2n+1)}{6} \quad \left(\text{recall } \Delta x = 4/n\right)$$

$$= \frac{32(n+1)}{n} - \frac{32(n+1)(2n+1)}{3n^2} \quad \text{(now simplify)}$$

$$= \frac{32}{3}\left(1 - \frac{1}{n^2}\right)$$

The result is an amazing, easy to use formula. To approximate the definite integral with 10 equally spaced subintervals and the Right Hand Rule, set $n = 10$ and compute

$$\int_0^4 (4x - x^2)\, dx \approx \frac{32}{3}\left(1 - \frac{1}{10^2}\right) = 10.56.$$

Recall how earlier we approximated the definite integral with 4 subintervals; with $n = 4$, the formula gives 10, our answer as before.

It is now easy to approximate the integral with 1,000,000 subintervals! Hand-held calculators will round off the answer a bit prematurely giving an answer of

Notes:

10.66666667. (The actual answer is 10.666666666656.)

We now take an important leap. Up to this point, our mathematics has been limited to geometry and algebra (finding areas and manipulating expressions). Now we apply *calculus*. For any *finite n*, we know that

$$\int_0^4 (4x - x^2)\, dx \approx \frac{32}{3}\left(1 - \frac{1}{n^2}\right).$$

Both common sense and high–level mathematics tell us that as *n* gets large, the approximation gets better. In fact, if we take the *limit* as $n \to \infty$, we get the *exact area* described by $\int_0^4 (4x - x^2)\, dx$. That is,

$$
\begin{aligned}
\int_0^4 (4x - x^2)\, dx &= \lim_{n\to\infty} \frac{32}{3}\left(1 - \frac{1}{n^2}\right) \\
&= \frac{32}{3}(1 - 0) \\
&= \frac{32}{3} = 10.\overline{6}
\end{aligned}
$$

This is a fantastic result. By considering *n* equally–spaced subintervals, we obtained a formula for an approximation of the definite integral that involved our variable *n*. As *n* grows large – without bound – the error shrinks to zero and we obtain the exact area.

This section started with a fundamental calculus technique: make an approximation, refine the approximation to make it better, then use limits in the refining process to get an exact answer. That is precisely what we just did.

Let's practice this again.

Example 124 **Approximating definite integrals with a formula, using sums**
Find a formula that approximates $\int_{-1}^5 x^3\, dx$ using the Right Hand Rule and *n* equally spaced subintervals, then take the limit as $n \to \infty$ to find the exact area.

SOLUTION Following Key Idea 8, we have $\Delta x = \frac{5-(-1)}{n} = 6/n$. We have $x_i = (-1) + (i - 1)\Delta x$; as the Right Hand Rule uses x_{i+1}, we have $x_{i+1} = (-1) + i\Delta x$.

The Riemann sum corresponding to the Right Hand Rule is (followed by sim-

Notes:

plifications):

$$\int_{-1}^{5} x^3\, dx \approx \sum_{i=1}^{n} f(x_{i+1})\,\Delta x$$

$$= \sum_{i=1}^{n} f(-1 + i\Delta x)\,\Delta x$$

$$= \sum_{i=1}^{n} (-1 + i\Delta x)^3\,\Delta x$$

$$= \sum_{i=1}^{n} \left((i\Delta x)^3 - 3(i\Delta x)^2 + 3i\Delta x - 1 \right)\Delta x \quad \text{(now distribute } \Delta x)$$

$$= \sum_{i=1}^{n} \left(i^3 \Delta x^4 - 3i^2 \Delta x^3 + 3i\Delta x^2 - \Delta x \right) \quad \text{(now split up summation)}$$

$$= \Delta x^4 \sum_{i=1}^{n} i^3 - 3\Delta x^3 \sum_{i=1}^{n} i^2 + 3\Delta x^2 \sum_{i=1}^{n} i - \sum_{i=1}^{n} \Delta x$$

$$= \Delta x^4 \left(\frac{n(n+1)}{2} \right)^2 - 3\Delta x^3 \frac{n(n+1)(2n+1)}{6} + 3\Delta x^2 \frac{n(n+1)}{2} - n\Delta x$$

(use $\Delta x = 6/n$)

$$= \frac{1296}{n^4} \cdot \frac{n^2(n+1)^2}{4} - 3\frac{216}{n^3} \cdot \frac{n(n+1)(2n+1)}{6} + 3\frac{36}{n^2} \frac{n(n+1)}{2} - 6$$

(now do a sizable amount of algebra to simplify)

$$= 156 + \frac{378}{n} + \frac{216}{n^2}$$

Once again, we have found a compact formula for approximating the definite integral with n equally spaced subintervals and the Right Hand Rule. Using 10 subintervals, we have an approximation of 195.96 (these rectangles are shown in Figure 5.21). Using $n = 100$ gives an approximation of 159.802.

Now find the exact answer using a limit:

$$\int_{-1}^{5} x^3\, dx = \lim_{n\to\infty} \left(156 + \frac{378}{n} + \frac{216}{n^2} \right) = 156.$$

Limits of Riemann Sums

We have used limits to evaluate exactly given definite limits. Will this always work? We will show, given not–very–restrictive conditions, that yes, it will always work.

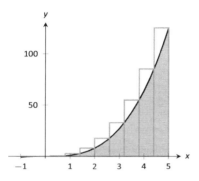

Figure 5.21: Approximating $\int_{-1}^{5} x^3\, dx$ using the Right Hand Rule and 10 evenly spaced subintervals.

Notes:

The previous two examples demonstrated how an expression such as

$$\sum_{i=1}^{n} f(x_{i+1})\Delta x$$

can be rewritten as an expression explicitly involving n, such as $32/3(1 - 1/n^2)$.

Viewed in this manner, we can think of the summation as a function of n. An n value is given (where n is a positive integer), and the sum of areas of n equally spaced rectangles is returned, using the Left Hand, Right Hand, or Midpoint Rules.

Given a definite integral $\int_a^b f(x)\, dx$, let:

- $S_L(n) = \sum_{i=1}^{n} f(x_i)\Delta x$, the sum of equally spaced rectangles formed using the Left Hand Rule,

- $S_R(n) = \sum_{i=1}^{n} f(x_{i+1})\Delta x$, the sum of equally spaced rectangles formed using the Right Hand Rule, and

- $S_M(n) = \sum_{i=1}^{n} f\left(\dfrac{x_i + x_{i+1}}{2}\right)\Delta x$, the sum of equally spaced rectangles formed using the Midpoint Rule.

Recall the definition of a limit as $n \to \infty$: $\lim\limits_{n\to\infty} S_L(n) = K$ if, given any $\varepsilon > 0$, there exists $N > 0$ such that

$$|S_L(n) - K| < \varepsilon \quad \text{when} \quad n \ge N.$$

The following theorem states that we can use any of our three rules to find the exact value of a definite integral $\int_a^b f(x)\, dx$. It also goes two steps further. The theorem states that the height of each rectangle doesn't have to be determined following a specific rule, but could be $f(c_i)$, where c_i is any point in the ith subinterval, as discussed before Riemann Sums where defined in Definition 21.

The theorem goes on to state that the rectangles do not need to be of the same width. Using the notation of Definition 21, let Δx_i denote the length of the ith subinterval in a partition of $[a, b]$. Now let $||\Delta x||$ represent the length of the largest subinterval in the partition: that is, $||\Delta x||$ is the largest of all the Δx_i's. If $||\Delta x||$ is small, then $[a, b]$ must be partitioned into many subintervals, since all subintervals must have small lengths. "Taking the limit as $||\Delta x||$ goes to zero" implies that the number n of subintervals in the partition is growing to

Notes:

infinity, as the largest subinterval length is becoming arbitrarily small. We then interpret the expression

$$\lim_{||\Delta x||\to 0} \sum_{i=1}^{n} f(c_i)\Delta x_i$$

as "the limit of the sum of rectangles, where the width of each rectangle can be different but getting small, and the height of each rectangle is not necessarily determined by a particular rule." The theorem states that this Riemann Sum also gives the value of the definite integral of f over $[a, b]$.

Theorem 38   Definite Integrals and the Limit of Riemann Sums

Let f be continuous on the closed interval $[a, b]$ and let $S_L(n)$, $S_R(n)$ and $S_M(n)$ be defined as before. Then:

1. $\displaystyle\lim_{n\to\infty} S_L(n) = \lim_{n\to\infty} S_R(n) = \lim_{n\to\infty} S_M(n) = \lim_{n\to\infty} \sum_{i=1}^{n} f(c_i)\Delta x,$

2. $\displaystyle\lim_{n\to\infty} \sum_{i=1}^{n} f(c_i)\Delta x = \int_a^b f(x)\, dx,$ and

3. $\displaystyle\lim_{||\Delta x||\to 0} \sum_{i=1}^{n} f(c_i)\Delta x_i = \int_a^b f(x)\, dx.$

We summarize what we have learned over the past few sections here.

- Knowing the "area under the curve" can be useful. One common example is: the area under a velocity curve is displacement.

- We have defined the definite integral, $\int_a^b f(x)\, dx$, to be the signed area under f on the interval $[a, b]$.

- While we can approximate a definite integral many ways, we have focused on using rectangles whose heights can be determined using: the Left Hand Rule, the Right Hand Rule and the Midpoint Rule.

- Sums of rectangles of this type are called Riemann sums.

- The exact value of the definite integral can be computed using the limit of a Riemann sum. We generally use one of the above methods as it makes the algebra simpler.

Notes:

We first learned of derivatives through limits then learned rules that made the process simpler. We know of a way to evaluate a definite integral using limits; in the next section we will see how the Fundamental Theorem of Calculus makes the process simpler. The key feature of this theorem is its connection between the indefinite integral and the definite integral.

Notes:

Exercises 5.3

Terms and Concepts

1. A fundamental calculus technique is to use _____ to refine approximations to get an exact answer.

2. What is the upper bound in the summation $\sum_{i=7}^{14}(48i - 201)$?

3. This section approximates definite integrals using what geometric shape?

4. T/F: A sum using the Right Hand Rule is an example of a Riemann Sum.

Problems

In Exercises 5 – 11, write out each term of the summation and compute the sum.

5. $\sum_{i=2}^{4} i^2$

6. $\sum_{i=-1}^{3}(4i - 2)$

7. $\sum_{i=-2}^{2} \sin(\pi i/2)$

8. $\sum_{i=1}^{5} \frac{1}{i}$

9. $\sum_{i=1}^{6}(-1)^i i$

10. $\sum_{i=1}^{4}\left(\frac{1}{i} - \frac{1}{i+1}\right)$

11. $\sum_{i=0}^{5}(-1)^i \cos(\pi i)$

In Exercises 12 – 15, write each sum in summation notation.

12. $3 + 6 + 9 + 12 + 15$

13. $-1 + 0 + 3 + 8 + 15 + 24 + 35 + 48 + 63$

14. $\frac{1}{2} + \frac{2}{3} + \frac{3}{4} + \frac{4}{5}$

15. $1 - e + e^2 - e^3 + e^4$

In Exercises 16 – 22, evaluate the summation using Theorem 37.

16. $\sum_{i=1}^{25} i$

17. $\sum_{i=1}^{10}(3i^2 - 2i)$

18. $\sum_{i=1}^{15}(2i^3 - 10)$

19. $\sum_{i=1}^{10}(-4i^3 + 10i^2 - 7i + 11)$

20. $\sum_{i=1}^{10}(i^3 - 3i^2 + 2i + 7)$

21. $1 + 2 + 3 + \ldots + 99 + 100$

22. $1 + 4 + 9 + \ldots + 361 + 400$

Theorem 37 states

$$\sum_{i=1}^{n} a_i = \sum_{i=1}^{k} a_i + \sum_{i=k+1}^{n} a_i \text{, so}$$

$$\sum_{i=k+1}^{n} a_i = \sum_{i=1}^{n} a_i - \sum_{i=1}^{k} a_i.$$

Use this fact, along with other parts of Theorem 37, to evaluate the summations given in Exercises 23 – 26.

23. $\sum_{i=11}^{20} i$

24. $\sum_{i=16}^{25} i^3$

25. $\sum_{i=7}^{12} 4$

26. $\sum_{i=5}^{10} 4i^3$

In Exercises 27 – 32, a definite integral
$$\int_a^b f(x)\, dx \text{ is given.}$$

(a) Graph $f(x)$ on $[a, b]$.

(b) Add to the sketch rectangles using the provided rule.

(c) Approximate $\int_a^b f(x)\, dx$ by summing the areas of the rectangles.

27. $\int_{-3}^{3} x^2\, dx$, with 6 rectangles using the Left Hand Rule.

28. $\int_{0}^{2} (5 - x^2)\, dx$, with 4 rectangles using the Midpoint Rule.

29. $\int_{0}^{\pi} \sin x\, dx$, with 6 rectangles using the Right Hand Rule.

30. $\int_{0}^{3} 2^x\, dx$, with 5 rectangles using the Left Hand Rule.

31. $\int_{1}^{2} \ln x\, dx$, with 3 rectangles using the Midpoint Rule.

32. $\int_{1}^{9} \frac{1}{x}\, dx$, with 4 rectangles using the Right Hand Rule.

In Exercises 33 – 38, a definite integral
$$\int_a^b f(x)\, dx \text{ is given. As demonstrated in Examples 123}$$
and 124, do the following.

(a) Find a formula to approximate $\int_a^b f(x)\, dx$ using n subintervals and the provided rule.

(b) Evaluate the formula using $n = 10$, 100 and 1, 000.

(c) Find the limit of the formula, as $n \to \infty$, to find the exact value of $\int_a^b f(x)\, dx$.

33. $\int_{0}^{1} x^3\, dx$, using the Right Hand Rule.

34. $\int_{-1}^{1} 3x^2\, dx$, using the Left Hand Rule.

35. $\int_{-1}^{3} (3x - 1)\, dx$, using the Midpoint Rule.

36. $\int_{1}^{4} (2x^2 - 3)\, dx$, using the Left Hand Rule.

37. $\int_{-10}^{10} (5 - x)\, dx$, using the Right Hand Rule.

38. $\int_{0}^{1} (x^3 - x^2)\, dx$, using the Right Hand Rule.

Review

In Exercises 39 – 44, find an antiderivative of the given function.

39. $f(x) = 5 \sec^2 x$

40. $f(x) = \dfrac{7}{x}$

41. $g(t) = 4t^5 - 5t^3 + 8$

42. $g(t) = 5 \cdot 8^t$

43. $g(t) = \cos t + \sin t$

44. $f(x) = \dfrac{1}{\sqrt{x}}$

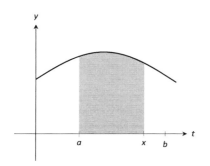

Figure 5.22: The area of the shaded region is $F(x) = \int_a^x f(t)\, dt$.

5.4 The Fundamental Theorem of Calculus

Let $f(t)$ be a continuous function defined on $[a, b]$. The definite integral $\int_a^b f(x)\, dx$ is the "area under f" on $[a, b]$. We can turn this concept into a function by letting the upper (or lower) bound vary.

Let $F(x) = \int_a^x f(t)\, dt$. It computes the area under f on $[a, x]$ as illustrated in Figure 5.22. We can study this function using our knowledge of the definite integral. For instance, $F(a) = 0$ since $\int_a^a f(t)\, dt = 0$.

We can also apply calculus ideas to $F(x)$; in particular, we can compute its derivative. While this may seem like an innocuous thing to do, it has far–reaching implications, as demonstrated by the fact that the result is given as an important theorem.

Theorem 39   **The Fundamental Theorem of Calculus, Part 1**

Let f be continuous on $[a, b]$ and let $F(x) = \int_a^x f(t)\, dt$. Then F is a differentiable function on (a, b), and

$$F'(x) = f(x).$$

Initially this seems simple, as demonstrated in the following example.

Example 125   **Using the Fundamental Theorem of Calculus, Part 1**

Let $F(x) = \int_{-5}^{x} (t^2 + \sin t)\, dt$. What is $F'(x)$?

SOLUTION   Using the Fundamental Theorem of Calculus, we have $F'(x) = x^2 + \sin x$.

This simple example reveals something incredible: $F(x)$ is an antiderivative of $x^2 + \sin x$! Therefore, $F(x) = \frac{1}{3}x^3 - \cos x + C$ for some value of C. (We can find C, but generally we do not care. We know that $F(-5) = 0$, which allows us to compute C. In this case, $C = \cos(-5) + \frac{125}{3}$.)

We have done more than found a complicated way of computing an antiderivative. Consider a function f defined on an open interval containing a, b and c. Suppose we want to compute $\int_a^b f(t)\, dt$. First, let $F(x) = \int_c^x f(t)\, dt$. Using

---

Notes:

the properties of the definite integral found in Theorem 36, we know

$$\int_a^b f(t)\, dt = \int_a^c f(t)\, dt + \int_c^b f(t)\, dt$$

$$= -\int_c^a f(t)\, dt + \int_c^b f(t)\, dt$$

$$= -F(a) + F(b)$$

$$= F(b) - F(a).$$

We now see how indefinite integrals and definite integrals are related: we can evaluate a definite integral using antiderivatives! This is the second part of the Fundamental Theorem of Calculus.

Theorem 40   The Fundamental Theorem of Calculus, Part 2

Let f be continuous on $[a, b]$ and let F be *any* antiderivative of f. Then

$$\int_a^b f(x)\, dx = F(b) - F(a).$$

Example 126   Using the Fundamental Theorem of Calculus, Part 2
We spent a great deal of time in the previous section studying $\int_0^4 (4x - x^2)\, dx$. Using the Fundamental Theorem of Calculus, evaluate this definite integral.

SOLUTION   We need an antiderivative of $f(x) = 4x - x^2$. All antiderivatives of f have the form $F(x) = 2x^2 - \frac{1}{3}x^3 + C$; for simplicity, choose $C = 0$.
The Fundamental Theorem of Calculus states

$$\int_0^4 (4x - x^2)\, dx = F(4) - F(0) = \left(2(4)^2 - \frac{1}{3}4^3\right) - (0 - 0) = 32 - \frac{64}{3} = 32/3.$$

This is the same answer we obtained using limits in the previous section, just with much less work.

Notation: A special notation is often used in the process of evaluating definite integrals using the Fundamental Theorem of Calculus. Instead of explicitly writing $F(b) - F(a)$, the notation $F(x)\Big|_a^b$ is used. Thus the solution to Example 126 would be written as:

$$\int_0^4 (4x - x^2)\, dx = \left(2x^2 - \frac{1}{3}x^3\right)\Big|_0^4 = \left(2(4)^2 - \frac{1}{3}4^3\right) - (0 - 0) = 32/3.$$

Notes:

The Constant C: *Any* antiderivative $F(x)$ can be chosen when using the Fundamental Theorem of Calculus to evaluate a definite integral, meaning any value of C can be picked. The constant *always* cancels out of the expression when evaluating $F(b) - F(a)$, so it does not matter what value is picked. This being the case, we might as well let $C = 0$.

Example 127   Using the Fundamental Theorem of Calculus, Part 2
Evaluate the following definite integrals.

1. $\displaystyle\int_{-2}^{2} x^3\,dx$ 2. $\displaystyle\int_{0}^{\pi} \sin x\,dx$ 3. $\displaystyle\int_{0}^{5} e^t\,dt$ 4. $\displaystyle\int_{4}^{9} \sqrt{u}\,du$ 5. $\displaystyle\int_{1}^{5} 2\,dx$

SOLUTION

1. $\displaystyle\int_{-2}^{2} x^3\,dx = \frac{1}{4}x^4\Big|_{-2}^{2} = \left(\frac{1}{4}2^4\right) - \left(\frac{1}{4}(-2)^4\right) = 0.$

2. $\displaystyle\int_{0}^{\pi} \sin x\,dx = -\cos x\Big|_{0}^{\pi} = -\cos\pi - \left(-\cos 0\right) = 1 + 1 = 2.$

   (This is interesting; it says that the area under one "hump" of a sine curve is 2.)

3. $\displaystyle\int_{0}^{5} e^t\,dt = e^t\Big|_{0}^{5} = e^5 - e^0 = e^5 - 1 \approx 147.41.$

4. $\displaystyle\int_{4}^{9} \sqrt{u}\,du = \int_{4}^{9} u^{\frac{1}{2}}\,du = \frac{2}{3}u^{\frac{3}{2}}\Big|_{4}^{9} = \frac{2}{3}\left(9^{\frac{3}{2}} - 4^{\frac{3}{2}}\right) = \frac{2}{3}(27 - 8) = \frac{38}{3}.$

5. $\displaystyle\int_{1}^{5} 2\,dx = 2x\Big|_{1}^{5} = 2(5) - 2 = 2(5 - 1) = 8.$

   This integral is interesting; the integrand is a constant function, hence we are finding the area of a rectangle with width $(5 - 1) = 4$ and height 2. Notice how the evaluation of the definite integral led to $2(4) = 8$.

   In general, if c is a constant, then $\int_a^b c\,dx = c(b - a)$.

Understanding Motion with the Fundamental Theorem of Calculus

We established, starting with Key Idea 1, that the derivative of a position function is a velocity function, and the derivative of a velocity function is an acceleration function. Now consider definite integrals of velocity and acceleration functions. Specifically, if $v(t)$ is a velocity function, what does $\displaystyle\int_a^b v(t)\,dt$ mean?

Notes:

The Fundamental Theorem of Calculus states that

$$\int_a^b v(t)\, dt = V(b) - V(a),$$

where $V(t)$ is any antiderivative of $v(t)$. Since $v(t)$ is a velocity function, $V(t)$ must be a position function, and $V(b) - V(a)$ measures a change in position, or **displacement**.

Example 128   Finding displacement
A ball is thrown straight up with velocity given by $v(t) = -32t + 20$ft/s, where t is measured in seconds. Find, and interpret, $\int_0^1 v(t)\, dt$.

SOLUTION   Using the Fundamental Theorem of Calculus, we have

$$\int_0^1 v(t)\, dt = \int_0^1 (-32t + 20)\, dt$$
$$= -16t^2 + 20t \Big|_0^1$$
$$= 4.$$

Thus if a ball is thrown straight up into the air with velocity $v(t) = -32t + 20$, the height of the ball, 1 second later, will be 4 feet above the initial height. (Note that the ball has *traveled* much farther. It has gone up to its peak and is falling down, but the difference between its height at $t = 0$ and $t = 1$ is 4ft.)

Integrating a rate of change function gives total change. Velocity is the rate of position change; integrating velocity gives the total change of position, i.e., displacement.

Integrating a speed function gives a similar, though different, result. Speed is also the rate of position change, but does not account for direction. So integrating a speed function gives total change of position, without the possibility of "negative position change." Hence the integral of a speed function gives *distance traveled.*

As acceleration is the rate of velocity change, integrating an acceleration function gives total change in velocity. We do not have a simple term for this analogous to displacement. If $a(t) = 5$miles/h^2 and t is measured in hours, then

$$\int_0^3 a(t)\, dt = 15$$

means the velocity has increased by 15m/h from $t = 0$ to $t = 3$.

Notes:

The Fundamental Theorem of Calculus and the Chain Rule

Part 1 of the Fundamental Theorem of Calculus (FTC) states that given $F(x) = \int_a^x f(t)\,dt$, $F'(x) = f(x)$. Using other notation, $\frac{d}{dx}(F(x)) = f(x)$. While we have just practiced evaluating definite integrals, sometimes finding antiderivatives is impossible and we need to rely on other techniques to approximate the value of a definite integral. Functions written as $F(x) = \int_a^x f(t)\,dt$ are useful in such situations.

It may be of further use to compose such a function with another. As an example, we may compose $F(x)$ with $g(x)$ to get

$$F\big(g(x)\big) = \int_a^{g(x)} f(t)\,dt.$$

What is the derivative of such a function? The Chain Rule can be employed to state

$$\frac{d}{dx}\Big(F(g(x))\Big) = F'(g(x))g'(x) = f(g(x))g'(x).$$

An example will help us understand this.

Example 129   The FTC, Part 1, and the Chain Rule

Find the derivative of $F(x) = \int_2^{x^2} \ln t\,dt$.

SOLUTION   We can view $F(x)$ as being the function $G(x) = \int_2^x \ln t\,dt$ composed with $g(x) = x^2$; that is, $F(x) = G\big(g(x)\big)$. The Fundamental Theorem of Calculus states that $G'(x) = \ln x$. The Chain Rule gives us

$$
\begin{aligned}
F'(x) &= G'\big(g(x)\big)g'(x) \\
&= \ln(g(x))g'(x) \\
&= \ln(x^2)2x \\
&= 2x \ln x^2
\end{aligned}
$$

Normally, the steps defining $G(x)$ and $g(x)$ are skipped.

Practice this once more.

Example 130   The FTC, Part 1, and the Chain Rule

Find the derivative of $F(x) = \int_{\cos x}^5 t^3\,dt$.

Notes:

SOLUTION   Note that $F(x) = -\int_5^{\cos x} t^3\, dt$. Viewed this way, the derivative of F is straightforward:

$$F'(x) = \sin x \cos^3 x.$$

Area Between Curves

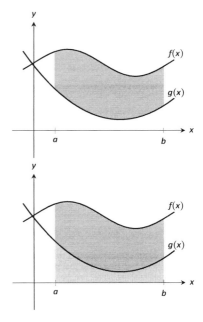

Consider continuous functions $f(x)$ and $g(x)$ defined on $[a, b]$, where $f(x) \geq g(x)$ for all x in $[a, b]$, as demonstrated in Figure 5.23. What is the area of the shaded region bounded by the two curves over $[a, b]$?

The area can be found by recognizing that this area is "the area under f − the area under g." Using mathematical notation, the area is

$$\int_a^b f(x)\, dx - \int_a^b g(x)\, dx.$$

Properties of the definite integral allow us to simplify this expression to

$$\int_a^b \big(f(x) - g(x)\big)\, dx.$$

Figure 5.23: Finding the area bounded by two functions on an interval; it is found by subtracting the area under g from the area under f.

Theorem 41   Area Between Curves

Let $f(x)$ and $g(x)$ be continuous functions defined on $[a, b]$ where $f(x) \geq g(x)$ for all x in $[a, b]$. The area of the region bounded by the curves $y = f(x)$, $y = g(x)$ and the lines $x = a$ and $x = b$ is

$$\int_a^b \big(f(x) - g(x)\big)\, dx.$$

Example 131   Finding area between curves
Find the area of the region enclosed by $y = x^2 + x - 5$ and $y = 3x - 2$.

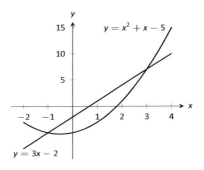

Figure 5.24: Sketching the region enclosed by $y = x^2 + x - 5$ and $y = 3x - 2$ in Example 131.

SOLUTION   It will help to sketch these two functions, as done in Figure 5.24. The region whose area we seek is completely bounded by these two functions; they seem to intersect at $x = -1$ and $x = 3$. To check, set $x^2 + x - 5 =$

Notes:

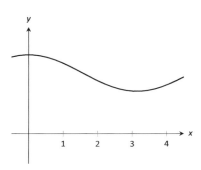

Figure 5.25: A graph of a function f to introduce the Mean Value Theorem.

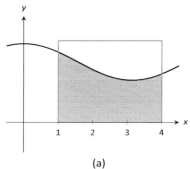

(a)

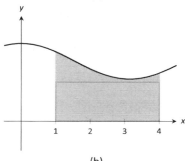

(b)

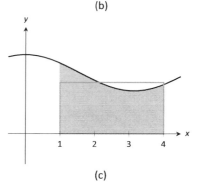

(c)

Figure 5.26: Differently sized rectangles give upper and lower bounds on $\int_1^4 f(x)\, dx$; the last rectangle matches the area exactly.

$3x - 2$ and solve for x:

$$x^2 + x - 5 = 3x - 2$$
$$(x^2 + x - 5) - (3x - 2) = 0$$
$$x^2 - 2x - 3 = 0$$
$$(x - 3)(x + 1) = 0$$
$$x = -1, 3.$$

Following Theorem 41, the area is

$$\int_{-1}^{3} \left(3x - 2 - (x^2 + x - 5)\right)\, dx = \int_{-1}^{3} (-x^2 + 2x + 3)\, dx$$
$$= \left(-\frac{1}{3}x^3 + x^2 + 3x\right)\Big|_{-1}^{3}$$
$$= -\frac{1}{3}(27) + 9 + 9 - \left(\frac{1}{3} + 1 - 3\right)$$
$$= 10\frac{2}{3} = 10.\overline{6}$$

The Mean Value Theorem and Average Value

Consider the graph of a function f in Figure 5.25 and the area defined by $\int_1^4 f(x)\, dx$. Three rectangles are drawn in Figure 5.26; in (a), the height of the rectangle is greater than f on $[1, 4]$, hence the area of this rectangle is is greater than $\int_0^4 f(x)\, dx$.

In (b), the height of the rectangle is smaller than f on $[1, 4]$, hence the area of this rectangle is less than $\int_1^4 f(x)\, dx$.

Finally, in (c) the height of the rectangle is such that the area of the rectangle is *exactly* that of $\int_0^4 f(x)\, dx$. Since rectangles that are "too big", as in (a), and rectangles that are "too little," as in (b), give areas greater/lesser than $\int_1^4 f(x)\, dx$, it makes sense that there is a rectangle, whose top intersects $f(x)$ somewhere on $[1, 4]$, whose area is *exactly* that of the definite integral.

We state this idea formally in a theorem.

Notes:

Theorem 42   **The Mean Value Theorem of Integration**

Let f be continuous on $[a, b]$. There exists a value c in $[a, b]$ such that

$$\int_a^b f(x)\, dx = f(c)(b - a).$$

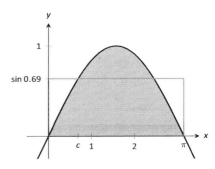

Figure 5.27: A graph of $y = \sin x$ on $[0, \pi]$ and the rectangle guaranteed by the Mean Value Theorem.

This is an *existential* statement; c exists, but we do not provide a method of finding it. Theorem 42 is directly connected to the Mean Value Theorem of Differentiation, given as Theorem 27; we leave it to the reader to see how.

We demonstrate the principles involved in this version of the Mean Value Theorem in the following example.

Example 132   **Using the Mean Value Theorem**
Consider $\int_0^\pi \sin x\, dx$. Find a value c guaranteed by the Mean Value Theorem.

SOLUTION   We first need to evaluate $\int_0^\pi \sin x\, dx$. (This was previously done in Example 127.)

$$\int_0^\pi \sin x\, dx = -\cos x \Big|_0^\pi = 2.$$

Thus we seek a value c in $[0, \pi]$ such that $\pi \sin c = 2$.

$$\pi \sin c = 2 \;\Rightarrow\; \sin c = 2/\pi \;\Rightarrow\; c = \arcsin(2/\pi) \approx 0.69.$$

In Figure 5.27 $\sin x$ is sketched along with a rectangle with height $\sin(0.69)$. The area of the rectangle is the same as the area under $\sin x$ on $[0, \pi]$.

Let f be a function on $[a, b]$ with c such that $f(c)(b-a) = \int_a^b f(x)\, dx$. Consider $\int_a^b \big(f(x) - f(c)\big)\, dx$:

$$\int_a^b \big(f(x) - f(c)\big)\, dx = \int_a^b f(x) - \int_a^b f(c)\, dx$$
$$= f(c)(b - a) - f(c)(b - a)$$
$$= 0.$$

When $f(x)$ is shifted by $-f(c)$, the amount of area under f above the x–axis on $[a, b]$ is the same as the amount of area below the x–axis above f; see Figure 5.28 for an illustration of this. In this sense, we can say that $f(c)$ is the *average value* of f on $[a, b]$.

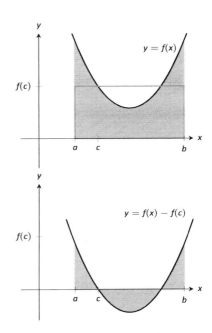

Figure 5.28: On top, a graph of $y = f(x)$ and the rectangle guaranteed by the Mean Value Theorem. Below, $y = f(x)$ is shifted down by $f(c)$; the resulting "area under the curve" is 0.

Notes:

The value $f(c)$ is the average value in another sense. First, recognize that the Mean Value Theorem can be rewritten as

$$f(c) = \frac{1}{b-a}\int_a^b f(x)\,dx,$$

for some value of c in $[a, b]$. Next, partition the interval $[a, b]$ into n equally spaced subintervals, $a = x_1 < x_2 < \ldots < x_{n+1} = b$ and choose any c_i in $[x_i, x_{i+1}]$. The average of the numbers $f(c_1), f(c_2), \ldots, f(c_n)$ is:

$$\frac{1}{n}\Big(f(c_1) + f(c_2) + \ldots + f(c_n)\Big) = \frac{1}{n}\sum_{i=1}^{n} f(c_i).$$

Multiply this last expression by 1 in the form of $\frac{(b-a)}{(b-a)}$:

$$\frac{1}{n}\sum_{i=1}^{n} f(c_i) = \sum_{i=1}^{n} f(c_i)\frac{1}{n}$$

$$= \sum_{i=1}^{n} f(c_i)\frac{1}{n}\frac{(b-a)}{(b-a)}$$

$$= \frac{1}{b-a}\sum_{i=1}^{n} f(c_i)\frac{b-a}{n}$$

$$= \frac{1}{b-a}\sum_{i=1}^{n} f(c_i)\Delta x \quad \text{(where } \Delta x = (b-a)/n\text{)}$$

Now take the limit as $n \to \infty$:

$$\lim_{n\to\infty} \frac{1}{b-a}\sum_{i=1}^{n} f(c_i)\Delta x \quad = \quad \frac{1}{b-a}\int_a^b f(x)\,dx \quad = \quad f(c).$$

This tells us this: when we evaluate f at n (somewhat) equally spaced points in $[a, b]$, the average value of these samples is $f(c)$ as $n \to \infty$.

This leads us to a definition.

Definition 22   The Average Value of f on $[a, b]$

Let f be continuous on $[a, b]$. The **average value of f on $[a, b]$** is $f(c)$, where c is a value in $[a, b]$ guaranteed by the Mean Value Theorem. I.e.,

$$\text{Average Value of } f \text{ on } [a, b] = \frac{1}{b-a}\int_a^b f(x)\,dx.$$

Notes:

An application of this definition is given in the following example.

Example 133   **Finding the average value of a function**

An object moves back and forth along a straight line with a velocity given by $v(t) = (t - 1)^2$ on $[0, 3]$, where t is measured in seconds and $v(t)$ is measured in ft/s.

What is the average velocity of the object?

SOLUTION   By our definition, the average velocity is:

$$\frac{1}{3 - 0} \int_0^3 (t - 1)^2 \, dt = \frac{1}{3} \int_0^3 (t^2 - 2t + 1) \, dt = \frac{1}{3} \left(\frac{1}{3}t^3 - t^2 + t \right) \Big|_0^3 = 1 \text{ ft/s}.$$

We can understand the above example through a simpler situation. Suppose you drove 100 miles in 2 hours. What was your average speed? The answer is simple: displacement/time = 100 miles/2 hours = 50 mph.

What was the displacement of the object in Example 133? We calculate this by integrating its velocity function: $\int_0^3 (t - 1)^2 \, dt = 3$ ft. Its final position was 3 feet from its initial position after 3 seconds: its average velocity was 1 ft/s.

This section has laid the groundwork for a lot of great mathematics to follow. The most important lesson is this: definite integrals can be evaluated using antiderivatives. Since the previous section established that definite integrals are the limit of Riemann sums, we can later create Riemann sums to approximate values other than "area under the curve," convert the sums to definite integrals, then evaluate these using the Fundamental Theorem of Calculus. This will allow us to compute the work done by a variable force, the volume of certain solids, the arc length of curves, and more.

The downside is this: generally speaking, computing antiderivatives is much more difficult than computing derivatives. The next chapter is devoted to techniques of finding antiderivatives so that a wide variety of definite integrals can be evaluated. Before that, the next section explores techniques of approximating the value of definite integrals beyond using the Left Hand, Right Hand and Midpoint Rules.

Notes:

Exercises 5.4

Terms and Concepts

1. How are definite and indefinite integrals related?

2. What constant of integration is most commonly used when evaluating definite integrals?

3. T/F: If f is a continuous function, then $F(x) = \int_a^x f(t)\, dt$ is also a continuous function.

4. The definite integral can be used to find "the area under a curve." Give two other uses for definite integrals.

Problems

In Exercises 5 – 28, evaluate the definite integral.

5. $\int_1^3 (3x^2 - 2x + 1)\, dx$

6. $\int_0^4 (x - 1)^2\, dx$

7. $\int_{-1}^1 (x^3 - x^5)\, dx$

8. $\int_{\pi/2}^{\pi} \cos x\, dx$

9. $\int_0^{\pi/4} \sec^2 x\, dx$

10. $\int_1^e \frac{1}{x}\, dx$

11. $\int_{-1}^1 5^x\, dx$

12. $\int_{-2}^{-1} (4 - 2x^3)\, dx$

13. $\int_0^{\pi} (2\cos x - 2\sin x)\, dx$

14. $\int_1^3 e^x\, dx$

15. $\int_0^4 \sqrt{t}\, dt$

16. $\int_9^{25} \frac{1}{\sqrt{t}}\, dt$

17. $\int_1^8 \sqrt[3]{x}\, dx$

18. $\int_1^2 \frac{1}{x}\, dx$

19. $\int_1^2 \frac{1}{x^2}\, dx$

20. $\int_1^2 \frac{1}{x^3}\, dx$

21. $\int_0^1 x\, dx$

22. $\int_0^1 x^2\, dx$

23. $\int_0^1 x^3\, dx$

24. $\int_0^1 x^{100}\, dx$

25. $\int_{-4}^4 dx$

26. $\int_{-10}^{-5} 3\, dx$

27. $\int_{-2}^2 0\, dx$

28. $\int_{\pi/6}^{\pi/3} \csc x \cot x\, dx$

29. Explain why:

   (a) $\int_{-1}^1 x^n\, dx = 0$, when n is a positive, odd integer, and

   (b) $\int_{-1}^1 x^n\, dx = 2\int_0^1 x^n\, dx$ when n is a positive, even integer.

In Exercises 30 – 33, find a value c guaranteed by the Mean Value Theorem.

30. $\int_0^2 x^2\, dx$

31. $\int_{-2}^2 x^2\, dx$

32. $\int_0^1 e^x\, dx$

33. $\int_0^{16} \sqrt{x}\, dx$

In Exercises 34 – 39, find the average value of the function on the given interval.

34. $f(x) = \sin x$ on $[0, \pi/2]$

35. $y = \sin x$ on $[0, \pi]$

36. $y = x$ on $[0, 4]$

37. $y = x^2$ on $[0, 4]$

38. $y = x^3$ on $[0, 4]$

39. $g(t) = 1/t$ on $[1, e]$

In Exercises 40 – 44, a velocity function of an object moving along a straight line is given. Find the displacement of the object over the given time interval.

40. $v(t) = -32t + 20$ft/s on $[0, 5]$

41. $v(t) = -32t + 200$ft/s on $[0, 10]$

42. $v(t) = 2^t$mph on $[-1, 1]$

43. $v(t) = \cos t$ ft/s on $[0, 3\pi/2]$

44. $v(t) = \sqrt[4]{t}$ ft/s on $[0, 16]$

In Exercises 45 – 48, an acceleration function of an object moving along a straight line is given. Find the change of the object's velocity over the given time interval.

45. $a(t) = -32$ft/s^2 on $[0, 2]$

46. $a(t) = 10$ft/s^2 on $[0, 5]$

47. $a(t) = t$ ft/s^2 on $[0, 2]$

48. $a(t) = \cos t$ ft/s^2 on $[0, \pi]$

In Exercises 49 – 52, sketch the given functions and find the area of the enclosed region.

49. $y = 2x, y = 5x$, and $x = 3$.

50. $y = -x + 1, y = 3x + 6, x = 2$ and $x = -1$.

51. $y = x^2 - 2x + 5, y = 5x - 5$.

52. $y = 2x^2 + 2x - 5, y = x^2 + 3x + 7$.

In Exercises 53 – 56, find $F'(x)$.

53. $F(x) = \int_2^{x^3 + x} \frac{1}{t} \, dt$

54. $F(x) = \int_{x^3}^0 t^3 \, dt$

55. $F(x) = \int_x^{x^2} (t + 2) \, dt$

56. $F(x) = \int_{\ln x}^{e^x} \sin t \, dt$

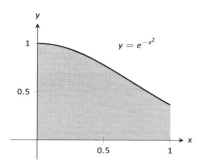

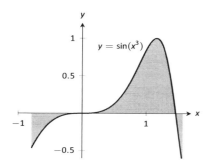

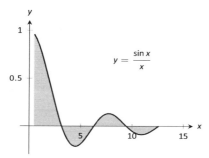

Figure 5.29: Graphically representing three definite integrals that cannot be evaluated using antiderivatives.

5.5 Numerical Integration

The Fundamental Theorem of Calculus gives a concrete technique for finding the exact value of a definite integral. That technique is based on computing antiderivatives. Despite the power of this theorem, there are still situations where we must *approximate* the value of the definite integral instead of finding its exact value. The first situation we explore is where we *cannot* compute the antiderivative of the integrand. The second case is when we actually do not know the integrand, but only its value when evaluated at certain points.

An **elementary function** is any function that is a combination of polynomials, n^{th} roots, rational, exponential, logarithmic and trigonometric functions. We can compute the derivative of any elementary function, but there are many elementary functions of which we cannot compute an antiderivative. For example, the following functions do not have antiderivatives that we can express with elementary functions:

$$e^{-x^2}, \quad \sin(x^3) \quad \text{and} \quad \frac{\sin x}{x}.$$

The simplest way to refer to the antiderivatives of e^{-x^2} is to simply write $\int e^{-x^2}\,dx$.

This section outlines three common methods of approximating the value of definite integrals. We describe each as a systematic method of approximating area under a curve. By approximating this area accurately, we find an accurate approximation of the corresponding definite integral.

We will apply the methods we learn in this section to the following definite integrals:

$$\int_0^1 e^{-x^2}\,dx, \quad \int_{-\frac{\pi}{4}}^{\frac{\pi}{2}} \sin(x^3)\,dx, \quad \text{and} \quad \int_{0.5}^{4\pi} \frac{\sin(x)}{x}\,dx,$$

as pictured in Figure 5.29.

The Left and Right Hand Rule Methods

In Section 5.3 we addressed the problem of evaluating definite integrals by approximating the area under the curve using rectangles. We revisit those ideas here before introducing other methods of approximating definite integrals.

We start with a review of notation. Let f be a continuous function on the interval $[a, b]$. We wish to approximate $\int_a^b f(x)\,dx$. We partition $[a, b]$ into n equally spaced subintervals, each of length $\Delta x = \dfrac{b-a}{n}$. The endpoints of these

---

Notes:

subintervals are labeled as

$$x_1 = a, \ x_2 = a + \Delta x, \ x_3 = a + 2\Delta x, \ \ldots, \ x_i = a + (i-1)\Delta x, \ \ldots, \ x_{n+1} = b.$$

Key Idea 8 states that to use the Left Hand Rule we use the summation $\sum_{i=1}^{n} f(x_i)\Delta x$ and to use the Right Hand Rule we use $\sum_{i=1}^{n} f(x_{i+1})\Delta x$. We review the use of these rules in the context of examples.

Example 134   **Approximating definite integrals with rectangles**

Approximate $\int_0^1 e^{-x^2} \, dx$ using the Left and Right Hand Rules with 5 equally spaced subintervals.

SOLUTION   We begin by partitioning the interval $[0,1]$ into 5 equally spaced intervals. We have $\Delta x = \frac{1-0}{5} = 1/5 = 0.2$, so

$$x_1 = 0, \ x_2 = 0.2, \ x_3 = 0.4, \ x_4 = 0.6, \ x_5 = 0.8, \text{ and } x_6 = 1.$$

Using the Left Hand Rule, we have:

$$\sum_{i=1}^{n} f(x_i)\Delta x = \big(f(x_1) + f(x_2) + f(x_3) + f(x_4) + f(x_5)\big)\Delta x$$

$$= \big(f(0) + f(0.2) + f(0.4) + f(0.6) + f(0.8)\big)\Delta x$$

$$\approx (1 + 0.961 + 0.852 + 0.698 + 0.527)(0.2)$$

$$\approx 0.808.$$

Using the Right Hand Rule, we have:

$$\sum_{i=1}^{n} f(x_{i+1})\Delta x = \big(f(x_2) + f(x_3) + f(x_4) + f(x_5) + f(x_6)\big)\Delta x$$

$$= \big(f(0.2) + f(0.4) + f(0.6) + f(0.8) + f(1)\big)\Delta x$$

$$\approx (0.961 + 0.852 + 0.698 + 0.527 + 0.368)(0.2)$$

$$\approx 0.681.$$

Figure 5.30 shows the rectangles used in each method to approximate the definite integral. These graphs show that in this particular case, the Left Hand Rule is an over approximation and the Right Hand Rule is an under approximation. To get a better approximation, we could use more rectangles, as we did in

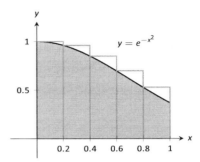

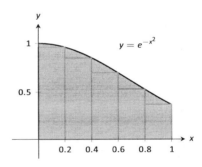

Figure 5.30: Approximating $\int_0^1 e^{-x^2} \, dx$ in Example 134.

Notes:

x_i	Exact	Approx.	$\sin(x_i^3)$
x_1	$-\pi/4$	-0.785	-0.466
x_2	$-7\pi/40$	-0.550	-0.165
x_3	$-\pi/10$	-0.314	-0.031
x_4	$-\pi/40$	-0.0785	0
x_5	$\pi/20$	0.157	0.004
x_6	$\pi/8$	0.393	0.061
x_7	$\pi/5$	0.628	0.246
x_8	$11\pi/40$	0.864	0.601
x_9	$7\pi/20$	1.10	0.971
x_{10}	$17\pi/40$	1.34	0.690
x_{11}	$\pi/2$	1.57	-0.670

Figure 5.31: Table of values used to approximate $\int_{-\frac{\pi}{4}}^{\frac{\pi}{2}} \sin(x^3)\, dx$ in Example 135.

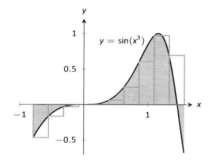

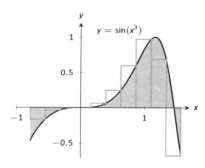

Figure 5.32: Approximating $\int_{-\frac{\pi}{4}}^{\frac{\pi}{2}} \sin(x^3)\, dx$ in Example 135.

Section 5.3. We could also average the Left and Right Hand Rule results together, giving

$$\frac{0.808 + 0.681}{2} = 0.7445.$$

The actual answer, accurate to 4 places after the decimal, is 0.7468, showing our average is a good approximation.

Example 135   **Approximating definite integrals with rectangles**

Approximate $\int_{-\frac{\pi}{4}}^{\frac{\pi}{2}} \sin(x^3)\, dx$ using the Left and Right Hand Rules with 10 equally spaced subintervals.

SOLUTION   We begin by finding Δx:

$$\frac{b-a}{n} = \frac{\pi/2 - (-\pi/4)}{10} = \frac{3\pi}{40} \approx 0.236.$$

It is useful to write out the endpoints of the subintervals in a table; in Figure 5.31, we give the exact values of the endpoints, their decimal approximations, and decimal approximations of $\sin(x^3)$ evaluated at these points.

Once this table is created, it is straightforward to approximate the definite integral using the Left and Right Hand Rules. (Note: the table itself is easy to create, especially with a standard spreadsheet program on a computer. The last two columns are all that are needed.) The Left Hand Rule sums the first 10 values of $\sin(x_i^3)$ and multiplies the sum by Δx; the Right Hand Rule sums the last 10 values of $\sin(x_i^3)$ and multiplies by Δx. Therefore we have:

Left Hand Rule: $\int_{-\frac{\pi}{4}}^{\frac{\pi}{2}} \sin(x^3)\, dx \approx (1.91)(0.236) = 0.451.$

Right Hand Rule: $\int_{-\frac{\pi}{4}}^{\frac{\pi}{2}} \sin(x^3)\, dx \approx (1.71)(0.236) = 0.404.$

Average of the Left and Right Hand Rules: 0.4275.

The actual answer, accurate to 3 places after the decimal, is 0.460. Our approximations were once again fairly good. The rectangles used in each approximation are shown in Figure 5.32. It is clear from the graphs that using more rectangles (and hence, narrower rectangles) should result in a more accurate approximation.

The Trapezoidal Rule

In Example 134 we approximated the value of $\int_0^1 e^{-x^2}\, dx$ with 5 rectangles of equal width. Figure 5.30 shows the rectangles used in the Left and Right Hand

Notes:

Rules. These graphs clearly show that rectangles do not match the shape of the graph all that well, and that accurate approximations will only come by using lots of rectangles.

Instead of using rectangles to approximate the area, we can instead use *trapezoids*. In Figure 5.33, we show the region under $f(x) = e^{-x^2}$ on $[0,1]$ approximated with 5 trapezoids of equal width; the top "corners" of each trapezoid lies on the graph of $f(x)$. It is clear from this figure that these trapezoids more accurately approximate the area under f and hence should give a better approximation of $\int_0^1 e^{-x^2}\, dx$. (In fact, these trapezoids seem to give a *great* approximation of the area!)

The formula for the area of a trapezoid is given in Figure 5.34. We approximate $\int_0^1 e^{-x^2}\, dx$ with these trapezoids in the following example.

Example 136   **Approximating definite integrals using trapezoids**

Use 5 trapezoids of equal width to approximate $\displaystyle\int_0^1 e^{-x^2}\, dx$.

SOLUTION   To compute the areas of the 5 trapezoids in Figure 5.33, it will again be useful to create a table of values as shown in Figure 5.35.

The leftmost trapezoid has legs of length 1 and 0.961 and a height of 0.2. Thus, by our formula, the area of the leftmost trapezoid is:

$$\frac{1 + 0.961}{2}(0.2) = 0.1961.$$

Moving right, the next trapezoid has legs of length 0.961 and 0.852 and a height of 0.2. Thus its area is:

$$\frac{0.961 + 0.852}{2}(0.2) = 0.1813.$$

The sum of the areas of all 5 trapezoids is:

$$\frac{1 + 0.961}{2}(0.2) + \frac{0.961 + 0.852}{2}(0.2) + \frac{0.852 + 0.698}{2}(0.2) +$$
$$\frac{0.698 + 0.527}{2}(0.2) + \frac{0.527 + 0.368}{2}(0.2) = 0.7445.$$

We approximate $\displaystyle\int_0^1 e^{-x^2}\, dx \approx 0.7445$.

There are many things to observe in this example. Note how each term in the final summation was multiplied by both 1/2 and by $\Delta x = 0.2$. We can factor these coefficients out, leaving a more concise summation as:

$$\frac{1}{2}(0.2)\Big[(1+0.961)+(0.961+0.852)+(0.852+0.698)+(0.698+0.527)+(0.527+0.368)\Big].$$

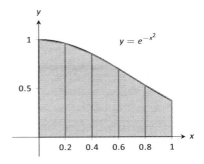

Figure 5.33: Approximating $\int_0^1 e^{-x^2}\, dx$ using 5 trapezoids of equal widths.

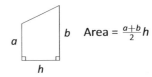

Figure 5.34: The area of a trapezoid.

x_i	$e^{-x_i^2}$
0	1
0.2	0.961
0.4	0.852
0.6	0.698
0.8	0.527
1	0.368

Figure 5.35: A table of values of e^{-x^2}.

Notes:

Now notice that all numbers except for the first and the last are added twice. Therefore we can write the summation even more concisely as

$$\frac{0.2}{2}\Big[1 + 2(0.961 + 0.852 + 0.698 + 0.527) + 0.368\Big].$$

This is the heart of the **Trapezoidal Rule**, wherein a definite integral $\int_a^b f(x)\,dx$ is approximated by using trapezoids of equal widths to approximate the corresponding area under f. Using n equally spaced subintervals with endpoints x_1, $x_2, \ldots, x_{n+1}$, we again have $\Delta x = \dfrac{b-a}{n}$. Thus:

$$\int_a^b f(x)\,dx \approx \sum_{i=1}^n \frac{f(x_i) + f(x_{i+1})}{2}\,\Delta x$$

$$= \frac{\Delta x}{2} \sum_{i=1}^n \big(f(x_i) + f(x_{i+1})\big)$$

$$= \frac{\Delta x}{2}\Big[f(x_1) + 2\sum_{i=2}^n f(x_i) + f(x_{n+1})\Big].$$

Example 137   Using the Trapezoidal Rule

Revisit Example 135 and approximate $\int_{-\frac{\pi}{4}}^{\frac{\pi}{2}} \sin(x^3)\,dx$ using the Trapezoidal Rule and 10 equally spaced subintervals.

SOLUTION   We refer back to Figure 5.31 for the table of values of $\sin(x^3)$. Recall that $\Delta x = 3\pi/40 \approx 0.236$. Thus we have:

$$\int_{-\frac{\pi}{4}}^{\frac{\pi}{2}} \sin(x^3)\,dx \approx \frac{0.236}{2}\Big[-0.466 + 2\Big(-0.165 + (-0.031) + \ldots + 0.69\Big) + (-0.67)\Big]$$

$$= 0.4275.$$

Notice how "quickly" the Trapezoidal Rule can be implemented once the table of values is created. This is true for all the methods explored in this section; the real work is creating a table of x_i and $f(x_i)$ values. Once this is completed, approximating the definite integral is not difficult. Again, using technology is wise. Spreadsheets can make quick work of these computations and make using lots of subintervals easy.

Also notice the approximations the Trapezoidal Rule gives. It is the average of the approximations given by the Left and Right Hand Rules! This effectively

Notes:

renders the Left and Right Hand Rules obsolete. They are useful when first learning about definite integrals, but if a real approximation is needed, one is generally better off using the Trapezoidal Rule instead of either the Left or Right Hand Rule.

How can we improve on the Trapezoidal Rule, apart from using more and more trapezoids? The answer is clear once we look back and consider what we have *really* done so far. The Left Hand Rule is not *really* about using rectangles to approximate area. Instead, it approximates a function f with constant functions on small subintervals and then computes the definite integral of these constant functions. The Trapezoidal Rule is really approximating a function f with a linear function on a small subinterval, then computes the definite integral of this linear function. In both of these cases the definite integrals are easy to compute in geometric terms.

So we have a progression: we start by approximating f with a constant function and then with a linear function. What is next? A quadratic function. By approximating the curve of a function with lots of parabolas, we generally get an even better approximation of the definite integral. We call this process **Simpson's Rule**, named after Thomas Simpson (1710-1761), even though others had used this rule as much as 100 years prior.

Simpson's Rule

Given one point, we can create a constant function that goes through that point. Given two points, we can create a linear function that goes through those points. Given three points, we can create a quadratic function that goes through those three points (given that no two have the same x–value).

Consider three points (x_1, y_1), (x_2, y_2) and (x_3, y_3) whose x–values are equally spaced and $x_1 < x_2 < x_3$. Let f be the quadratic function that goes through these three points. It is not hard to show that

$$\int_{x_1}^{x_3} f(x)\, dx = \frac{x_3 - x_1}{6} \left(y_1 + 4y_2 + y_3 \right). \tag{5.4}$$

Consider Figure 5.36. A function f goes through the 3 points shown and the parabola g that also goes through those points is graphed with a dashed line. Using our equation from above, we know exactly that

$$\int_{1}^{3} g(x)\, dx = \frac{3 - 1}{6} \left(3 + 4(1) + 2 \right) = 3.$$

Since g is a good approximation for f on $[1, 3]$, we can state that

$$\int_{1}^{3} f(x)\, dx \approx 3.$$

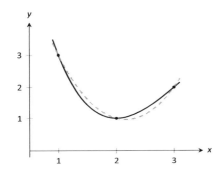

Figure 5.36: A graph of a function f and a parabola that approximates it well on $[1, 3]$.

Notes:

x_i	$e^{-x_i^2}$
0	1
0.25	0.939
0.5	0.779
0.75	0.570
1	0.368

(a)

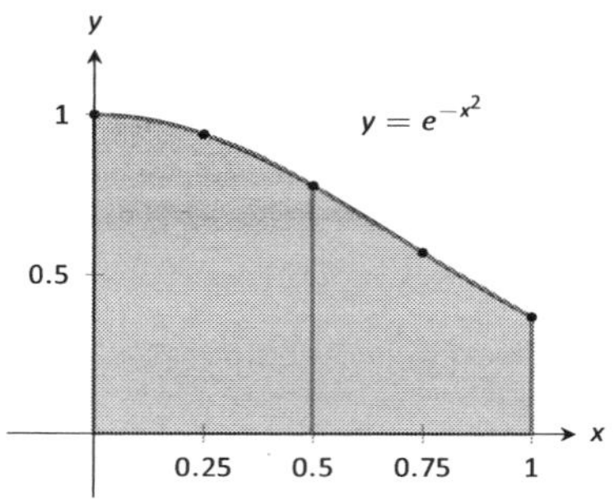

(b)

Figure 5.37: A table of values to approximate $\int_0^1 e^{-x^2}\,dx$, along with a graph of the function.

x_i	$\sin(x_i^3)$
−0.785	−0.466
−0.550	−0.165
−0.314	−0.031
−0.0785	0
0.157	0.004
0.393	0.061
0.628	0.246
0.864	0.601
1.10	0.971
1.34	0.690
1.57	−0.670

Figure 5.38: Table of values used to approximate $\int_{-\frac{\pi}{4}}^{\frac{\pi}{2}} \sin(x^3)\,dx$ in Example 139.

Notice how the interval $[1, 3]$ was split into two subintervals as we needed 3 points. Because of this, whenever we use Simpson's Rule, we need to break the interval into an even number of subintervals.

In general, to approximate $\int_a^b f(x)\,dx$ using Simpson's Rule, subdivide $[a, b]$ into n subintervals, where n is even and each subinterval has width $\Delta x = (b - a)/n$. We approximate f with $n/2$ parabolic curves, using Equation (5.4) to compute the area under these parabolas. Adding up these areas gives the formula:

$$\int_a^b f(x)\,dx \approx \frac{\Delta x}{3}\Big[f(x_1)+4f(x_2)+2f(x_3)+4f(x_4)+\ldots+2f(x_{n-1})+4f(x_n)+f(x_{n+1})\Big].$$

Note how the coefficients of the terms in the summation have the pattern 1, 4, 2, 4, 2, 4, . . ., 2, 4, 1.

Let's demonstrate Simpson's Rule with a concrete example.

Example 138   Using Simpson's Rule

Approximate $\int_0^1 e^{-x^2}\,dx$ using Simpson's Rule and 4 equally spaced subintervals.

SOLUTION   We begin by making a table of values as we have in the past, as shown in Figure 5.37(a). Simpson's Rule states that

$$\int_0^1 e^{-x^2}\,dx \approx \frac{0.25}{3}\Big[1 + 4(0.939) + 2(0.779) + 4(0.570) + 0.368\Big] = 0.746\overline{83}.$$

Recall in Example 134 we stated that the correct answer, accurate to 4 places after the decimal, was 0.7468. Our approximation with Simpson's Rule, with 4 subintervals, is better than our approximation with the Trapezoidal Rule using 5!

Figure 5.37(b) shows $f(x) = e^{-x^2}$ along with its approximating parabolas, demonstrating how good our approximation is. The approximating curves are nearly indistinguishable from the actual function.

Example 139   Using Simpson's Rule

Approximate $\int_{-\frac{\pi}{4}}^{\frac{\pi}{2}} \sin(x^3)\,dx$ using Simpson's Rule and 10 equally spaced intervals.

SOLUTION   Figure 5.38 shows the table of values that we used in the past for this problem, shown here again for convenience. Again, $\Delta x = (\pi/2 + \pi/4)/10 \approx 0.236$.

Notes:

Simpson's Rule states that

$$\int_{-\frac{\pi}{4}}^{\frac{\pi}{2}} \sin(x^3)\,dx \approx \frac{0.236}{3}\Big[(-0.466) + 4(-0.165) + 2(-0.031) + \ldots$$

$$\ldots + 2(0.971) + 4(0.69) + (-0.67)\Big]$$

$$= 0.4701$$

Recall that the actual value, accurate to 3 decimal places, is 0.460. Our approximation is within one $1/100^{\text{th}}$ of the correct value. The graph in Figure 5.39 shows how closely the parabolas match the shape of the graph.

Summary and Error Analysis

We summarize the key concepts of this section thus far in the following Key Idea.

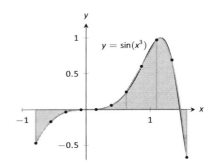

Figure 5.39: Approximating $\int_{-\frac{\pi}{4}}^{\frac{\pi}{2}} \sin(x^3)\,dx$ in Example 139 with Simpson's Rule and 10 equally spaced intervals.

Key Idea 9   Numerical Integration

Let f be a continuous function on $[a, b]$, let n be a positive integer, and let $\Delta x = \dfrac{b - a}{n}$.

Set $x_1 = a$, $x_2 = a + \Delta x$, $\ldots$, $x_i = a + (i - 1)\Delta x$, $x_{n+1} = b$.

Consider $\displaystyle\int_a^b f(x)\,dx$.

Left Hand Rule: $\displaystyle\int_a^b f(x)\,dx \approx \Delta x\Big[f(x_1) + f(x_2) + \ldots + f(x_n)\Big]$.

Right Hand Rule: $\displaystyle\int_a^b f(x)\,dx \approx \Delta x\Big[f(x_2) + f(x_3) + \ldots + f(x_{n+1})\Big]$.

Trapezoidal Rule: $\displaystyle\int_a^b f(x)\,dx \approx \dfrac{\Delta x}{2}\Big[f(x_1) + 2f(x_2) + 2f(x_3) + \ldots + 2f(x_n) + f(x_{n+1})\Big]$.

Simpson's Rule: $\displaystyle\int_a^b f(x)\,dx \approx \dfrac{\Delta x}{3}\Big[f(x_1) + 4f(x_2) + 2f(x_3) + \ldots + 4f(x_n) + f(x_{n+1})\Big]$ (n even).

In our examples, we approximated the value of a definite integral using a given method then compared it to the "right" answer. This should have raised several questions in the reader's mind, such as:

1. How was the "right" answer computed?

2. If the right answer can be found, what is the point of approximating?

3. If there is value to approximating, how are we supposed to know if the approximation is any good?

Notes:

These are good questions, and their answers are educational. In the examples, *the* right answer was never computed. Rather, an approximation accurate to a certain number of places after the decimal was given. In Example 134, we do not know the *exact* answer, but we know it starts with 0.7468. These more accurate approximations were computed using numerical integration but with more precision (i.e., more subintervals and the help of a computer).

Since the exact answer cannot be found, approximation still has its place. How are we to tell if the approximation is any good?

"Trial and error" provides one way. Using technology, make an approximation with, say, 10, 100, and 200 subintervals. This likely will not take much time at all, and a trend should emerge. If a trend does not emerge, try using yet more subintervals. Keep in mind that trial and error is never foolproof; you might stumble upon a problem in which a trend will not emerge.

A second method is to use Error Analysis. While the details are beyond the scope of this text, there are some formulas that give *bounds* for how good your approximation will be. For instance, the formula might state that the approximation is within 0.1 of the correct answer. If the approximation is 1.58, then one knows that the correct answer is between 1.48 and 1.68. By using lots of subintervals, one can get an approximation as accurate as one likes. Theorem 43 states what these bounds are.

Theorem 43   Error Bounds in the Trapezoidal and Simpson's Rules

1. Let E_T be the error in approximating $\int_a^b f(x)\, dx$ using the Trapezoidal Rule.

   If f has a continuous 2nd derivative on $[a, b]$ and M is any upper bound of $\left|f''(x)\right|$ on $[a, b]$, then

   $$E_T \leq \frac{(b-a)^3}{12n^2} M.$$

2. Let E_S be the error in approximating $\int_a^b f(x)\, dx$ using Simpson's Rule.

   If f has a continuous 4th derivative on $[a, b]$ and M is any upper bound of $\left|f^{(4)}\right|$ on $[a, b]$, then

   $$E_S \leq \frac{(b-a)^5}{180n^4} M.$$

Notes:

There are some key things to note about this theorem.

1. The larger the interval, the larger the error. This should make sense intuitively.

2. The error shrinks as more subintervals are used (i.e., as n gets larger).

3. The error in Simpson's Rule has a term relating to the 4th derivative of f. Consider a cubic polynomial: it's 4th derivative is 0. Therefore, the error in approximating the definite integral of a cubic polynomial with Simpson's Rule is 0 – Simpson's Rule computes the exact answer!

We revisit Examples 136 and 138 and compute the error bounds using Theorem 43 in the following example.

Example 140   **Computing error bounds**

Find the error bounds when approximating $\int_0^1 e^{-x^2}\ dx$ using the Trapezoidal Rule and 5 subintervals, and using Simpson's Rule with 4 subintervals.

SOLUTION

Trapezoidal Rule with $n = 5$:

We start by computing the 2nd derivative of $f(x) = e^{-x^2}$:

$$f''(x) = e^{-x^2}(4x^2 - 2).$$

Figure 5.40 shows a graph of $f''(x)$ on $[0, 1]$. It is clear that the largest value of f'', in absolute value, is 2. Thus we let $M = 2$ and apply the error formula from Theorem 43.

$$E_T = \frac{(1-0)^3}{12 \cdot 5^2} \cdot 2 = 0.00\overline{6}.$$

Our error estimation formula states that our approximation of 0.7445 found in Example 136 is within 0.0067 of the correct answer, hence we know that

$$0.7445 - 0.0067 = .7378 \leq \int_0^1 e^{-x^2}\ dx \leq 0.7512 = 0.7445 + 0.0067.$$

We had earlier computed the exact answer, correct to 4 decimal places, to be 0.7468, affirming the validity of Theorem 43.

Simpson's Rule with $n = 4$:

We start by computing the 4th derivative of $f(x) = e^{-x^2}$:

$$f^{(4)}(x) = e^{-x^2}(16x^4 - 48x^2 + 12).$$

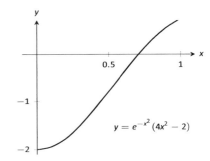

Figure 5.40: Graphing $f''(x)$ in Example 140 to help establish error bounds.

Notes:

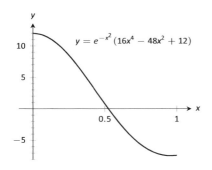

$$y = e^{-x^2}(16x^4 - 48x^2 + 12)$$

Figure 5.41: Graphing $f^{(4)}(x)$ in Example 140 to help establish error bounds.

Time	Speed (mph)
0	0
1	25
2	22
3	19
4	39
5	0
6	43
7	59
8	54
9	51
10	43
11	35
12	40
13	43
14	30
15	0
16	0
17	28
18	40
19	42
20	40
21	39
22	40
23	23
24	0

Figure 5.42: Speed data collected at 30 second intervals for Example 141.

Figure 5.41 shows a graph of $f^{(4)}(x)$ on $[0, 1]$. It is clear that the largest value of $f^{(4)}$, in absolute value, is 12. Thus we let $M = 12$ and apply the error formula from Theorem 43.

$$E_s = \frac{(1-0)^5}{180 \cdot 4^4} \cdot 12 = 0.00026.$$

Our error estimation formula states that our approximation of $0.74683\overline{3}$ found in Example 138 is within 0.00026 of the correct answer, hence we know that

$$0.74683 - 0.00026 = .74657 \leq \int_0^1 e^{-x^2}\, dx \leq 0.74709 = 0.74683 + 0.00026.$$

Once again we affirm the validity of Theorem 43.

At the beginning of this section we mentioned two main situations where numerical integration was desirable. We have considered the case where an antiderivative of the integrand cannot be computed. We now investigate the situation where the integrand is not known. This is, in fact, the most widely used application of Numerical Integration methods. "Most of the time" we observe behavior but do not know "the" function that describes it. We instead collect data about the behavior and make approximations based off of this data. We demonstrate this in an example.

Example 141   Approximating distance traveled

One of the authors drove his daughter home from school while she recorded their speed every 30 seconds. The data is given in Figure 5.42. Approximate the distance they traveled.

SOLUTION   Recall that by integrating a speed function we get distance traveled. We have information about $v(t)$; we will use Simpson's Rule to approximate $\int_a^b v(t)\, dt$.

The most difficult aspect of this problem is converting the given data into the form we need it to be in. The speed is measured in miles per hour, whereas the time is measured in 30 second increments.

We need to compute $\Delta x = (b-a)/n$. Clearly, $n = 24$. What are a and b? Since we start at time $t = 0$, we have that $a = 0$. The final recorded time came after 24 periods of 30 seconds, which is 12 minutes or 1/5 of an hour. Thus we have

$$\Delta x = \frac{b-a}{n} = \frac{1/5 - 0}{24} = \frac{1}{120}; \quad \frac{\Delta x}{3} = \frac{1}{360}.$$

Notes:

Thus the distance traveled is approximately:

$$\int_0^{0.2} v(t)\,dt \approx \frac{1}{360}\Big[f(x_1) + 4f(x_2) + 2f(x_3) + \cdots + 4f(x_n) + f(x_{n+1})\Big]$$

$$= \frac{1}{360}\Big[0 + 4 \cdot 25 + 2 \cdot 22 + \cdots + 2 \cdot 40 + 4 \cdot 23 + 0\Big]$$

$$\approx 6.2167 \text{ miles.}$$

We approximate the author drove 6.2 miles. (Because we are sure the reader wants to know, the author's odometer recorded the distance as about 6.05 miles.)

We started this chapter learning about antiderivatives and indefinite integrals. We then seemed to change focus by looking at areas between the graph of a function and the x-axis. We defined these areas as the definite integral of the function, using a notation very similar to the notation of the indefinite integral. The Fundamental Theorem of Calculus tied these two seemingly separate concepts together: we can find areas under a curve, i.e., we can evaluate a definite integral, using antiderivatives.

We ended the chapter by noting that antiderivatives are sometimes more than difficult to find: they are impossible. Therefore we developed numerical techniques that gave us good approximations of definite integrals.

We used the definite integral to compute areas, and also to compute displacements and distances traveled. There is far more we can do than that. In Chapter 7 we'll see more applications of the definite integral. Before that, in Chapter 6 we'll learn advanced techniques of integration, analogous to learning rules like the Product, Quotient and Chain Rules of differentiation.

Notes:

Exercises 5.5

Terms and Concepts

1. T/F: Simpson's Rule is a method of approximating antiderivatives.

2. What are the two basic situations where approximating the value of a definite integral is necessary?

3. Why are the Left and Right Hand Rules rarely used?

Problems

In Exercises 4 – 11, a definite integral is given.

    (a) Approximate the definite integral with the Trapezoidal Rule and $n = 4$.

    (b) Approximate the definite integral with Simpson's Rule and $n = 4$.

    (c) Find the exact value of the integral.

4. $\int_{-1}^{1} x^2 \, dx$

5. $\int_{0}^{10} 5x \, dx$

6. $\int_{0}^{\pi} \sin x \, dx$

7. $\int_{0}^{4} \sqrt{x} \, dx$

8. $\int_{0}^{3} (x^3 + 2x^2 - 5x + 7) \, dx$

9. $\int_{0}^{1} x^4 \, dx$

10. $\int_{0}^{2\pi} \cos x \, dx$

11. $\int_{-3}^{3} \sqrt{9 - x^2} \, dx$

In Exercises 12 – 19, approximate the definite integral with the Trapezoidal Rule and Simpson's Rule, with $n = 6$.

12. $\int_{0}^{1} \cos\left(x^2\right) dx$

13. $\int_{-1}^{1} e^{x^2} \, dx$

14. $\int_{0}^{5} \sqrt{x^2 + 1} \, dx$

15. $\int_{0}^{\pi} x \sin x \, dx$

16. $\int_{0}^{\pi/2} \sqrt{\cos x} \, dx$

17. $\int_{1}^{4} \ln x \, dx$

18. $\int_{-1}^{1} \frac{1}{\sin x + 2} \, dx$

19. $\int_{0}^{6} \frac{1}{\sin x + 2} \, dx$

In Exercises 20 – 23, find n such that the error in approximating the given definite integral is less than 0.0001 when using:

    (a) the Trapezoidal Rule

    (b) Simpson's Rule

20. $\int_{0}^{\pi} \sin x \, dx$

21. $\int_{1}^{4} \frac{1}{\sqrt{x}} \, dx$

22. $\int_{0}^{\pi} \cos\left(x^2\right) dx$

23. $\int_{0}^{5} x^4 \, dx$

In Exercises 24 – 25, a region is given. Find the area of the region using Simpson's Rule:

    (a) where the measurements are in centimeters, taken in 1 cm increments, and

    (b) where the measurements are in hundreds of yards, taken in 100 yd increments.

24.

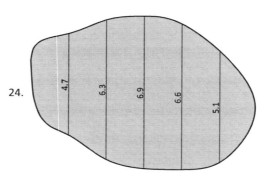

25.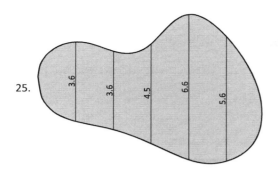

6: TECHNIQUES OF ANTIDIFFERENTIATION

The previous chapter introduced the antiderivative and connected it to signed areas under a curve through the Fundamental Theorem of Calculus. The next chapter explores more applications of definite integrals than just area. As evaluating definite integrals will become important, we will want to find antiderivatives of a variety of functions.

This chapter is devoted to exploring techniques of antidifferentiation. While not every function has an antiderivative in terms of elementary functions (a concept introduced in the section on Numerical Integration), we can still find antiderivatives of a wide variety of functions.

6.1 Substitution

We motivate this section with an example. Let $f(x) = (x^2 + 3x - 5)^{10}$. We can compute $f'(x)$ using the Chain Rule. It is:

$$f'(x) = 10(x^2 + 3x - 5)^9 \cdot (2x + 3) = (20x + 30)(x^2 + 3x - 5)^9.$$

Now consider this: What is $\int (20x + 30)(x^2 + 3x - 5)^9 \, dx$? We have the answer in front of us;

$$\int (20x + 30)(x^2 + 3x - 5)^9 \, dx = (x^2 + 3x - 5)^{10} + C.$$

How would we have evaluated this indefinite integral without starting with $f(x)$ as we did?

This section explores *integration by substitution*. It allows us to "undo the Chain Rule." Substitution allows us to evaluate the above integral without knowing the original function first.

The underlying principle is to rewrite a "complicated" integral of the form $\int f(x) \, dx$ as a not–so–complicated integral $\int h(u) \, du$. We'll formally establish later how this is done. First, consider again our introductory indefinite integral, $\int (20x + 30)(x^2 + 3x - 5)^9 \, dx$. Arguably the most "complicated" part of the integrand is $(x^2 + 3x - 5)^9$. We wish to make this simpler; we do so through a substitution. Let $u = x^2 + 3x - 5$. Thus

$$(x^2 + 3x - 5)^9 = u^9.$$

We have established u as a function of x, so now consider the differential of u:

$$du = (2x + 3)dx.$$

Keep in mind that $(2x+3)$ and dx are multiplied; the dx is not "just sitting there."

Return to the original integral and do some substitutions through algebra:

$$\int (20x + 30)(x^2 + 3x - 5)^9 \, dx = \int 10(2x + 3)(x^2 + 3x - 5)^9 \, dx$$

$$= \int 10\underbrace{(x^2 + 3x - 5)}_{u}{}^9 \, \underbrace{(2x + 3) \, dx}_{du}$$

$$= \int 10u^9 \, du$$

$$= u^{10} + C \quad \text{(replace } u \text{ with } x^2 + 3x - 5)$$

$$= (x^2 + 3x - 5)^{10} + C$$

One might well look at this and think "I (sort of) followed how that worked, but I could never come up with that on my own," but the process is learnable. This section contains numerous examples through which the reader will gain understanding and mathematical maturity enabling them to regard substitution as a natural tool when evaluating integrals.

We stated before that integration by substitution "undoes" the Chain Rule. Specifically, let $F(x)$ and $g(x)$ be differentiable functions and consider the derivative of their composition:

$$\frac{d}{dx}\Big(F\big(g(x)\big)\Big) = F'(g(x))g'(x).$$

Thus

$$\int F'(g(x))g'(x) \, dx = F(g(x)) + C.$$

Integration by substitution works by recognizing the "inside" function $g(x)$ and replacing it with a variable. By setting $u = g(x)$, we can rewrite the derivative as

$$\frac{d}{dx}\Big(F(u)\Big) = F'(u)u'.$$

Since $du = g'(x)dx$, we can rewrite the above integral as

$$\int F'(g(x))g'(x) \, dx = \int F'(u)du = F(u) + C = F(g(x)) + C.$$

This concept is important so we restate it in the context of a theorem.

Notes:

Theorem 44 **Integration by Substitution**

Let F and g be differentiable functions, where the range of g is an interval I contained in the domain of F. Then

$$\int F'(g(x))g'(x)\,dx = F(g(x)) + C.$$

If $u = g(x)$, then $du = g'(x)dx$ and

$$\int F'(g(x))g'(x)\,dx = \int F'(u)\,du = F(u) + C = F(g(x)) + C.$$

The point of substitution is to make the integration step easy. Indeed, the step $\int F'(u)\,du = F(u) + C$ looks easy, as the antiderivative of the derivative of F is just F, plus a constant. The "work" involved is making the proper substitution. There is not a step–by–step process that one can memorize; rather, experience will be one's guide. To gain experience, we now embark on many examples.

Example 142 **Integrating by substitution**

Evaluate $\int x\sin(x^2 + 5)\,dx.$

SOLUTION Knowing that substitution is related to the Chain Rule, we choose to let u be the "inside" function of $\sin(x^2 + 5)$. (This is not *always* a good choice, but it is often the best place to start.)

Let $u = x^2 + 5$, hence $du = 2x\,dx$. The integrand has an $x\,dx$ term, but not a $2x\,dx$ term. (Recall that multiplication is commutative, so the x does not physically have to be next to dx for there to be an $x\,dx$ term.) We can divide both sides of the du expression by 2:

$$du = 2x\,dx \quad\Rightarrow\quad \frac{1}{2}du = x\,dx.$$

We can now substitute.

$$\int x\sin(x^2 + 5)\,dx = \int \sin(\underbrace{x^2 + 5}_{u})\,\underbrace{x\,dx}_{\frac{1}{2}du}$$

$$= \int \frac{1}{2}\sin u\,du$$

Notes:

$$= -\frac{1}{2}\cos u + C \quad \text{(now replace } u \text{ with } x^2 + 5\text{)}$$

$$= -\frac{1}{2}\cos(x^2 + 5) + C.$$

Thus $\int x\sin(x^2 + 5)\,dx = -\frac{1}{2}\cos(x^2 + 5) + C$. We can check our work by evaluating the derivative of the right hand side.

Example 143   **Integrating by substitution**

Evaluate $\displaystyle\int \cos(5x)\,dx$.

SOLUTION   Again let u replace the "inside" function. Letting $u = 5x$, we have $du = 5dx$. Since our integrand does not have a $5dx$ term, we can divide the previous equation by 5 to obtain $\frac{1}{5}du = dx$. We can now substitute.

$$\int \cos(5x)\,dx = \int \cos(\underbrace{5x}_{u})\underbrace{dx}_{\frac{1}{5}du}$$

$$= \int \frac{1}{5}\cos u\,du$$

$$= \frac{1}{5}\sin u + C$$

$$= \frac{1}{5}\sin(5x) + C.$$

We can again check our work through differentiation.

The previous example exhibited a common, and simple, type of substitution. The "inside" function was a linear function (in this case, $y = 5x$). When the inside function is linear, the resulting integration is very predictable, outlined here.

Key Idea 10   **Substitution With A Linear Function**

Consider $\int F'(ax + b)\,dx$, where $a \neq 0$ and b are constants. Letting $u = ax + b$ gives $du = a \cdot dx$, leading to the result

$$\int F'(ax + b)\,dx = \frac{1}{a}F(ax + b) + C.$$

Thus $\int \sin(7x - 4)\,dx = -\frac{1}{7}\cos(7x - 4) + C$. Our next example can use Key Idea 10, but we will only employ it after going through all of the steps.

Notes:

Example 144   **Integrating by substituting a linear function**

Evaluate $\displaystyle\int \frac{7}{-3x+1}\, dx$.

   SOLUTION   View this a composition of functions $f(g(x))$, where $f(x) = 7/x$ and $g(x) = -3x + 1$. Employing our understanding of substitution, we let $u = -3x + 1$, the inside function. Thus $du = -3dx$. The integrand lacks a -3; hence divide the previous equation by -3 to obtain $-du/3 = dx$. We can now evaluate the integral through substitution.

$$\int \frac{7}{-3x+1}\, dx = \int \frac{7}{u} \frac{du}{-3}$$
$$= \frac{-7}{3} \int \frac{du}{u}$$
$$= \frac{-7}{3} \ln|u| + C$$
$$= -\frac{7}{3} \ln|-3x+1| + C.$$

Using Key Idea 10 is faster, recognizing that u is linear and $a = -3$. One may want to continue writing out all the steps until they are comfortable with this particular shortcut.

   Not all integrals that benefit from substitution have a clear "inside" function. Several of the following examples will demonstrate ways in which this occurs.

Example 145   **Integrating by substitution**

Evaluate $\displaystyle\int \sin x \cos x\, dx$.

   SOLUTION   There is not a composition of function here to exploit; rather, just a product of functions. Do not be afraid to experiment; when given an integral to evaluate, it is often beneficial to think "If I let u be *this*, then du must be *that* …" and see if this helps simplify the integral at all.

   In this example, let's set $u = \sin x$. Then $du = \cos x\, dx$, which we have as part of the integrand! The substitution becomes very straightforward:

$$\int \sin x \cos x\, dx = \int u\, du$$
$$= \frac{1}{2}u^2 + C$$
$$= \frac{1}{2}\sin^2 x + C.$$

Notes:

One would do well to ask "What would happen if we let $u = \cos x$?" The result is just as easy to find, yet looks very different. The challenge to the reader is to evaluate the integral letting $u = \cos x$ and discover why the answer is the same, yet looks different.

Our examples so far have required "basic substitution." The next example demonstrates how substitutions can be made that often strike the new learner as being "nonstandard."

Example 146   **Integrating by substitution**

Evaluate $\int x\sqrt{x+3}\,dx$.

Solution   Recognizing the composition of functions, set $u = x + 3$. Then $du = dx$, giving what seems initially to be a simple substitution. But at this stage, we have:

$$\int x\sqrt{x+3}\,dx = \int x\sqrt{u}\,du.$$

We cannot evaluate an integral that has both an x and an u in it. We need to convert the x to an expression involving just u.

Since we set $u = x + 3$, we can also state that $u - 3 = x$. Thus we can replace x in the integrand with $u - 3$. It will also be helpful to rewrite $\sqrt{u}$ as $u^{\frac{1}{2}}$.

$$\int x\sqrt{x+3}\,dx = \int (u-3)u^{\frac{1}{2}}\,du$$

$$= \int \left(u^{\frac{3}{2}} - 3u^{\frac{1}{2}} \right) du$$

$$= \frac{2}{5}u^{\frac{5}{2}} - 2u^{\frac{3}{2}} + C$$

$$= \frac{2}{5}(x+3)^{\frac{5}{2}} - 2(x+3)^{\frac{3}{2}} + C.$$

Checking your work is always a good idea. In this particular case, some algebra will be needed to make one's answer match the integrand in the original problem.

Example 147   **Integrating by substitution**

Evaluate $\int \dfrac{1}{x \ln x}\,dx$.

Solution   This is another example where there does not seem to be an obvious composition of functions. The line of thinking used in Example 146 is useful here: choose something for u and consider what this implies du must

Notes:

be. If u can be chosen such that du also appears in the integrand, then we have chosen well.

Choosing $u = 1/x$ makes $du = -1/x^2\ dx$; that does not seem helpful. However, setting $u = \ln x$ makes $du = 1/x\ dx$, which is part of the integrand. Thus:

$$\int \frac{1}{x \ln x}\ dx = \int \underbrace{\frac{1}{\ln x}}_{1/u}\ \underbrace{\frac{1}{x}\ dx}_{du}$$

$$= \int \frac{1}{u}\ du$$

$$= \ln|u| + C$$

$$= \ln|\ln x| + C.$$

The final answer is interesting; the natural log of the natural log. Take the derivative to confirm this answer is indeed correct.

Integrals Involving Trigonometric Functions

Section 6.3 delves deeper into integrals of a variety of trigonometric functions; here we use substitution to establish a foundation that we will build upon.

The next three examples will help fill in some missing pieces of our antiderivative knowledge. We know the antiderivatives of the sine and cosine functions; what about the other standard functions tangent, cotangent, secant and cosecant? We discover these next.

Example 148   **Integration by substitution: antiderivatives of $\tan x$**

Evaluate $\displaystyle\int \tan x\ dx$.

SOLUTION   The previous paragraph established that we did not know the antiderivatives of tangent, hence we must assume that we have learned something in this section that can help us evaluate this indefinite integral.

Rewrite $\tan x$ as $\sin x/\cos x$. While the presence of a composition of functions may not be immediately obvious, recognize that $\cos x$ is "inside" the $1/x$ function. Therefore, we see if setting $u = \cos x$ returns usable results. We have

Notes:

that $du = -\sin x \, dx$, hence $-du = \sin x \, dx$. We can integrate:

$$\int \tan x \, dx = \int \frac{\sin x}{\cos x} \, dx$$
$$= \int \underbrace{\frac{1}{\cos x}}_{u} \underbrace{\sin x \, dx}_{-du}$$
$$= \int \frac{-1}{u} \, du$$
$$= -\ln|u| + C$$
$$= -\ln|\cos x| + C.$$

Some texts prefer to bring the -1 inside the logarithm as a power of $\cos x$, as in:

$$-\ln|\cos x| + C = \ln|(\cos x)^{-1}| + C$$
$$= \ln\left|\frac{1}{\cos x}\right| + C$$
$$= \ln|\sec x| + C.$$

Thus the result they give is $\int \tan x \, dx = \ln|\sec x| + C$. These two answers are equivalent.

Example 149   **Integrating by substitution: antiderivatives of** $\sec x$

Evaluate $\displaystyle\int \sec x \, dx$.

SOLUTION   This example employs a wonderful trick: multiply the integrand by "1" so that we see how to integrate more clearly. In this case, we write "1" as

$$1 = \frac{\sec x + \tan x}{\sec x + \tan x}.$$

This may seem like it came out of left field, but it works beautifully. Consider:

$$\int \sec x \, dx = \int \sec x \cdot \frac{\sec x + \tan x}{\sec x + \tan x} \, dx$$
$$= \int \frac{\sec^2 x + \sec x \tan x}{\sec x + \tan x} \, dx.$$

Notes:

Now let $u = \sec x + \tan x$; this means $du = (\sec x \tan x + \sec^2 x)\,dx$, which is our numerator. Thus:

$$= \int \frac{du}{u}$$
$$= \ln|u| + C$$
$$= \ln|\sec x + \tan x| + C.$$

We can use similar techniques to those used in Examples 148 and 149 to find antiderivatives of $\cot x$ and $\csc x$ (which the reader can explore in the exercises.) We summarize our results here.

Theorem 45   Antiderivatives of Trigonometric Functions

1. $\displaystyle\int \sin x\,dx = -\cos x + C$

2. $\displaystyle\int \cos x\,dx = \sin x + C$

3. $\displaystyle\int \tan x\,dx = -\ln|\cos x| + C$

4. $\displaystyle\int \csc x\,dx = -\ln|\csc x + \cot x| + C$

5. $\displaystyle\int \sec x\,dx = \ln|\sec x + \tan x| + C$

6. $\displaystyle\int \cot x\,dx = \ln|\sin x| + C$

We explore one more common trigonometric integral.

Example 150   Integration by substitution: powers of $\cos x$ and $\sin x$

Evaluate $\displaystyle\int \cos^2 x\,dx.$

SOLUTION   We have a composition of functions as $\cos^2 x = (\cos x)^2$. However, setting $u = \cos x$ means $du = -\sin x\,dx$, which we do not have in the integral. Another technique is needed.

The process we'll employ is to use a Power Reducing formula for $\cos^2 x$ (perhaps consult the back of this text for this formula), which states

$$\cos^2 x = \frac{1 + \cos(2x)}{2}.$$

The right hand side of this equation is not difficult to integrate. We have:

$$\int \cos^2 x\,dx = \int \frac{1 + \cos(2x)}{2}\,dx$$
$$= \int \left(\frac{1}{2} + \frac{1}{2}\cos(2x)\right)\,dx.$$

Notes:

Now use Key Idea 10:

$$= \frac{1}{2}x + \frac{1}{2}\frac{\sin(2x)}{2} + C$$

$$= \frac{1}{2}x + \frac{\sin(2x)}{4} + C.$$

We'll make significant use of this power–reducing technique in future sections.

Simplifying the Integrand

It is common to be reluctant to manipulate the integrand of an integral; at first, our grasp of integration is tenuous and one may think that working with the integrand will improperly change the results. Integration by substitution works using a different logic: as long as *equality* is maintained, the integrand can be manipulated so that its *form* is easier to deal with. The next two examples demonstrate common ways in which using algebra first makes the integration easier to perform.

Example 151   **Integration by substitution: simplifying first**
Evaluate $\displaystyle\int \frac{x^3 + 4x^2 + 8x + 5}{x^2 + 2x + 1}\,dx.$

SOLUTION   One may try to start by setting u equal to either the numerator or denominator; in each instance, the result is not workable.

When dealing with rational functions (i.e., quotients made up of polynomial functions), it is an almost universal rule that everything works better when the degree of the numerator is less than the degree of the denominator. Hence we use polynomial division.

We skip the specifics of the steps, but note that when $x^2 + 2x + 1$ is divided into $x^3 + 4x^2 + 8x + 5$, it goes in $x + 2$ times with a remainder of $3x + 3$. Thus

$$\frac{x^3 + 4x^2 + 8x + 5}{x^2 + 2x + 1} = x + 2 + \frac{3x + 3}{x^2 + 2x + 1}.$$

Integrating $x + 2$ is simple. The fraction can be integrated by setting $u = x^2 + 2x + 1$, giving $du = (2x + 2)\,dx$. This is very similar to the numerator. Note that

Notes:

$du/2 = (x + 1) \, dx$ and then consider the following:

$$\int \frac{x^3 + 4x^2 + 8x + 5}{x^2 + 2x + 1} \, dx = \int \left(x + 2 + \frac{3x + 3}{x^2 + 2x + 1} \right) dx$$

$$= \int (x + 2) \, dx + \int \frac{3(x + 1)}{x^2 + 2x + 1} \, dx$$

$$= \frac{1}{2}x^2 + 2x + C_1 + \int \frac{3}{u} \frac{du}{2}$$

$$= \frac{1}{2}x^2 + 2x + C_1 + \frac{3}{2} \ln |u| + C_2$$

$$= \frac{1}{2}x^2 + 2x + \frac{3}{2} \ln |x^2 + 2x + 1| + C.$$

In some ways, we "lucked out" in that after dividing, substitution was able to be done. In later sections we'll develop techniques for handling rational functions where substitution is not directly feasible.

Example 152   **Integration by alternate methods**

Evaluate $\displaystyle\int \frac{x^2 + 2x + 3}{\sqrt{x}} \, dx$ with, and without, substitution.

    SOLUTION   We already know how to integrate this particular example. Rewrite $\sqrt{x}$ as $x^{\frac{1}{2}}$ and simplify the fraction:

$$\frac{x^2 + 2x + 3}{x^{1/2}} = x^{\frac{3}{2}} + 2x^{\frac{1}{2}} + 3x^{-\frac{1}{2}}.$$

We can now integrate using the Power Rule:

$$\int \frac{x^2 + 2x + 3}{x^{1/2}} \, dx = \int \left(x^{\frac{3}{2}} + 2x^{\frac{1}{2}} + 3x^{-\frac{1}{2}} \right) dx$$

$$= \frac{2}{5}x^{\frac{5}{2}} + \frac{4}{3}x^{\frac{3}{2}} + 6x^{\frac{1}{2}} + C$$

This is a perfectly fine approach. We demonstrate how this can also be solved using substitution as its implementation is rather clever.

    Let $u = \sqrt{x} = x^{\frac{1}{2}}$; therefore

$$du = \frac{1}{2}x^{-\frac{1}{2}} dx = \frac{1}{2\sqrt{x}} \, dx \quad \Rightarrow \quad 2du = \frac{1}{\sqrt{x}} \, dx.$$

This gives us $\displaystyle\int \frac{x^2 + 2x + 3}{\sqrt{x}} \, dx = \int (x^2 + 2x + 3) \cdot 2 \, du$. What are we to do with the other x terms? Since $u = x^{\frac{1}{2}}$, $u^2 = x$, etc. We can then replace x^2 and

---

Notes:

x with appropriate powers of *u*. We thus have

$$\int \frac{x^2 + 2x + 3}{\sqrt{x}}\, dx = \int (x^2 + 2x + 3) \cdot 2\, du$$

$$= \int 2(u^4 + 2u^2 + 3)\, du$$

$$= \frac{2}{5}u^5 + \frac{4}{3}u^3 + 6u + C$$

$$= \frac{2}{5}x^{\frac{5}{2}} + \frac{4}{3}x^{\frac{3}{2}} + 6x^{\frac{1}{2}} + C,$$

which is obviously the same answer we obtained before. In this situation, substitution is arguably more work than our other method. The fantastic thing is that it works. It demonstrates how flexible integration is.

Substitution and Inverse Trigonometric Functions

When studying derivatives of inverse functions, we learned that

$$\frac{d}{dx}\left(\tan^{-1} x \right) = \frac{1}{1 + x^2}.$$

Applying the Chain Rule to this is not difficult; for instance,

$$\frac{d}{dx}\left(\tan^{-1} 5x \right) = \frac{5}{1 + 25x^2}.$$

We now explore how Substitution can be used to "undo" certain derivatives that are the result of the Chain Rule applied to Inverse Trigonometric functions. We begin with an example.

Example 153   **Integrating by substitution: inverse trigonometric functions**

Evaluate $\displaystyle\int \frac{1}{25 + x^2}\, dx.$

    SOLUTION   The integrand looks similar to the derivative of the arctangent function. Note:

$$\frac{1}{25 + x^2} = \frac{1}{25\left(1 + \frac{x^2}{25}\right)}$$

$$= \frac{1}{25\left(1 + \left(\frac{x}{5}\right)^2\right)}$$

$$= \frac{1}{25} \frac{1}{1 + \left(\frac{x}{5}\right)^2}.$$

Notes:

Thus

$$\int \frac{1}{25 + x^2}\, dx = \frac{1}{25} \int \frac{1}{1 + \left(\frac{x}{5}\right)^2}\, dx.$$

This can be integrated using Substitution. Set $u = x/5$, hence $du = dx/5$ or $dx = 5\,du$. Thus

$$\begin{aligned}
\int \frac{1}{25 + x^2}\, dx &= \frac{1}{25} \int \frac{1}{1 + \left(\frac{x}{5}\right)^2}\, dx \\
&= \frac{1}{5} \int \frac{1}{1 + u^2}\, du \\
&= \frac{1}{5} \tan^{-1} u + C \\
&= \frac{1}{5} \tan^{-1} \left(\frac{x}{5}\right) + C
\end{aligned}$$

Example 153 demonstrates a general technique that can be applied to other integrands that result in inverse trigonometric functions. The results are summarized here.

Theorem 46   Integrals Involving Inverse Trigonomentric Functions

Let $a > 0$.

1. $\displaystyle \int \frac{1}{a^2 + x^2}\, dx = \frac{1}{a} \tan^{-1} \left(\frac{x}{a}\right) + C$

2. $\displaystyle \int \frac{1}{\sqrt{a^2 - x^2}}\, dx = \sin^{-1} \left(\frac{x}{a}\right) + C$

3. $\displaystyle \int \frac{1}{x\sqrt{x^2 - a^2}}\, dx = \frac{1}{a} \sec^{-1} \left(\frac{|x|}{a}\right) + C$

Let's practice using Theorem 46.

Example 154   Integrating by substitution: inverse trigonometric functions
Evaluate the given indefinite integrals.

$$\int \frac{1}{9 + x^2}\, dx, \quad \int \frac{1}{x\sqrt{x^2 - \frac{1}{100}}}\, dx \quad \text{and} \quad \int \frac{1}{\sqrt{5 - x^2}}\, dx.$$

Notes:

SOLUTION Each can be answered using a straightforward application of Theorem 46.

$$\int \frac{1}{9 + x^2}\, dx = \frac{1}{3} \tan^{-1} \frac{x}{3} + C, \text{ as } a = 3.$$

$$\int \frac{1}{x\sqrt{x^2 - \frac{1}{100}}}\, dx = 10 \sec^{-1} 10x + C, \text{ as } a = \frac{1}{10}.$$

$$\int \frac{1}{\sqrt{5 - x^2}} = \sin^{-1} \frac{x}{\sqrt{5}} + C, \text{ as } a = \sqrt{5}.$$

Most applications of Theorem 46 are not as straightforward. The next examples show some common integrals that can still be approached with this theorem.

Example 155 **Integrating by substitution: completing the square**

Evaluate $\displaystyle\int \frac{1}{x^2 - 4x + 13}\, dx$.

SOLUTION Initially, this integral seems to have nothing in common with the integrals in Theorem 46. As it lacks a square root, it almost certainly is not related to arcsine or arcsecant. It is, however, related to the arctangent function.

We see this by *completing the square* in the denominator. We give a brief reminder of the process here.

Start with a quadratic with a leading coefficient of 1. It will have the form of $x^2 + bx + c$. Take 1/2 of b, square it, and add/subtract it back into the expression. I.e.,

$$x^2 + bx + c = \underbrace{x^2 + bx + \frac{b^2}{4}}_{(x+b/2)^2} - \frac{b^2}{4} + c$$

$$= \left(x + \frac{b}{2}\right)^2 + c - \frac{b^2}{4}$$

In our example, we take half of -4 and square it, getting 4. We add/subtract it into the denominator as follows:

$$\frac{1}{x^2 - 4x + 13} = \frac{1}{\underbrace{x^2 - 4x + 4}_{(x-2)^2} - 4 + 13}$$

$$= \frac{1}{(x - 2)^2 + 9}$$

Notes:

We can now integrate this using the arctangent rule. Technically, we need to substitute first with $u = x - 2$, but we can employ Key Idea 10 instead. Thus we have

$$\int \frac{1}{x^2 - 4x + 13}\, dx = \int \frac{1}{(x-2)^2 + 9}\, dx = \frac{1}{3}\tan^{-1}\frac{x-2}{3} + C.$$

Example 156   **Integrals requiring multiple methods**

Evaluate $\displaystyle\int \frac{4-x}{\sqrt{16-x^2}}\, dx$.

    SOLUTION   This integral requires two different methods to evaluate it. We get to those methods by splitting up the integral:

$$\int \frac{4-x}{\sqrt{16-x^2}}\, dx = \int \frac{4}{\sqrt{16-x^2}}\, dx - \int \frac{x}{\sqrt{16-x^2}}\, dx.$$

The first integral is handled using a straightforward application of Theorem 46; the second integral is handled by substitution, with $u = 16 - x^2$. We handle each separately.

$$\int \frac{4}{\sqrt{16-x^2}}\, dx = 4\sin^{-1}\frac{x}{4} + C.$$

$\displaystyle\int \frac{x}{\sqrt{16-x^2}}\, dx$: Set $u = 16 - x^2$, so $du = -2x\,dx$ and $x\,dx = -du/2$. We have

$$\int \frac{x}{\sqrt{16-x^2}}\, dx = \int \frac{-du/2}{\sqrt{u}}$$
$$= -\frac{1}{2}\int \frac{1}{\sqrt{u}}\, du$$
$$= -\sqrt{u} + C$$
$$= -\sqrt{16-x^2} + C.$$

Combining these together, we have

$$\int \frac{4-x}{\sqrt{16-x^2}}\, dx = 4\sin^{-1}\frac{x}{4} + \sqrt{16-x^2} + C.$$

Substitution and Definite Integration

    This section has focused on evaluating indefinite integrals as we are learning a new technique for finding antiderivatives. However, much of the time integration is used in the context of a definite integral. Definite integrals that require substitution can be calculated using the following workflow:

Notes:

1. Start with a definite integral $\displaystyle\int_a^b f(x)\,dx$ that requires substitution.

2. Ignore the bounds; use substitution to evaluate $\displaystyle\int f(x)\,dx$ and find an antiderivative $F(x)$.

3. Evaluate $F(x)$ at the bounds; that is, evaluate $F(x)\Big|_a^b = F(b) - F(a)$.

This workflow works fine, but substitution offers an alternative that is powerful and amazing (and a little time saving).

At its heart, (using the notation of Theorem 44) substitution converts integrals of the form $\int F'(g(x))g'(x)\,dx$ into an integral of the form $\int F'(u)\,du$ with the substitution of $u = g(x)$. The following theorem states how the bounds of a definite integral can be changed as the substitution is performed.

Theorem 47   Substitution with Definite Integrals

Let F and g be differentiable functions, where the range of g is an interval I that is contained in the domain of F. Then

$$\int_a^b F'\big(g(x)\big)g'(x)\,dx = \int_{g(a)}^{g(b)} F'(u)\,du.$$

In effect, Theorem 47 states that once you convert to integrating with respect to u, you do not need to switch back to evaluating with respect to x. A few examples will help one understand.

Example 157   Definite integrals and substitution: changing the bounds
Evaluate $\displaystyle\int_0^2 \cos(3x - 1)\,dx$ using Theorem 47.

SOLUTION   Observing the composition of functions, let $u = 3x - 1$, hence $du = 3dx$. As $3dx$ does not appear in the integrand, divide the latter equation by 3 to get $du/3 = dx$.

By setting $u = 3x - 1$, we are implicitly stating that $g(x) = 3x - 1$. Theorem 47 states that the new lower bound is $g(0) = -1$; the new upper bound is

---

Notes:

$g(2) = 5$. We now evaluate the definite integral:

$$\int_{1}^{2} \cos(3x - 1)\, dx = \int_{-1}^{5} \cos u \, \frac{du}{3}$$
$$= \frac{1}{3} \sin u \Big|_{-1}^{5}$$
$$= \frac{1}{3}\big(\sin 5 - \sin(-1)\big) \approx -0.039.$$

Notice how once we converted the integral to be in terms of u, we never went back to using x.

The graphs in Figure 6.1 tell more of the story. In (a) the area defined by the original integrand is shaded, whereas in (b) the area defined by the new integrand is shaded. In this particular situation, the areas look very similar; the new region is "shorter" but "wider," giving the same area.

Example 158   **Definite integrals and substitution: changing the bounds**

Evaluate $\int_{0}^{\pi/2} \sin x \cos x \, dx$ using Theorem 47.

SOLUTION   We saw the corresponding indefinite integral in Example 145. In that example we set $u = \sin x$ but stated that we could have let $u = \cos x$. For variety, we do the latter here.

Let $u = g(x) = \cos x$, giving $du = -\sin x \, dx$ and hence $\sin x \, dx = -du$. The new upper bound is $g(\pi/2) = 0$; the new lower bound is $g(0) = 1$. Note how the lower bound is actually larger than the upper bound now. We have

$$\int_{0}^{\pi/2} \sin x \cos x \, dx = \int_{1}^{0} -u \, du \quad \text{(switch bounds \& change sign)}$$
$$= \int_{0}^{1} u \, du$$
$$= \frac{1}{2} u^2 \Big|_{0}^{1} = 1/2.$$

In Figure 6.2 we have again graphed the two regions defined by our definite integrals. Unlike the previous example, they bear no resemblance to each other. However, Theorem 47 guarantees that they have the same area.

Integration by substitution is a powerful and useful integration technique. The next section introduces another technique, called Integration by Parts. As substitution "undoes" the Chain Rule, integration by parts "undoes" the Product Rule. Together, these two techniques provide a strong foundation on which most other integration techniques are based.

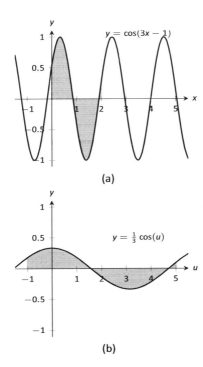

Figure 6.1: Graphing the areas defined by the definite integrals of Example 157.

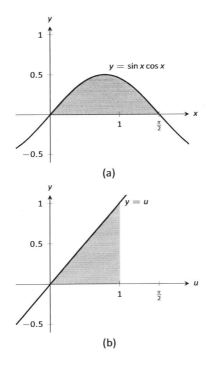

Figure 6.2: Graphing the areas defined by the definite integrals of Example 158.

Notes:

Exercises 6.1

Terms and Concepts

1. Substitution "undoes" what derivative rule?

2. T/F: One can use algebra to rewrite the integrand of an integral to make it easier to evaluate.

Problems

In Exercises 3 – 14, evaluate the indefinite integral to develop an understanding of Substitution.

3. $\int 3x^2 \left(x^3 - 5\right)^7 dx$

4. $\int (2x - 5) \left(x^2 - 5x + 7\right)^3 dx$

5. $\int x \left(x^2 + 1\right)^8 dx$

6. $\int (12x + 14) \left(3x^2 + 7x - 1\right)^5 dx$

7. $\int \frac{1}{2x + 7} dx$

8. $\int \frac{1}{\sqrt{2x + 3}} dx$

9. $\int \frac{x}{\sqrt{x + 3}} dx$

10. $\int \frac{x^3 - x}{\sqrt{x}} dx$

11. $\int \frac{e^{\sqrt{x}}}{\sqrt{x}} dx$

12. $\int \frac{x^4}{\sqrt{x^5 + 1}} dx$

13. $\int \frac{\frac{1}{x} + 1}{x^2} dx$

14. $\int \frac{\ln(x)}{x} dx$

In Exercises 15 – 23, use Substitution to evaluate the indefinite integral involving trigonometric functions.

15. $\int \sin^2(x) \cos(x) dx$

16. $\int \cos(3 - 6x) dx$

17. $\int \sec^2(4 - x) dx$

18. $\int \sec(2x) dx$

19. $\int \tan^2(x) \sec^2(x) dx$

20. $\int x \cos \left(x^2\right) dx$

21. $\int \tan^2(x) dx$

22. $\int \cot x\, dx$. Do not just refer to Theorem 45 for the answer; justify it through Substitution.

23. $\int \csc x\, dx$. Do not just refer to Theorem 45 for the answer; justify it through Substitution.

In Exercises 24 – 30, use Substitution to evaluate the indefinite integral involving exponential functions.

24. $\int e^{3x - 1} dx$

25. $\int e^{x^3} x^2 dx$

26. $\int e^{x^2 - 2x + 1}(x - 1) dx$

27. $\int \frac{e^x + 1}{e^x} dx$

28. $\int \frac{e^x - e^{-x}}{e^{2x}} dx$

29. $\int 3^{3x} dx$

30. $\int 4^{2x} dx$

In Exercises 31 – 34, use Substitution to evaluate the indefinite integral involving logarithmic functions.

31. $\int \frac{\ln x}{x} dx$

32. $\int \frac{\left(\ln x\right)^2}{x} dx$

33. $\int \frac{\ln \left(x^3\right)}{x} dx$

34. $\displaystyle\int \frac{1}{x\ln\left(x^2\right)}\,dx$

52. $\displaystyle\int \left(3x^2 + 2x\right)\left(5x^3 + 5x^2 + 2\right)^8 dx$

In Exercises 35 – 40, use Substitution to evaluate the indefinite integral involving rational functions.

53. $\displaystyle\int \frac{x}{\sqrt{1-x^2}}\,dx$

35. $\displaystyle\int \frac{x^2 + 3x + 1}{x}\,dx$

54. $\displaystyle\int x^2\csc^2\left(x^3 + 1\right)dx$

36. $\displaystyle\int \frac{x^3 + x^2 + x + 1}{x}\,dx$

55. $\displaystyle\int \sin(x)\sqrt{\cos(x)}\,dx$

37. $\displaystyle\int \frac{x^3 - 1}{x + 1}\,dx$

56. $\displaystyle\int \frac{1}{x - 5}\,dx$

38. $\displaystyle\int \frac{x^2 + 2x - 5}{x - 3}\,dx$

57. $\displaystyle\int \frac{7}{3x + 2}\,dx$

39. $\displaystyle\int \frac{3x^2 - 5x + 7}{x + 1}\,dx$

58. $\displaystyle\int \frac{3x^3 + 4x^2 + 2x - 22}{x^2 + 3x + 5}\,dx$

40. $\displaystyle\int \frac{x^2 + 2x + 1}{x^3 + 3x^2 + 3x}\,dx$

59. $\displaystyle\int \frac{2x + 7}{x^2 + 7x + 3}\,dx$

In Exercises 41 – 50, use Substitution to evaluate the indefinite integral involving inverse trigonometric functions.

60. $\displaystyle\int \frac{9(2x + 3)}{3x^2 + 9x + 7}\,dx$

41. $\displaystyle\int \frac{7}{x^2 + 7}\,dx$

61. $\displaystyle\int \frac{-x^3 + 14x^2 - 46x - 7}{x^2 - 7x + 1}\,dx$

42. $\displaystyle\int \frac{3}{\sqrt{9 - x^2}}\,dx$

62. $\displaystyle\int \frac{x}{x^4 + 81}\,dx$

43. $\displaystyle\int \frac{14}{\sqrt{5 - x^2}}\,dx$

63. $\displaystyle\int \frac{2}{4x^2 + 1}\,dx$

44. $\displaystyle\int \frac{2}{x\sqrt{x^2 - 9}}\,dx$

64. $\displaystyle\int \frac{1}{x\sqrt{4x^2 - 1}}\,dx$

45. $\displaystyle\int \frac{5}{\sqrt{x^4 - 16x^2}}\,dx$

65. $\displaystyle\int \frac{1}{\sqrt{16 - 9x^2}}\,dx$

46. $\displaystyle\int \frac{x}{\sqrt{1 - x^4}}\,dx$

66. $\displaystyle\int \frac{3x - 2}{x^2 - 2x + 10}\,dx$

47. $\displaystyle\int \frac{1}{x^2 - 2x + 8}\,dx$

67. $\displaystyle\int \frac{7 - 2x}{x^2 + 12x + 61}\,dx$

48. $\displaystyle\int \frac{2}{\sqrt{-x^2 + 6x + 7}}\,dx$

68. $\displaystyle\int \frac{x^2 + 5x - 2}{x^2 - 10x + 32}\,dx$

49. $\displaystyle\int \frac{3}{\sqrt{-x^2 + 8x + 9}}\,dx$

69. $\displaystyle\int \frac{x^3}{x^2 + 9}\,dx$

50. $\displaystyle\int \frac{5}{x^2 + 6x + 34}\,dx$

70. $\displaystyle\int \frac{x^3 - x}{x^2 + 4x + 9}\,dx$

In Exercises 51 – 75, evaluate the indefinite integral.

71. $\displaystyle\int \frac{\sin(x)}{\cos^2(x) + 1}\,dx$

51. $\displaystyle\int \frac{x^2}{\left(x^3 + 3\right)^2}\,dx$

72. $\displaystyle\int \frac{\cos(x)}{\sin^2(x) + 1}\,dx$

73. $\displaystyle\int \frac{\cos(x)}{1 - \sin^2(x)}\,dx$

74. $\displaystyle\int \frac{3x - 3}{\sqrt{x^2 - 2x - 6}}\,dx$

75. $\displaystyle\int \frac{x - 3}{\sqrt{x^2 - 6x + 8}}\,dx$

In Exercises 76 – 83, evaluate the definite integral.

76. $\displaystyle\int_1^3 \frac{1}{x - 5}\,dx$

77. $\displaystyle\int_2^6 x\sqrt{x - 2}\,dx$

78. $\displaystyle\int_{-\pi/2}^{\pi/2} \sin^2 x \cos x\,dx$

79. $\displaystyle\int_0^1 2x(1 - x^2)^4\,dx$

80. $\displaystyle\int_{-2}^{-1} (x + 1)e^{x^2 + 2x + 1}\,dx$

81. $\displaystyle\int_{-1}^1 \frac{1}{1 + x^2}\,dx$

82. $\displaystyle\int_2^4 \frac{1}{x^2 - 6x + 10}\,dx$

83. $\displaystyle\int_1^{\sqrt{3}} \frac{1}{\sqrt{4 - x^2}}\,dx$

A: Solutions To Selected Problems

Chapter 1

Section 1.1

1. Answers will vary.

3. F

5. Answers will vary.

7. -5

9. 2

11. Limit does not exist.

13. 7

15. Limit does not exist.

17.

h	$\frac{f(a+h)-f(a)}{h}$
-0.1	9
-0.01	9
0.01	9
0.1	9

The limit seems to be exactly 9.

19.

h	$\frac{f(a+h)-f(a)}{h}$
-0.1	-0.114943
-0.01	-0.111483
0.01	-0.110742
0.1	-0.107527

The limit is approx. -0.11.

21.

h	$\frac{f(a+h)-f(a)}{h}$
-0.1	0.202027
-0.01	0.2002
0.01	0.1998
0.1	0.198026

The limit is approx. 0.2.

23.

h	$\frac{f(a+h)-f(a)}{h}$
-0.1	-0.0499583
-0.01	-0.00499996
0.01	0.00499996
0.1	0.0499583

The limit is approx. 0.005.

Section 1.2

1. ε should be given first, and the restriction $|x - a| < \delta$ implies $|f(x) - K| < \varepsilon$, not the other way around.

3. T

5. Let $\varepsilon > 0$ be given. We wish to find $\delta > 0$ such that when $|x - 5| < \delta$, $|f(x) - (-2)| < \varepsilon$.
Consider $|f(x) - (-2)| < \varepsilon$:
$$|f(x) + 2| < \varepsilon$$
$$|(3 - x) + 2| < \varepsilon$$
$$|5 - x| < \varepsilon$$
$$-\varepsilon < 5 - x < \varepsilon$$
$$-\varepsilon < x - 5 < \varepsilon.$$

This implies we can let $\delta = \varepsilon$. Then:
$$|x - 5| < \delta$$
$$-\delta < x - 5 < \delta$$
$$-\varepsilon < x - 5 < \varepsilon$$
$$-\varepsilon < (x - 3) - 2 < \varepsilon$$
$$-\varepsilon < (-x + 3) - (-2) < \varepsilon$$
$$|3 - x - (-2)| < \varepsilon,$$

which is what we wanted to prove.

7. Let $\varepsilon > 0$ be given. We wish to find $\delta > 0$ such that when $|x - 4| < \delta$, $|f(x) - 15| < \varepsilon$.
Consider $|f(x) - 15| < \varepsilon$, keeping in mind we want to make a statement about $|x - 4|$:
$$|f(x) - 15| < \varepsilon$$
$$|x^2 + x - 5 - 15| < \varepsilon$$
$$|x^2 + x - 20| < \varepsilon$$
$$|x - 4| \cdot |x + 5| < \varepsilon$$
$$|x - 4| < \varepsilon/|x + 5|$$

Since x is near 4, we can safely assume that, for instance, $3 < x < 5$. Thus
$$3 + 5 < x + 5 < 5 + 5$$
$$8 < x + 5 < 10$$
$$\frac{1}{10} < \frac{1}{x + 5} < \frac{1}{8}$$
$$\frac{\varepsilon}{10} < \frac{\varepsilon}{x + 5} < \frac{\varepsilon}{8}$$

Let $\delta = \frac{\varepsilon}{10}$. Then:
$$|x - 4| < \delta$$
$$|x - 4| < \frac{\varepsilon}{10}$$
$$|x - 4| < \frac{\varepsilon}{x + 5}$$
$$|x - 4| \cdot |x + 5| < \frac{\varepsilon}{x + 5} \cdot |x + 5|$$

Assuming x is near 4, $x + 5$ is positive and we can drop the absolute value signs on the right.
$$|x - 4| \cdot |x + 5| < \frac{\varepsilon}{x + 5} \cdot (x + 5)$$
$$|x^2 + x - 20| < \varepsilon$$
$$|(x^2 + x - 5) - 15| < \varepsilon,$$

which is what we wanted to prove.

9. Let $\varepsilon > 0$ be given. We wish to find $\delta > 0$ such that when $|x - 2| < \delta$, $|f(x) - 5| < \varepsilon$. However, since $f(x) = 5$, a constant function, the latter inequality is simply $|5 - 5| < \varepsilon$, which is always true. Thus we can choose any δ we like; we arbitrarily choose $\delta = \varepsilon$.

11. Let $\varepsilon > 0$ be given. We wish to find $\delta > 0$ such that when $|x - 0| < \delta$, $|f(x) - 0| < \varepsilon$. In simpler terms, we want to show that when $|x| < \delta$, $|\sin x| < \varepsilon$.
Set $\delta = \varepsilon$. We start with assuming that $|x| < \delta$. Using the hint, we have that $|\sin x| < |x| < \delta = \varepsilon$. Hence if $|x| < \delta$, we know immediately that $|\sin x| < \varepsilon$.

Section 1.3

1. Answers will vary.

3. Answers will vary.

5. As x is near 1, both f and g are near 0, but f is approximately twice the size of g. (I.e., $f(x) \approx 2g(x)$.)

7. 6

9. Limit does not exist.

11. Not possible to know.

13. -45

15. -1

17. π

19. $-0.000000015 \approx 0$

21. Limit does not exist

23. 2

25. $\frac{\pi^2 + 3\pi + 5}{5\pi^2 - 2\pi - 3} \approx 0.6064$

27. -8

29. 10

31. $-3/2$

33. 0

35. 1

37. 3

39. 1

Section 1.4

1. The function approaches different values from the left and right; the function grows without bound; the function oscillates.

3. F

5.   (a) 2
     (b) 2
     (c) 2
     (d) 1
     (e) As f is not defined for $x < 0$, this limit is not defined.
     (f) 1

7.   (a) Does not exist.
     (b) Does not exist.
     (c) Does not exist.
     (d) Not defined.
     (e) 0
     (f) 0

9.   (a) 2
     (b) 2
     (c) 2
     (d) 2

11. (a) 2
     (b) 2
     (c) 2
     (d) 0
     (e) 2
     (f) 2
     (g) 2
     (h) Not defined

13. (a) 2
     (b) -4
     (c) Does not exist.
     (d) 2

15. (a) 0
     (b) 0
     (c) 0
     (d) 0
     (e) 2
     (f) 2
     (g) 2
     (h) 2

17. (a) $1 - \cos^2 a = \sin^2 a$
     (b) $\sin^2 a$
     (c) $\sin^2 a$
     (d) $\sin^2 a$

19. (a) 4
     (b) 4
     (c) 4
     (d) 3

21. (a) -1
     (b) 1
     (c) Does not exist
     (d) 0

23. $2/3$

25. -9

Section 1.5

1. Answers will vary.

3. A root of a function f is a value c such that $f(c) = 0$.

5. F

7. T

9. F

11. No; $\lim\limits_{x \to 1} f(x) = 2$, while $f(1) = 1$.

13. No; $f(1)$ does not exist.

15. Yes

17. (a) No; $\lim\limits_{x \to -2} f(x) \neq f(-2)$
     (b) Yes
     (c) No; $f(2)$ is not defined.

19. (a) Yes
     (b) No; the left and right hand limits at 1 are not equal.

21. (a) Yes
     (b) No. $\lim_{x \to 8} f(x) = 16/5 \neq f(8) = 5$.

23. $(-\infty, -2] \cup [2, \infty)$

25. $(-\infty, -\sqrt{6}] \cup [\sqrt{6}, \infty)$

27. $(-\infty, \infty)$

29. $(0, \infty)$

31. $(-\infty, 0]$

33. Yes, by the Intermediate Value Theorem.

35. We cannot say; the Intermediate Value Theorem only applies to function values between -10 and 10; as 11 is outside this range, we do not know.

37. Approximate root is $x = 1.23$. The intervals used are:
    $[1, 1.5]$ $[1, 1.25]$ $[1.125, 1.25]$
    $[1.1875, 1.25]$ $[1.21875, 1.25]$ $[1.234375, 1.25]$
    $[1.234375, 1.2421875]$ $[1.234375, 1.2382813]$

39. Approximate root is $x = 0.69$. The intervals used are:
$[0.65, 0.7]$ $[0.675, 0.7]$ $[0.6875, 0.7]$
$[0.6875, 0.69375]$ $[0.690625, 0.69375]$

41. (a) 20

 (b) 25

 (c) Limit does not exist

 (d) 25

43. Answers will vary.

Section 1.6

1. F

3. F

5. T

7. Answers will vary.

9. (a) ∞

 (b) ∞

11. (a) 1

 (b) 0

 (c) $1/2$

 (d) $1/2$

13. (a) Limit does not exist

 (b) Limit does not exist

15. Tables will vary.

(a)

x	$f(x)$
2.9	-15.1224
2.99	-159.12
2.999	-1599.12

It seems $\lim_{x\to 3^-} f(x) = -\infty$.

(b)

x	$f(x)$
3.1	16.8824
3.01	160.88
3.001	1600.88

It seems $\lim_{x\to 3^+} f(x) = \infty$.

 (c) It seems $\lim_{x\to 3} f(x)$ does not exist.

17. Tables will vary.

(a)

x	$f(x)$
2.9	132.857
2.99	12124.4

It seems $\lim_{x\to 3^-} f(x) = \infty$.

(b)

x	$f(x)$
3.1	108.039
3.01	11876.4

It seems $\lim_{x\to 3^+} f(x) = \infty$.

 (c) It seems $\lim_{x\to 3} f(x) = \infty$.

19. Horizontal asymptote at $y = 2$; vertical asymptotes at $x = -5, 4$.

21. Horizontal asymptote at $y = 0$; vertical asymptotes at $x = -1, 0$.

23. No horizontal or vertical asymptotes.

25. ∞

27. $-\infty$

29. Solution omitted.

31. Yes. The only "questionable" place is at $x = 3$, but the left and right limits agree.

Chapter 2

Section 2.1

1. T

3. Answers will vary.

5. Answers will vary.

7. $f'(x) = 2$

9. $g'(x) = 2x$

11. $r'(x) = \frac{-1}{x^2}$

13. (a) $y = 6$

 (b) $x = -2$

15. (a) $y = -3x + 4$

 (b) $y = 1/3(x - 7) - 17$

17. (a) $y = -7(x + 1) + 8$

 (b) $y = 1/7(x + 1) + 8$

19. (a) $y = -1(x - 3) + 1$

 (b) $y = 1(x - 3) + 1$

21. $y = -0.099(x - 9) + 1$

23. $y = -0.05x + 1$

25. (a) Approximations will vary; they should match (c) closely.

 (b) $f'(x) = -1/(x + 1)^2$

 (c) At $(0, 1)$, slope is -1. At $(1, 0.5)$, slope is $-1/4$.

27.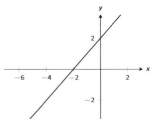

29.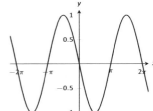

31. Approximately 24.

33. (a) $(-\infty, \infty)$

 (b) $(-\infty, -1) \cup (-1, 1) \cup (1, \infty)$

 (c) $(-\infty, 5]$

 (d) $[-5, 5]$

Section 2.2

1. Velocity

3. Linear functions.

5. -17

7. $f(10.1)$ is likely most accurate, as accuracy is lost the farther from $x = 10$ we go.

9. 6

11. ft/s^2

13. (a) thousands of dollars per car

 (b) It is likely that $P(0) < 0$. That is, negative profit for not producing any cars.

15. $f(x) = g'(x)$

17. Either $g(x) = f'(x)$ or $f(x) = g'(x)$ is acceptable. The actual answer is $g(x) = f'(x)$, but is very hard to show that $f(x) \neq g'(x)$ given the level of detail given in the graph.

19. $f'(x) = 10x$

21. $f'(\pi) \approx 0$.

Section 2.3

1. Power Rule.

3. One answer is $f(x) = 10e^x$.

5. $g(x)$ and $h(x)$

7. One possible answer is $f(x) = 17x - 205$.

9. $f'(x)$ is a velocity function, and $f''(x)$ is acceleration.

11. $f'(x) = 14x - 5$

13. $m'(t) = 45t^4 - \frac{3}{8}t^2 + 3$

15. $f'(r) = 6e^r$

17. $f'(x) = \frac{2}{x} - 1$

19. $h'(t) = e^t - \cos t + \sin t$

21. $f'(t) = 0$

23. $g'(x) = 24x^2 - 120x + 150$

25. $f'(x) = 18x - 12$

27. $f'(x) = 6x^5 \; f''(x) = 30x^4 \; f'''(x) = 120x^3 \; f^{(4)}(x) = 360x^2$

29. $h'(t) = 2t - e^t \; h''(t) = 2 - e^t \; h'''(t) = -e^t \; h^{(4)}(t) = -e^t$

31. $f'(\theta) = \cos\theta + \sin\theta \; f''(\theta) = -\sin\theta + \cos\theta$
    $f'''(\theta) = -\cos\theta - \sin\theta \; f^{(4)}(\theta) = \sin\theta - \cos\theta$

33. Tangent line: $y = 2(x - 1)$
    Normal line: $y = -1/2(x - 1)$

35. Tangent line: $y = x - 1$
    Normal line: $y = -x + 1$

37. Tangent line: $y = \frac{\sqrt{2}}{2}(x - \frac{\pi}{4}) - \sqrt{2}$
    Normal line: $y = \frac{-2}{\sqrt{2}}(x - \frac{\pi}{4}) - \sqrt{2}$

39. The tangent line to $f(x) = e^x$ at $x = 0$ is $y = x + 1$; thus $e^{0.1} \approx y(0.1) = 1.1$.

Section 2.4

1. F

3. T

5. F

7. (a) $f'(x) = (x^2 + 3x) + x(2x + 3)$
   (b) $f'(x) = 3x^2 + 6x$
   (c) They are equal.

9. (a) $h'(s) = 2(s + 4) + (2s - 1)(1)$
   (b) $h'(s) = 4s + 7$
   (c) They are equal.

11. (a) $f'(x) = \frac{x(2x) - (x^2 + 3)1}{x^2}$
    (b) $f'(x) = 1 - \frac{3}{x^2}$
    (c) They are equal.

13. (a) $h'(s) = \frac{4s^3(0) - 3(12s^2)}{16s^6}$
    (b) $h'(s) = -9/4s^{-4}$
    (c) They are equal.

15. $f'(x) = \sin x + x \cos x$

17. $g'(x) = \frac{-12}{(x-5)^2}$

19. $h'(x) = -\csc^2 x - e^x$

21. (a) $f'(x) = \frac{(x+2)(4x^3 + 6x^2) - (x^4 + 2x^3)(1)}{(x+2)^2}$
    (b) $f(x) = x^3$ when $x \neq -2$, so $f'(x) = 3x^2$.
    (c) They are equal.

23. $f'(t) = 5t^4(\sec t + e^t) + t^5(\sec t \tan t + e^t)$

25. $g'(x) = 0$

27. $f'(x) = \frac{(t^2 \cos t + 2)(2t \sin t + t^2 \cos t) - (t^2 \sin t + 3)(2t \cos t - t^2 \sin t)}{(t^2 \cos t + 2)^2}$

29. $g'(x) = 2\sin x \sec x + 2x \cos x \sec x + 2x \sin x \sec x \tan x = 2\tan x + 2x + 2x \tan^2 x = 2\tan x + 2x \sec^2 x$

31. Tangent line: $y = -(x - \frac{3\pi}{2}) - \frac{3\pi}{2} = -x$
    Normal line: $y = (x - \frac{3\pi}{2}) - \frac{3\pi}{2} = x - 3\pi$

33. Tangent line: $y = -9x - 5$
    Normal line: $y = 1/9x - 5$

35. $x = 0$

37. $x = -2, 0$

39. $f^{(4)}(x) = -4\cos x + x \sin x$

41. $f^{(8)} = 0$

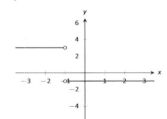

43.

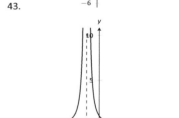

45.

Section 2.5

1. T

3. F

5. T

7. $f'(x) = 10(4x^3 - x)^9 \cdot (12x^2 - 1) = (120x^2 - 10)(4x^3 - x)^9$

9. $g'(\theta) = 3(\sin\theta + \cos\theta)^2(\cos\theta - \sin\theta)$

11. $f'(x) = 4(x + \frac{1}{x})^3(1 - \frac{1}{x^2})$

13. $g'(x) = 5\sec^2(5x)$

15. $p'(t) = -3\cos^2(t^2 + 3t + 1)\sin(t^2 + 3t + 1)(2t + 3)$

17. $f'(x) = 2/x$

19. $g'(r) = \ln 4 \cdot 4^r$

21. $g'(t) = 0$

23. $f'(x) = \frac{(3^t + 2)((\ln 2)2^t) - (2^t + 3)((\ln 3)3^t)}{(3^t + 2)^2}$

25. $f'(x) = \frac{2^{x^2}(\ln 3 \cdot 3^x x^2 2x + 1) - (3^{x^2} + x)(\ln 2 \cdot 2^{x^2} 2x)}{2^{2x^2}}$

27. $g'(t) = 5\cos(t^2 + 3t)\cos(5t - 7) - (2t + 3)\sin(t^2 + 3t)\sin(5t - 7)$

29. Tangent line: $y = 0$

   Normal line: $x = 0$

31. Tangent line: $y = -3(\theta - \pi/2) + 1$

   Normal line: $y = 1/3(\theta - \pi/2) + 1$

33. In both cases the derivative is the same: $1/x$.

35. (a) $^\circ$ F/mph

   (b) The sign would be negative; when the wind is blowing at 10 mph, any increase in wind speed will make it feel colder, i.e., a lower number on the Fahrenheit scale.

Section 2.6

1. Answers will vary.

3. T

5. $f'(x) = \frac{1}{2}x^{-1/2} - \frac{1}{2}x^{-3/2} = \frac{1}{2\sqrt{x}} - \frac{1}{2\sqrt{x^3}}$

7. $f'(t) = \frac{-t}{\sqrt{1-t^2}}$

9. $h'(x) = 1.5x^{0.5} = 1.5\sqrt{x}$

11. $g'(x) = \frac{\sqrt{x}(1) - (x+7)(1/2x^{-1/2})}{x} = \frac{1}{2\sqrt{x}} - \frac{7}{2\sqrt{x^3}}$

13. $\frac{dy}{dx} = \frac{-4x^3}{2y+1}$

15. $\frac{dy}{dx} = \sin(x)\sec(y)$

17. $\frac{dy}{dx} = \frac{y}{x}$

19. $-\frac{2\sin(y)\cos(y)}{x}$

21. $\frac{1}{2y+2}$

23. $\frac{-\cos(x)(x+\cos(y))+\sin(x)+y}{\sin(y)(\sin(x)+y)+x+\cos(y)}$

25. $-\frac{2x+y}{2y+x}$

27. (a) $y = 0$

   (b) $y = -1.859(x - 0.1) + 0.281$

29. (a) $y = 4$

   (b) $y = 0.93(x - 2) + \sqrt[4]{108}$

31. (a) $y = -\frac{1}{\sqrt{3}}(x - \frac{7}{2}) + \frac{6+3\sqrt{3}}{2}$

   (b) $y = \sqrt{3}(x - \frac{4+3\sqrt{3}}{2}) + \frac{3}{2}$

33. $\frac{d^2y}{dx^2} = \frac{3}{5}\frac{y^{3/5}}{x^{8/5}} + \frac{3}{5}\frac{1}{yx^{6/5}}$

35. $\frac{d^2y}{dx^2} = 0$

37. $y' = (2x)^{x^2}(2x\ln(2x) + x)$

   Tangent line: $y = (2 + 4\ln 2)(x - 1) + 2$

39. $y' = x^{\sin(x)+2}(\cos x \ln x + \frac{\sin x + 2}{x})$

   Tangent line: $y = (3\pi^2/4)(x - \pi/2) + (\pi/2)^3$

41. $y' = \frac{(x+1)(x+2)}{(x+3)(x+4)}(\frac{1}{x+1} + \frac{1}{x+2} - \frac{1}{x+3} - \frac{1}{x+4})$

   Tangent line: $y = 11/72x + 1/6$

Section 2.7

1. F

3. The point $(10, 1)$ lies on the graph of $y = f^{-1}(x)$ (assuming f is invertible).

5. Compose $f(g(x))$ and $g(f(x))$ to confirm that each equals x.

7. Compose $f(g(x))$ and $g(f(x))$ to confirm that each equals x.

9. $(f^{-1})'(20) = \frac{1}{f'(2)} = 1/5$

11. $(f^{-1})'(\sqrt{3}/2) = \frac{1}{f'(\pi/6)} = 1$

13. $(f^{-1})'(1/2) = \frac{1}{f'(1)} = -2$

15. $h'(t) = \frac{2}{\sqrt{1-4t^2}}$

17. $g'(x) = \frac{2}{1+4x^2}$

19. $g'(t) = \cos^{-1}(t)\cos(t) - \frac{\sin(t)}{\sqrt{1-t^2}}$

21. $h'(x) = \frac{\sin^{-1}(x)+\cos^{-1}(x)}{\sqrt{1-x^2}\cos^{-1}(x)^2}$

23. $f'(x) = -\frac{1}{\sqrt{1-x^2}}$

25. (a) $f(x) = x$, so $f'(x) = 1$

   (b) $f'(x) = \cos(\sin^{-1}x)\frac{1}{\sqrt{1-x^2}} = 1$.

27. (a) $f(x) = \sqrt{1-x^2}$, so $f'(x) = \frac{-x}{\sqrt{1-x^2}}$

   (b) $f'(x) = \cos(\cos^{-1}x)(\frac{1}{\sqrt{1-x^2}}) = \frac{-x}{\sqrt{1-x^2}}$

29. $y = -4(x - \sqrt{3}/4) + \pi/6$

31. $y = -4/5(x - 1) + 2$

Chapter 3

Section 3.1

1. Answers will vary.

3. Answers will vary.

5. F

7. A: abs. min B: none C: abs. max D: none E: none

9. $f'(0) = 0\ f'(2) = 0$

11. $f'(0) = 0\ f'(3.2) = 0\ f'(4)$ is undefined

13. $f'(0)$ is not defined

15. min: $(-0.5, 3.75)$

   max: $(2, 10)$

17. min: $(\pi/4, 3\sqrt{2}/2)$

   max: $(\pi/2, 3)$

19. min: $(\sqrt{3}, 2\sqrt{3})$

   max: $(5, 28/5)$

21. min: $(\pi, -e^\pi)$

   max: $(\pi/4, \frac{\sqrt{2}e^{\pi/4}}{2})$

23. min: $(1, 0)$

   max: $(e, 1/e)$

25. $\frac{dy}{dx} = \frac{y(y-2x)}{x(x-2y)}$

27. $3x^2 + 1$

Section 3.2

1. Answers will vary.

3. Any c in $[-1, 1]$ is valid.

5. $c = -1/2$

7. Rolle's Thm. does not apply.

9. Rolle's Thm. does not apply.

11. $c = 0$

13. $c = 3/\sqrt{2}$

15. The Mean Value Theorem does not apply.

17. $c = \pm \sec^{-1}(2/\sqrt{\pi})$

19. $c = \frac{5 \pm 7\sqrt{7}}{6}$

21. Max value of 19 at $x = -2$ and $x = 5$; min value of 6.75 at $x = 1.5$.

23. They are the odd, integer valued multiples of $\pi/2$ (such as $0, \pm\pi/2, \pm3\pi/2, \pm5\pi/2$, etc.)

Section 3.3

1. Answers will vary.

3. Answers will vary.

5. Increasing

7. Graph and verify.

9. Graph and verify.

11. Graph and verify.

13. Graph and verify.

15. domain=$(-\infty, \infty)$
    c.p. at $c = -2, 0$;
    increasing on $(-\infty, -2) \cup (0, \infty)$;
    decreasing on $(-2, 0)$;
    rel. min at $x = 0$;
    rel. max at $x = -2$.

17. domain=$(-\infty, \infty)$
    c.p. at $c = 1$;
    increasing on $(-\infty, \infty)$;

19. domain=$(-\infty, -1) \cup (-1, 1) \cup (1, \infty)$
    c.p. at $c = 0$;
    decreasing on $(-\infty, -1) \cup (-1, 0)$;
    increasing on $(0, 1) \cup (1, \infty)$;
    rel. min at $x = 0$;

21. domain=$(-\infty, 0) \cup (0, \infty)$;
    c.p. at $c = 2, 6$;
    decreasing on $(-\infty, 0) \cup (0, 2) \cup (6, \infty)$;
    increasing on $(2, 6)$;
    rel. min at $x = 2$; rel. max at $x = 6$.

23. domain = $(-\infty, \infty)$;
    c.p. at $c = -1, 1$;
    decreasing on $(-1, 1)$;
    increasing on $(-\infty, -1) \cup (1, \infty)$;
    rel. min at $x = 1$;
    rel. max at $x = -1$

25. $c = \pm \cos^{-1}(2/\pi)$

Section 3.4

1. Answers will vary.

3. Yes; Answers will vary.

5. Graph and verify.

7. Graph and verify.

9. Graph and verify.

11. Graph and verify.

13. Graph and verify.

15. Graph and verify.

17. Possible points of inflection: none; concave down on $(-\infty, \infty)$

19. Possible points of inflection: $x = 1/2$; concave down on $(-\infty, 1/2)$; concave up on $(1/2, \infty)$

21. Possible points of inflection: $x = (1/3)(2 \pm \sqrt{7})$; concave up on $((1/3)(2 - \sqrt{7}), (1/3)(2 + \sqrt{7}))$; concave down on $(-\infty, (1/3)(2 - \sqrt{7})) \cup ((1/3)(2 + \sqrt{7}), \infty)$

23. Possible points of inflection: $x = \pm1/\sqrt{3}$; concave down on $(-1/\sqrt{3}, 1/\sqrt{3})$; concave up on $(-\infty, -1/\sqrt{3}) \cup (1/\sqrt{3}, \infty)$

25. Possible points of inflection: $x = -\pi/4, 3\pi/4$; concave down on $(-\pi/4, 3\pi/4)$ concave up on $(-\pi, -\pi/4) \cup (3\pi/4, \pi)$

27. Possible points of inflection: $x = 1/e^{3/2}$; concave down on $(0, 1/e^{3/2})$ concave up on $(1/e^{3/2}, \infty)$

29. min: $x = 1$

31. max: $x = -1/\sqrt{3}$ min: $x = 1/\sqrt{3}$

33. min: $x = 1$

35. min: $x = 1$

37. critical values: $x = -1, 1$; no max/min

39. max: $x = -2$; min: $x = 0$

41. max: $x = 0$

43. f' has no maximal or minimal value

45. f' has a minimal value at $x = 1/2$

47. f' has a relative max at: $x = (1/3)(2 + \sqrt{7})$ relative min at: $x = (1/3)(2 - \sqrt{7})$

49. f' has a relative max at $x = -1/\sqrt{3}$; relative min at $x = 1/\sqrt{3}$

51. f' has a relative min at $x = 3\pi/4$; relative max at $x = -\pi/4$

53. f' has a relative min at $x = 1/\sqrt{e^3} = e^{-3/2}$

Section 3.5

1. Answers will vary.

3. T

5. T

7. A good sketch will include the x and y intercepts..

9. Use technology to verify sketch.

11. Use technology to verify sketch.

13. Use technology to verify sketch.

15. Use technology to verify sketch.

17. Use technology to verify sketch.

19. Use technology to verify sketch.

21. Use technology to verify sketch.

23. Use technology to verify sketch.

25. Use technology to verify sketch.

27. Critical points: $x = \frac{n\pi/2 - b}{a}$, where n is an odd integer Points of inflection: $(n\pi - b)/a$, where n is an integer.

29. $\frac{dy}{dx} = -x/y$, so the function is increasing in second and fourth quadrants, decreasing in the first and third quadrants.
    $\frac{d^2y}{dx^2} = -1/y - x^2/y^3$, which is positive when $y < 0$ and is negative when $y > 0$. Hence the function is concave down in the first and second quadrants and concave up in the third and fourth quadrants.

Chapter 4

Section 4.1

1. F

3. $x_0 = 1.5, x_1 = 1.5709148, x_2 = 1.5707963, x_3 = 1.5707963,$
   $x_4 = 1.5707963, x_5 = 1.5707963$

5. $x_0 = 0, x_1 = 2, x_2 = 1.2, x_3 = 1.0117647, x_4 = 1.0000458,$
   $x_5 = 1$

7. $x_0 = 2, x_1 = 0.6137056389, x_2 = 0.9133412072,$
   $x_3 = 0.9961317034, x_4 = 0.9999925085, x_5 = 1$

9. roots are: $x = -3.714, x = -0.857, x = 1$ and $x = 1.571$

11. roots are: $x = -2.165, x = 0, x = 0.525$ and $x = 1.813$

13. $x = -0.637, x = 1.410$

15. $x = \pm 4.493, x = 0$

17. The approximations alternate between $x = 1, x = 2$ and $x = 3$.

Section 4.2

1. T

3. (a) $5/(2\pi) \approx 0.796$cm/s

   (b) $1/(4\pi) \approx 0.0796$ cm/s

   (c) $1/(40\pi) \approx 0.00796$ cm/s

5. 63.14mph

7. Due to the height of the plane, the gun does not have to rotate very fast.

   (a) 0.0573 rad/s

   (b) 0.0725 rad/s

   (c) In the limit, rate goes to 0.0733 rad/s

9. (a) 0.04 ft/s

   (b) 0.458 ft/s

   (c) 3.35 ft/s

   (d) Not defined; as the distance approaches 24, the rates approaches ∞.

11. (a) 50.92 ft/min

    (b) 0.509 ft/min

    (c) 0.141 ft/min

    As the tank holds about 523.6ft^3, it will take about 52.36 minutes.

13. (a) The rope is 80ft long.

    (b) 1.71 ft/sec

    (c) 1.87 ft/sec

    (d) About 34 feet.

15. The cone is rising at a rate of 0.003ft/s.

Section 4.3

1. T

3. 2500; the two numbers are each 50.

5. There is no maximum sum; the fundamental equation has only 1 critical value that corresponds to a minimum.

7. Area = 1/4, with sides of length $1/\sqrt{2}$.

9. The radius should be about 3.84cm and the height should be $2r = 7.67$cm. No, this is not the size of the standard can.

11. The height and width should be 18 and the length should be 36, giving a volume of 11, 664in^3.

13. $5 - 10/\sqrt{39} \approx 3.4$ miles should be run underground, giving a minimum cost of \$374,899.96.

15. The dog should run about 19 feet along the shore before starting to swim.

17. The largest area is 2 formed by a square with sides of length $\sqrt{2}$.

Section 4.4

1. T

3. F

5. Answers will vary.

7. Use $y = x^2; dy = 2x \cdot dx$ with $x = 6$ and $dx = -0.07$. Thus $dy = -0.84$; knowing $6^2 = 36$, we have $5.93^2 \approx 35.16$.

9. Use $y = x^3; dy = 3x^2 \cdot dx$ with $x = 7$ and $dx = -0.2$. Thus $dy = -29.4$; knowing $7^3 = 343$, we have $6.8^3 \approx 313.6$.

11. Use $y = \sqrt{x}; dy = 1/(2\sqrt{x}) \cdot dx$ with $x = 25$ and $dx = -1$. Thus $dy = -0.1$; knowing $\sqrt{25} = 5$, we have $\sqrt{24} \approx 4.9$.

13. Use $y = \sqrt[3]{x}; dy = 1/(3\sqrt[3]{x^2}) \cdot dx$ with $x = 8$ and $dx = 0.5$. Thus $dy = 1/24 \approx 1/25 = 0.04$; knowing $\sqrt[3]{8} = 2$, we have $\sqrt[3]{8.5} \approx 2.04$.

15. Use $y = \cos x; dy = -\sin x \cdot dx$ with $x = \pi/2 \approx 1.57$ and $dx \approx -0.07$. Thus $dy = 0.07$; knowing $\cos \pi/2 = 0$, we have $\cos 1.5 \approx 0.07$.

17. $dy = (2x + 3)dx$

19. $dy = \frac{-2}{4x^3} dx$

21. $dy = \left(2xe^{3x} + 3x^2 e^{3x}\right) dx$

23. $dy = \frac{2(\tan x + 1) - 2x \sec^2 x}{(\tan x + 1)^2} dx$

25. $dy = (e^x \sin x + e^x \cos x)dx$

27. $dy = \frac{1}{(x+2)^2} dx$

29. $dy = (\ln x)dx$

31. (a) ± 12.8 feet

    (b) ± 32 feet

33. ± 48in^2, or 1/3ft^2

35. (a) 298.8 feet

    (b) ± 17.3 ft

    (c) ± 5.8%

37. The isosceles triangle setup works the best with the smallest percent error.

Chapter 5

Section 5.1

1. Answers will vary.

3. Answers will vary.

5. Answers will vary.

7. velocity

9. $1/9x^9 + C$

11. $t + C$

13. $-1/(3t) + C$

15. $2\sqrt{x} + C$

17. $-\cos \theta + C$

19. $5e^\theta + C$

21. $\frac{5^t}{2 \ln 5} + C$

23. $t^6/6 + t^4/4 - 3t^2 + C$

25. $e^\pi x + C$

27. (a) $x > 0$

(b) $1/x$

(c) $x < 0$

(d) $1/x$

(e) $\ln |x| + C$. Explanations will vary.

29. $5e^x + 5$

31. $\tan x + 4$

33. $5/2x^2 + 7x + 3$

35. $5e^x - 2x$

37. $\frac{2x^4 \ln^2(2) + 2^x + x \ln 2)(\ln 32 - 1) + \ln^2(2) \cos(x) - 1 - \ln^2(2)}{\ln^2(2)}$

39. No answer provided.

Section 5.2

1. Answers will vary.

3. 0

5. (a) 3
   (b) 4
   (c) 3
   (d) 0
   (e) −4
   (f) 9

7. (a) 4
   (b) 2
   (c) 4
   (d) 2
   (e) 1
   (f) 2

9. (a) π
   (b) π
   (c) 2π
   (d) 10π

11. (a) $4/\pi$
    (b) $-4/\pi$
    (c) 0
    (d) $2/\pi$

13. (a) $40/3$
    (b) $26/3$
    (c) $8/3$
    (d) $38/3$

15. (a) 3ft/s
    (b) 9.5ft
    (c) 9.5ft

17. (a) 96ft/s
    (b) 6 seconds
    (c) 6 seconds
    (d) Never; the maximum height is 208ft.

19. 5

21. Answers can vary; one solution is $a = -2, b = 7$

23. -7

25. Answers can vary; one solution is $a = -11, b = 18$

27. $-\cos x - \sin x + \tan x + C$

29. $\ln |x| + \csc x + C$

Section 5.3

1. limits

3. Rectangles.

5. $2^2 + 3^2 + 4^2 = 29$

7. $0 - 1 + 0 + 1 + 0 = 0$

9. $-1 + 2 - 3 + 4 - 5 + 6 = 3$

11. $1 + 1 + 1 + 1 + 1 + 1 = 6$

13. Answers may vary; $\sum_{i=0}^{8}(i^2 - 1)$

15. Answers may vary; $\sum_{i=0}^{4}(-1)^i e^i$

17. 1045

19. -8525

21. 5050

23. 155

25. 24

27. 19

29. $\pi/3 + \pi/(2\sqrt{3}) \approx 1.954$

31. 0.388584

33. (a) Exact expressions will vary; $\frac{(1+n)^2}{4n^2}$.
    (b) $121/400, 10201/40000, 1002001/4000000$
    (c) $1/4$

35. (a) 8.
    (b) 8, 8, 8
    (c) 8

37. (a) Exact expressions will vary; $100 - 200/n$.
    (b) 80, 98, 499/5
    (c) 100

39. $F(x) = 5\tan x + 4$

41. $G(t) = 4/6t^6 - 5/4t^4 + 8t + 9$

43. $G(t) = \sin t - \cos t - 78$

Section 5.4

1. Answers will vary.

3. T

5. 20

7. 0

9. 1

11. $(5 - 1/5)/\ln 5$

13. -4

15. $16/3$

17. $45/4$

19. $1/2$

21. $1/2$

23. $1/4$

25. 8

27. 0

29. Explanations will vary. A sketch will help.

31. $c = \pm 2/\sqrt{3}$

33. $c = 64/9 \approx 7.1$

35. $2/pi$

37. $16/3$

39. $1/(e - 1)$

41. 400ft

43. -1ft

45. -64ft/s

47. 2ft/s

49. $27/2$

51. $9/2$

53. $F'(x) = (3x^2 + 1)\frac{1}{x^3 + x}$

55. $F'(x) = 2x(x^2 + 2) - (x + 2)$

Section 5.5

1. F

3. They are superseded by the Trapezoidal Rule; it takes an equal amount of work and is generally more accurate.

5.   (a) 250

    (b) 250

    (c) 250

7.   (a) $2 + \sqrt{2} + \sqrt{3} \approx 5.15$

    (b) $2/3(3 + \sqrt{2} + 2\sqrt{3}) \approx 5.25$

    (c) $16/3 \approx 5.33$

9.   (a) 0.2207

    (b) 0.2005

    (c) $1/5$

11.   (a) $9/2(1 + \sqrt{3}) \approx 12.294$

    (b) $3 + 6\sqrt{3} \approx 13.392$

    (c) $9\pi/2 \approx 14.137$

13. Trapezoidal Rule: 3.0241
    Simpson's Rule: 2.9315

15. Trapezoidal Rule: 3.0695
    Simpson's Rule: 3.14295

17. Trapezoidal Rule: 2.52971
    Simpson's Rule: 2.5447

19. Trapezoidal Rule: 3.5472
    Simpson's Rule: 3.6133

21.   (a) $n = 150$ (using max $\left(f''(x)\right) = 1$)

    (b) $n = 18$ (using max $\left(f^{(4)}(x)\right) = 7$)

23.   (a) $n = 5591$ (using max $\left(f''(x)\right) = 300$)

    (b) $n = 46$ (using max $\left(f^{(4)}(x)\right) = 24$)

25.   (a) Area is 25.0667 cm^2

    (b) Area is 250,667 yd^2

Chapter 6

Section 6.1

1. Chain Rule.

3. $\frac{1}{8}(x^3 - 5)^8 + C$

5. $\frac{1}{18}\left(x^2 + 1\right)^9 + C$

7. $\frac{1}{2}\ln|2x + 7| + C$

9. $\frac{2}{3}(x + 3)^{3/2} - 6(x + 3)^{1/2} + C = \frac{2}{3}(x - 6)\sqrt{x + 3} + C$

11. $2e^{\sqrt{x}} + C$

13. $-\frac{1}{2x^2} - \frac{1}{x} + C$

15. $\frac{\sin^3(x)}{3} + C$

17. $-\tan(4 - x) + C$

19. $\frac{\tan^3(x)}{3} + C$

21. $\tan(x) - x + C$

23. The key is to multiply $\csc x$ by 1 in the form $(\csc x + \cot x)/(\csc x + \cot x)$.

25. $\frac{e^{x^3}}{3} + C$

27. $x - e^{-x} + C$

29. $\frac{27^x}{\ln 27} + C$

31. $\frac{1}{2}\ln^2(x) + C$

33. $\frac{1}{6}\ln^2\left(x^3\right) + C$

35. $\frac{x^2}{2} + 3x + \ln|x| + C$

37. $\frac{x^3}{3} - \frac{x^2}{2} + x - 2\ln|x + 1| + C$

39. $\frac{3}{2}x^2 - 8x + 15\ln|x + 1| + C$

41. $\sqrt{7}\tan^{-1}\left(\frac{x}{\sqrt{7}}\right) + C$

43. $14\sin^{-1}\left(\frac{x}{\sqrt{5}}\right) + C$

45. $\frac{5}{4}\sec^{-1}(|x|/4) + C$

47. $\frac{\tan^{-1}\left(\frac{x-1}{\sqrt{7}}\right)}{\sqrt{7}} + C$

49. $3\sin^{-1}\left(\frac{x-4}{5}\right) + C$

51. $-\frac{1}{3\left(x^3 + 3\right)} + C$

53. $-\sqrt{1 - x^2} + C$

55. $-\frac{2}{3}\cos^{\frac{3}{2}}(x) + C$

57. $\frac{7}{3}\ln|3x + 2| + C$

59. $\ln\left|x^2 + 7x + 3\right| + C$

61. $-\frac{x^2}{2} + 2\ln\left|x^2 - 7x + 1\right| + 7x + C$

63. $\tan^{-1}(2x) + C$

65. $\frac{1}{3}\sin^{-1}\left(\frac{3x}{4}\right) + C$

67. $\frac{19}{5}\tan^{-1}\left(\frac{x+6}{5}\right) - \ln\left|x^2 + 12x + 61\right| + C$

69. $\frac{x^2}{2} - \frac{9}{2}\ln\left|x^2 + 9\right| + C$

71. $-\tan^{-1}(\cos(x)) + C$

73. $\ln|\sec x + \tan x| + C$ (integrand simplifies to sec x)

75. $\sqrt{x^2 - 6x + 8} + C$

77. $352/15$

79. $1/5$

81. $\pi/2$

83. $\pi/6$

Index

!, 397
Absolute Convergence Theorem, 448
absolute maximum, 123
absolute minimum, 123
Absolute Value Theorem, 401
acceleration, 73, 642
Alternating Harmonic Series, 419, 446, 459
Alternating Series Test
    for series, 442
a_N, 660, 670
analytic function, 480
angle of elevation, 647
antiderivative, 189
arc length, 370, 519, 545, 639, 664
arc length parameter, 664, 666
asymptote
    horizontal, 49
    vertical, 47
a_T, 660, 670
average rate of change, 627
average value of a function, 769
average value of function, 236

Binomial Series, 480
Bisection Method, 42
boundary point, 682
bounded sequence, 404
    convergence, 405
bounded set, 682

center of mass, 783–785, 787, 814
Chain Rule, 97
    multivariable, 713, 716
    notation, 103
circle of curvature, 669
closed, 682
closed disk, 682
concave down, 144
concave up, 144
concavity, 144, 516
    inflection point, 145
    test for, 145
conic sections, 490
    degenerate, 490
    ellipse, 493
    hyperbola, 496
    parabola, 490
Constant Multiple Rule
    of derivatives, 80
    of integration, 193
    of series, 419

constrained optimization, 745
continuous function, 37, 688
    properties, 40, 689
    vector–valued, 630
contour lines, 676
convergence
    absolute, 446, 448
    Alternating Series Test, 442
    conditional, 446
    Direct Comparison Test, 429
      for integration, 339
    Integral Test, 426
    interval of, 454
    Limit Comparison Test, 430
      for integration, 341
    n^{th}–term test, 422
    of geometric series, 414
    of improper int., 334, 339, 341
    of monotonic sequences, 407
    of p-series, 415
    of power series, 453
    of sequence, 400, 405
    of series, 411
    radius of, 454
    Ratio Comparison Test, 435
    Root Comparison Test, 438
critical number, 125
critical point, 125, 740–742
cross product
    and derivatives, 635
    applications, 597
      area of parallelogram, 598
      torque, 600
      volume of parallelepiped, 599
    definition, 593
    properties, 595, 596
curvature, 666
    and motion, 670
    equations for, 668
    of circle, 668, 669
    radius of, 669
curve
    parametrically defined, 503
    rectangular equation, 503
    smooth, 509
curve sketching, 152
cusp, 509
cycloid, 625
cylinder, 555

decreasing function, 136

finding intervals, 137
strictly, 136
definite integral, 201
and substitution, 270
properties, 203
derivative
acceleration, 74
as a function, 64
at a point, 60
basic rules, 78
Chain Rule, 97, 103, 713, 716
Constant Multiple Rule, 80
Constant Rule, 78
differential, 181
directional, 720, 722, 723, 726, 727
exponential functions, 103
First Deriv. Test, 139
Generalized Power Rule, 98
higher order, 81
interpretation, 82
hyperbolic funct., 316
implicit, 106, 718
interpretation, 71
inverse function, 117
inverse hyper., 319
inverse trig., 120
Mean Value Theorem, 132
mixed partial, 696
motion, 74
multivariable differentiability, 705, 710
normal line, 61
notation, 64, 81
parametric equations, 513
partial, 692, 700
Power Rule, 78, 91, 111
power series, 457
Product Rule, 85
Quotient Rule, 88
Second Deriv. Test, 148
Sum/Difference Rule, 80
tangent line, 60
trigonometric functions, 90
vector–valued functions, 631, 632, 635
velocity, 74
differentiable, 60, 705, 710
differential, 181
notation, 181
Direct Comparison Test
for integration, 339
for series, 429
directional derivative, 720, 722, 723, 726, 727
directrix, 490, 555
Disk Method, 355
displacement, 230, 626, 639
distance
between lines, 611
between point and line, 611
between point and plane, 620
between points in space, 552
traveled, 650

divergence
Alternating Series Test, 442
Direct Comparison Test, 429
for integration, 339
Integral Test, 426
Limit Comparison Test, 430
for integration, 341
n^{th}–term test, 422
of geometric series, 414
of improper int., 334, 339, 341
of p-series, 415
of sequence, 400
of series, 411
Ratio Comparison Test, 435
Root Comparison Test, 438
dot product
and derivatives, 635
definition, 580
properties, 581, 582
double integral, 762, 763
in polar, 773
properties, 766

eccentricity, 495, 499
elementary function, 240
ellipse
definition, 493
eccentricity, 495
parametric equations, 509
reflective property, 496
standard equation, 494
extrema
absolute, 123, 740
and First Deriv. Test, 139
and Second Deriv. Test, 148
finding, 126
relative, 124, 740, 741
Extreme Value Theorem, 124, 745
extreme values, 123

factorial, 397
First Derivative Test, 139
floor function, 38
fluid pressure/force, 388, 390
focus, 490, 493, 496
Fubini's Theorem, 763
function
of three variables, 679
of two variables, 675
vector–valued, 623
Fundamental Theorem of Calculus, 228, 229
and Chain Rule, 232

Gabriel's Horn, 376
Generalized Power Rule, 98
geometric series, 413, 414
gradient, 722, 723, 726, 727, 737
and level curves, 723
and level surfaces, 737

Harmonic Series, 419

Head To Tail Rule, 570
Hooke's Law, 381
hyperbola
    definition, 496
    eccentricity, 499
    parametric equations, 509
    reflective property, 499
    standard equation, 497
hyperbolic function
    definition, 313
    derivatives, 316
    identities, 316
    integrals, 316
    inverse, 317
        derivative, 319
        integration, 319
        logarithmic def., 318

implicit differentiation, 106, 718
improper integration, 334, 337
increasing function, 136
    finding intervals, 137
    strictly, 136
indefinite integral, 189
indeterminate form, 2, 48, 327, 328
inflection point, 145
initial point, 566
initial value problem, 194
Integral Test, 426
integration
    arc length, 370
    area, 201, 754, 755
    area between curves, 233, 346
    average value, 236
    by parts, 275
    by substitution, 257
    definite, 201
        and substitution, 270
        properties, 203
        Riemann Sums, 224
    displacement, 230
    distance traveled, 650
    double, 762
    fluid force, 388, 390
    Fun. Thm. of Calc., 228, 229
    general application technique, 345
    hyperbolic funct., 316
    improper, 334, 337, 339, 341
    indefinite, 189
    inverse hyper., 319
    iterated, 753
    Mean Value Theorem, 235
    multiple, 753
    notation, 190, 201, 229, 753
    numerical, 240
        Left/Right Hand Rule, 240, 247
        Simpson's Rule, 245, 247, 248
        Trapezoidal Rule, 243, 247, 248
    of multivariable functions, 751
    of power series, 457

of trig. functions, 263
of trig. powers, 286, 291
of vector–valued functions, 637
partial fraction decomp., 306
Power Rule, 194
Sum/Difference Rule, 194
surface area, 374, 521, 546
trig. subst., 297
triple, 800, 811, 813
volume
    cross-sectional area, 353
    Disk Method, 355
    Shell Method, 362, 366
    Washer Method, 357, 366
work, 378
interior point, 682
Intermediate Value Theorem, 42
interval of convergence, 454
iterated integration, 753, 762, 763, 800, 811, 813
    changing order, 757
    properties, 766, 807

L'Hôpital's Rule, 324, 326
lamina, 779
Left Hand Rule, 210, 215, 240
Left/Right Hand Rule, 247
level curves, 676, 723
level surface, 680, 737
limit
    Absolute Value Theorem, 401
    at infinity, 49
    definition, 10
    difference quotient, 6
    does not exist, 4, 32
    indeterminate form, 2, 48, 327, 328
    L'Hôpital's Rule, 324, 326
    left handed, 30
    of infinity, 46
    of multivariable function, 683, 684, 690
    of sequence, 400
    of vector–valued functions, 629
    one sided, 30
    properties, 18, 684
    pseudo-definition, 2
    right handed, 30
    Squeeze Theorem, 22
Limit Comparison Test
    for integration, 341
    for series, 430
lines, 604
    distances between, 611
    equations for, 606
    intersecting, 607
    parallel, 607
    skew, 607
logarithmic differentiation, 113

Maclaurin Polynomial, *see* Taylor Polynomial
    definition, 466
Maclaurin Series, *see* Taylor Series

definition, 477
magnitude of vector, 566
mass, 779, 780, 814
    center of, 783
maximum
    absolute, 123, 740
    and First Deriv. Test, 139
    and Second Deriv. Test, 148
    relative/local, 124, 740, 743
Mean Value Theorem
    of differentiation, 132
    of integration, 235
Midpoint Rule, 210, 215
minimum
    absolute, 123, 740
    and First Deriv. Test, 139, 148
    relative/local, 124, 740, 743
moment, 785, 787, 814
monotonic sequence, 405
multiple integration, *see* iterated integration
multivariable function, 675, 679
    continuity, 688–690, 706, 711
    differentiability, 705, 706, 710, 711
    domain, 675, 679
    level curves, 676
    level surface, 680
    limit, 683, 684, 690
    range, 675, 679

Newton's Method, 160
norm, 566
normal line, 61, 513, 733
normal vector, 615
n^{th}–term test, 422
numerical integration, 240
    Left/Right Hand Rule, 240, 247
    Simpson's Rule, 245, 247
      error bounds, 248
    Trapezoidal Rule, 243, 247
      error bounds, 248

open, 682
open ball, 690
open disk, 682
optimization, 173
    constrained, 745
orthogonal, 584, 733
    decomposition, 588
orthogonal decomposition of vectors, 588
orthogonal projection, 586
osculating circle, 669

p-series, 415
parabola
    definition, 490
    general equation, 491
    reflective property, 493
parallel vectors, 574
Parallelogram Law, 570
parametric equations
    arc length, 519

concavity, 516
definition, 503
finding $\frac{d^2y}{dx^2}$, 517
finding $\frac{dy}{dx}$, 513
normal line, 513
surface area, 521
tangent line, 513
partial derivative, 692, 700
    high order, 700
    meaning, 694
    mixed, 696
    second derivative, 696
    total differential, 704, 710
perpendicular, *see* orthogonal
planes
    coordinate plane, 554
    distance between point and plane, 620
    equations of, 616
    introduction, 554
    normal vector, 615
    tangent, 736
point of inflection, 145
polar
    coordinates, 525
    function
      arc length, 545
      gallery of graphs, 532
      surface area, 546
    functions, 528
      area, 541
      area between curves, 543
      finding $\frac{dy}{dx}$, 538
      graphing, 528
polar coordinates, 525
    plotting points, 525
Power Rule
    differentiation, 78, 85, 91, 111
    integration, 194
power series, 452
    algebra of, 482
    convergence, 453
    derivatives and integrals, 457
projectile motion, 647, 648, 661

quadric surface
    definition, 558
    ellipsoid, 560
    elliptic cone, 559
    elliptic paraboloid, 559
    gallery, 559–561
    hyperbolic paraboloid, 561
    hyperboloid of one sheet, 560
    hyperboloid of two sheets, 561
    sphere, 560
    trace, 558
Quotient Rule, 88

$\mathbb{R}$, 566
radius of convergence, 454
radius of curvature, 669

Ratio Comparison Test
    for series, 435
rearrangements of series, 447, 448
related rates, 166
Riemann Sum, 210, 214, 217
    and definite integral, 224
Right Hand Rule, 210, 215, 240
right hand rule
    of Cartesian coordinates, 552
Rolle's Theorem, 132
Root Comparison Test
    for series, 438

saddle point, 742, 743
Second Derivative Test, 148, 743
sensitivity analysis, 709
sequence
    Absolute Value Theorem, 401
    positive, 429
sequences
    boundedness, 404
    convergent, 400, 405, 407
    definition, 397
    divergent, 400
    limit, 400
    limit properties, 403
    monotonic, 405
series
    absolute convergence, 446
    Absolute Convergence Theorem, 448
    alternating, 441
      Approximation Theorem, 444
    Alternating Series Test, 442
    Binomial, 480
    conditional convergence, 446
    convergent, 411
    definition, 411
    Direct Comparison Test, 429
    divergent, 411
    geometric, 413, 414
    Integral Test, 426
    interval of convergence, 454
    Limit Comparison Test, 430
    Maclaurin, 477
    n^{th}–term test, 422
    p-series, 415
    partial sums, 411
    power, 452, 453
      derivatives and integrals, 457
    properties, 419
    radius of convergence, 454
    Ratio Comparison Test, 435
    rearrangements, 447, 448
    Root Comparison Test, 438
    Taylor, 477
    telescoping, 416, 417
Shell Method, 362, 366
signed area, 201
signed volume, 762, 763
Simpson's Rule, 245, 247

error bounds, 248
smooth, 634
smooth curve, 509
speed, 642
sphere, 553
Squeeze Theorem, 22
Sum/Difference Rule
    of derivatives, 80
    of integration, 194
    of series, 419
summation
    notation, 211
    properties, 213
surface area, 792
    solid of revolution, 374, 521, 546
surface of revolution, 556, 557

tangent line, 60, 513, 538, 633
    directional, 730
tangent plane, 736
Taylor Polynomial
    definition, 466
    Taylor's Theorem, 469
Taylor Series
    common series, 482
    definition, 477
    equality with generating function, 479
Taylor's Theorem, 469
telescoping series, 416, 417
terminal point, 566
total differential, 704, 710
    sensitivity analysis, 709
total signed area, 201
trace, 558
Trapezoidal Rule, 243, 247
    error bounds, 248
triple integral, 800, 811, 813
    properties, 807

unbounded sequence, 404
unbounded set, 682
unit normal vector
    a_{N}, 660
    and acceleration, 659, 660
    and curvature, 670
    definition, 657
    in $\mathbb{R}^2$, 659
unit tangent vector
    and acceleration, 659, 660
    and curvature, 666, 670
    a_{T}, 660
    definition, 655
    in $\mathbb{R}^2$, 659
unit vector, 572
    properties, 574
    standard unit vector, 576
    unit normal vector, 657
    unit tangent vector, 655

vector–valued function
    algebra of, 624

  arc length, 639
  average rate of change, 627
  continuity, 630
  definition, 623
  derivatives, 631, 632, 635
  describing motion, 642
  displacement, 626
  distance traveled, 650
  graphing, 623
  integration, 637
  limits, 629
  of constant length, 637, 646, 647, 656
  projectile motion, 647, 648
  smooth, 634
  tangent line, 633
vectors, 566
  algebra of, 569
  algebraic properties, 572
  component form, 567
  cross product, 593, 595, 596
  definition, 566
  dot product, 580–582
  Head To Tail Rule, 570
  magnitude, 566
  norm, 566
  normal vector, 615
  orthogonal, 584
  orthogonal decomposition, 588
  orthogonal projection, 586
  parallel, 574
  Parallelogram Law, 570
  resultant, 570
  standard unit vector, 576
  unit vector, 572, 574
  zero vector, 570
velocity, 73, 642
volume, 762, 763, 798

Washer Method, 357, 366
work, 378, 590

Differentiation Rules

1. $\dfrac{d}{dx}(cx) = c$

2. $\dfrac{d}{dx}(u \pm v) = u' \pm v'$

3. $\dfrac{d}{dx}(u \cdot v) = uv' + u'v$

4. $\dfrac{d}{dx}\left(\dfrac{u}{v}\right) = \dfrac{vu' - uv'}{v^2}$

5. $\dfrac{d}{dx}(u(v)) = u'(v)v'$

6. $\dfrac{d}{dx}(c) = 0$

7. $\dfrac{d}{dx}(x) = 1$

8. $\dfrac{d}{dx}(x^n) = nx^{n-1}$

9. $\dfrac{d}{dx}(e^x) = e^x$

10. $\dfrac{d}{dx}(a^x) = \ln a \cdot a^x$

11. $\dfrac{d}{dx}(\ln x) = \dfrac{1}{x}$

12. $\dfrac{d}{dx}(\log_a x) = \dfrac{1}{\ln a} \cdot \dfrac{1}{x}$

13. $\dfrac{d}{dx}(\sin x) = \cos x$

14. $\dfrac{d}{dx}(\cos x) = -\sin x$

15. $\dfrac{d}{dx}(\csc x) = -\csc x \cot x$

16. $\dfrac{d}{dx}(\sec x) = \sec x \tan x$

17. $\dfrac{d}{dx}(\tan x) = \sec^2 x$

18. $\dfrac{d}{dx}(\cot x) = -\csc^2 x$

19. $\dfrac{d}{dx}(\sin^{-1} x) = \dfrac{1}{\sqrt{1-x^2}}$

20. $\dfrac{d}{dx}(\cos^{-1} x) = \dfrac{-1}{\sqrt{1-x^2}}$

21. $\dfrac{d}{dx}(\csc^{-1} x) = \dfrac{-1}{|x|\sqrt{x^2-1}}$

22. $\dfrac{d}{dx}(\sec^{-1} x) = \dfrac{1}{|x|\sqrt{x^2-1}}$

23. $\dfrac{d}{dx}(\tan^{-1} x) = \dfrac{1}{1+x^2}$

24. $\dfrac{d}{dx}(\cot^{-1} x) = \dfrac{-1}{1+x^2}$

25. $\dfrac{d}{dx}(\cosh x) = \sinh x$

26. $\dfrac{d}{dx}(\sinh x) = \cosh x$

27. $\dfrac{d}{dx}(\tanh x) = \operatorname{sech}^2 x$

28. $\dfrac{d}{dx}(\operatorname{sech} x) = -\operatorname{sech} x \tanh x$

29. $\dfrac{d}{dx}(\operatorname{csch} x) = -\operatorname{csch} x \coth x$

30. $\dfrac{d}{dx}(\coth x) = -\operatorname{csch}^2 x$

31. $\dfrac{d}{dx}(\cosh^{-1} x) = \dfrac{1}{\sqrt{x^2-1}}$

32. $\dfrac{d}{dx}(\sinh^{-1} x) = \dfrac{1}{\sqrt{x^2+1}}$

33. $\dfrac{d}{dx}(\operatorname{sech}^{-1} x) = \dfrac{-1}{x\sqrt{1-x^2}}$

34. $\dfrac{d}{dx}(\operatorname{csch}^{-1} x) = \dfrac{-1}{|x|\sqrt{1+x^2}}$

35. $\dfrac{d}{dx}(\tanh^{-1} x) = \dfrac{1}{1-x^2}$

36. $\dfrac{d}{dx}(\coth^{-1} x) = \dfrac{1}{1-x^2}$

Integration Rules

1. $\displaystyle\int c \cdot f(x)\, dx = c \int f(x)\, dx$

2. $\displaystyle\int f(x) \pm g(x)\, dx = $
$\displaystyle\int f(x)\, dx \pm \int g(x)\, dx$

3. $\displaystyle\int 0\, dx = C$

4. $\displaystyle\int 1\, dx = x + C$

5. $\displaystyle\int x^n\, dx = \dfrac{1}{n+1}x^{n+1} + C,\ n \neq -1$
$n \neq -1$

6. $\displaystyle\int e^x\, dx = e^x + C$

7. $\displaystyle\int a^x\, dx = \dfrac{1}{\ln a} \cdot a^x + C$

8. $\displaystyle\int \dfrac{1}{x}\, dx = \ln|x| + C$

9. $\displaystyle\int \cos x\, dx = \sin x + C$

10. $\displaystyle\int \sin x\, dx = -\cos x + C$

11. $\displaystyle\int \tan x\, dx = -\ln|\cos x| + C$

12. $\displaystyle\int \sec x\, dx = \ln|\sec x + \tan x| + C$

13. $\displaystyle\int \csc x\, dx = -\ln|\csc x + \cot x| + C$

14. $\displaystyle\int \cot x\, dx = \ln|\sin x| + C$

15. $\displaystyle\int \sec^2 x\, dx = \tan x + C$

16. $\displaystyle\int \csc^2 x\, dx = -\cot x + C$

17. $\displaystyle\int \sec x \tan x\, dx = \sec x + C$

18. $\displaystyle\int \csc x \cot x\, dx = -\csc x + C$

19. $\displaystyle\int \cos^2 x\, dx = \dfrac{1}{2}x + \dfrac{1}{4}\sin(2x) + C$

20. $\displaystyle\int \sin^2 x\, dx = \dfrac{1}{2}x - \dfrac{1}{4}\sin(2x) + C$

21. $\displaystyle\int \dfrac{1}{x^2+a^2}\, dx = \dfrac{1}{a}\tan^{-1}\left(\dfrac{x}{a}\right) + C$

22. $\displaystyle\int \dfrac{1}{\sqrt{a^2-x^2}}\, dx = \sin^{-1}\left(\dfrac{x}{a}\right) + C$

23. $\displaystyle\int \dfrac{1}{x\sqrt{x^2-a^2}}\, dx = \dfrac{1}{a}\sec^{-1}\left(\dfrac{|x|}{a}\right) + C$

24. $\displaystyle\int \cosh x\, dx = \sinh x + C$

25. $\displaystyle\int \sinh x\, dx = \cosh x + C$

26. $\displaystyle\int \tanh x\, dx = \ln(\cosh x) + C$

27. $\displaystyle\int \coth x\, dx = \ln|\sinh x| + C$

28. $\displaystyle\int \dfrac{1}{\sqrt{x^2-a^2}}\, dx = \ln\left|x + \sqrt{x^2-a^2}\right| + C$

29. $\displaystyle\int \dfrac{1}{\sqrt{x^2+a^2}}\, dx = \ln\left|x + \sqrt{x^2+a^2}\right| + C$

30. $\displaystyle\int \dfrac{1}{a^2-x^2}\, dx = \dfrac{1}{2}\ln\left|\dfrac{a+x}{a-x}\right| + C$

31. $\displaystyle\int \dfrac{1}{x\sqrt{a^2-x^2}}\, dx = \dfrac{1}{a}\ln\left(\dfrac{x}{a+\sqrt{a^2-x^2}}\right) + C$

32. $\displaystyle\int \dfrac{1}{x\sqrt{x^2+a^2}}\, dx = \dfrac{1}{a}\ln\left|\dfrac{x}{a+\sqrt{x^2+a^2}}\right| + C$

The Unit Circle

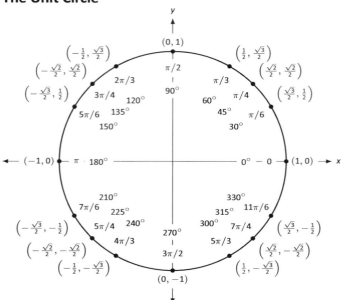

Definitions of the Trigonometric Functions

Unit Circle Definition

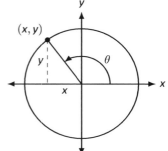

$$\sin\theta = y \qquad \cos\theta = x$$

$$\csc\theta = \frac{1}{y} \qquad \sec\theta = \frac{1}{x}$$

$$\tan\theta = \frac{y}{x} \qquad \cot\theta = \frac{x}{y}$$

Right Triangle Definition

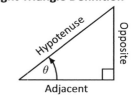

$$\sin\theta = \frac{O}{H} \qquad \csc\theta = \frac{H}{O}$$

$$\cos\theta = \frac{A}{H} \qquad \sec\theta = \frac{H}{A}$$

$$\tan\theta = \frac{O}{A} \qquad \cot\theta = \frac{A}{O}$$

Common Trigonometric Identities

Pythagorean Identities

$$\sin^2 x + \cos^2 x = 1$$

$$\tan^2 x + 1 = \sec^2 x$$

$$1 + \cot^2 x = \csc^2 x$$

Cofunction Identities

$$\sin\left(\frac{\pi}{2} - x\right) = \cos x \qquad \csc\left(\frac{\pi}{2} - x\right) = \sec x$$

$$\cos\left(\frac{\pi}{2} - x\right) = \sin x \qquad \sec\left(\frac{\pi}{2} - x\right) = \csc x$$

$$\tan\left(\frac{\pi}{2} - x\right) = \cot x \qquad \cot\left(\frac{\pi}{2} - x\right) = \tan x$$

Double Angle Formulas

$$\sin 2x = 2\sin x \cos x$$

$$\cos 2x = \cos^2 x - \sin^2 x$$

$$= 2\cos^2 x - 1$$

$$= 1 - 2\sin^2 x$$

$$\tan 2x = \frac{2\tan x}{1 - \tan^2 x}$$

Sum to Product Formulas

$$\sin x + \sin y = 2\sin\left(\frac{x+y}{2}\right)\cos\left(\frac{x-y}{2}\right)$$

$$\sin x - \sin y = 2\sin\left(\frac{x-y}{2}\right)\cos\left(\frac{x+y}{2}\right)$$

$$\cos x + \cos y = 2\cos\left(\frac{x+y}{2}\right)\cos\left(\frac{x-y}{2}\right)$$

$$\cos x - \cos y = -2\sin\left(\frac{x+y}{2}\right)\sin\left(\frac{x-y}{2}\right)$$

Power–Reducing Formulas

$$\sin^2 x = \frac{1 - \cos 2x}{2}$$

$$\cos^2 x = \frac{1 + \cos 2x}{2}$$

$$\tan^2 x = \frac{1 - \cos 2x}{1 + \cos 2x}$$

Even/Odd Identities

$$\sin(-x) = -\sin x$$

$$\cos(-x) = \cos x$$

$$\tan(-x) = -\tan x$$

$$\csc(-x) = -\csc x$$

$$\sec(-x) = \sec x$$

$$\cot(-x) = -\cot x$$

Product to Sum Formulas

$$\sin x \sin y = \frac{1}{2}\left(\cos(x-y) - \cos(x+y)\right)$$

$$\cos x \cos y = \frac{1}{2}\left(\cos(x-y) + \cos(x+y)\right)$$

$$\sin x \cos y = \frac{1}{2}\left(\sin(x+y) + \sin(x-y)\right)$$

Angle Sum/Difference Formulas

$$\sin(x \pm y) = \sin x \cos y \pm \cos x \sin y$$

$$\cos(x \pm y) = \cos x \cos y \mp \sin x \sin y$$

$$\tan(x \pm y) = \frac{\tan x \pm \tan y}{1 \mp \tan x \tan y}$$

Areas and Volumes

Triangles

$h = a \sin \theta$

Area $= \frac{1}{2} bh$

Law of Cosines:
$c^2 = a^2 + b^2 - 2ab \cos \theta$

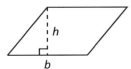

Right Circular Cone

Volume $= \frac{1}{3} \pi r^2 h$

Surface Area =
$\pi r \sqrt{r^2 + h^2} + \pi r^2$

Parallelograms

Area $= bh$

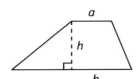

Right Circular Cylinder

Volume $= \pi r^2 h$

Surface Area =
$2\pi rh + 2\pi r^2$

Trapezoids

Area $= \frac{1}{2}(a + b)h$

Sphere

Volume $= \frac{4}{3} \pi r^3$

Surface Area $= 4\pi r^2$

Circles

Area $= \pi r^2$

Circumference $= 2\pi r$

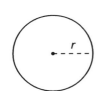

General Cone

Area of Base $= A$

Volume $= \frac{1}{3} Ah$

Sectors of Circles

θ in radians

Area $= \frac{1}{2} \theta r^2$

$s = r\theta$

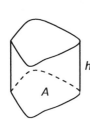

General Right Cylinder

Area of Base $= A$

Volume $= Ah$

Algebra

Factors and Zeros of Polynomials
Let $p(x) = a_n x^n + a_{n-1} x^{n-1} + \cdots + a_1 x + a_0$ be a polynomial. If $p(a) = 0$, then a is a *zero* of the polynomial and a solution of the equation $p(x) = 0$. Furthermore, $(x - a)$ is a *factor* of the polynomial.

Fundamental Theorem of Algebra
An nth degree polynomial has n (not necessarily distinct) zeros. Although all of these zeros may be imaginary, a real polynomial of odd degree must have at least one real zero.

Quadratic Formula
If $p(x) = ax^2 + bx + c$, and $0 \le b^2 - 4ac$, then the real zeros of p are $x = (-b \pm \sqrt{b^2 - 4ac})/2a$

Special Factors
$$x^2 - a^2 = (x - a)(x + a) \qquad\qquad x^3 - a^3 = (x - a)(x^2 + ax + a^2)$$
$$x^3 + a^3 = (x + a)(x^2 - ax + a^2) \qquad x^4 - a^4 = (x^2 - a^2)(x^2 + a^2)$$
$$(x + y)^n = x^n + nx^{n-1}y + \tfrac{n(n-1)}{2!}x^{n-2}y^2 + \cdots + nxy^{n-1} + y^n$$
$$(x - y)^n = x^n - nx^{n-1}y + \tfrac{n(n-1)}{2!}x^{n-2}y^2 - \cdots \pm nxy^{n-1} \mp y^n$$

Binomial Theorem
$$(x + y)^2 = x^2 + 2xy + y^2 \qquad\qquad (x - y)^2 = x^2 - 2xy + y^2$$
$$(x + y)^3 = x^3 + 3x^2y + 3xy^2 + y^3 \qquad (x - y)^3 = x^3 - 3x^2y + 3xy^2 - y^3$$
$$(x + y)^4 = x^4 + 4x^3y + 6x^2y^2 + 4xy^3 + y^4 \qquad (x - y)^4 = x^4 - 4x^3y + 6x^2y^2 - 4xy^3 + y^4$$

Rational Zero Theorem
If $p(x) = a_n x^n + a_{n-1} x^{n-1} + \cdots + a_1 x + a_0$ has integer coefficients, then every *rational zero* of p is of the form $x = r/s$, where r is a factor of a_0 and s is a factor of a_n.

Factoring by Grouping
$$acx^3 + adx^2 + bcx + bd = ax^2(cs + d) + b(cx + d) = (ax^2 + b)(cx + d)$$

Arithmetic Operations
$$ab + ac = a(b + c) \qquad \frac{a}{b} + \frac{c}{d} = \frac{ad + bc}{bd} \qquad \frac{a + b}{c} = \frac{a}{c} + \frac{b}{c}$$

$$\frac{\left(\frac{a}{b}\right)}{\left(\frac{c}{d}\right)} = \left(\frac{a}{b}\right)\left(\frac{d}{c}\right) = \frac{ad}{bc} \qquad \frac{\left(\frac{a}{b}\right)}{c} = \frac{a}{bc} \qquad \frac{a}{\left(\frac{b}{c}\right)} = \frac{ac}{b}$$

$$a\left(\frac{b}{c}\right) = \frac{ab}{c} \qquad \frac{a - b}{c - d} = \frac{b - a}{d - c} \qquad \frac{ab + ac}{a} = b + c$$

Exponents and Radicals
$$a^0 = 1, \ a \neq 0 \qquad (ab)^x = a^x b^x \qquad a^x a^y = a^{x+y} \qquad \sqrt{a} = a^{1/2} \qquad \frac{a^x}{a^y} = a^{x-y} \qquad \sqrt[n]{a} = a^{1/n}$$

$$\left(\frac{a}{b}\right)^x = \frac{a^x}{b^x} \qquad \sqrt[n]{a^m} = a^{m/n} \qquad a^{-x} = \frac{1}{a^x} \qquad \sqrt[n]{ab} = \sqrt[n]{a}\sqrt[n]{b} \qquad (a^x)^y = a^{xy} \qquad \sqrt[n]{\frac{a}{b}} = \frac{\sqrt[n]{a}}{\sqrt[n]{b}}$$

Additional Formulas

Summation Formulas:

$$\sum_{i=1}^{n} c = cn \qquad\qquad \sum_{i=1}^{n} i = \frac{n(n+1)}{2}$$

$$\sum_{i=1}^{n} i^2 = \frac{n(n+1)(2n+1)}{6} \qquad\qquad \sum_{i=1}^{n} i^3 = \left(\frac{n(n+1)}{2}\right)^2$$

Trapezoidal Rule:

$$\int_a^b f(x)\,dx \approx \frac{\Delta x}{2}\left[f(x_1) + 2f(x_2) + 2f(x_3) + \ldots + 2f(x_n) + f(x_{n+1})\right]$$

with Error $\leq \dfrac{(b-a)^3}{12n^2}\left[\max|f''(x)|\right]$

Simpson's Rule:

$$\int_a^b f(x)\,dx \approx \frac{\Delta x}{3}\left[f(x_1) + 4f(x_2) + 2f(x_3) + 4f(x_4) + \ldots + 2f(x_{n-1}) + 4f(x_n) + f(x_{n+1})\right]$$

with Error $\leq \dfrac{(b-a)^5}{180n^4}\left[\max|f^{(4)}(x)|\right]$

Arc Length:

$$L = \int_a^b \sqrt{1 + f'(x)^2}\ dx$$

Surface of Revolution:

$$S = 2\pi \int_a^b f(x)\sqrt{1 + f'(x)^2}\ dx$$

(where $f(x) \geq 0$)

$$S = 2\pi \int_a^b x\sqrt{1 + f'(x)^2}\ dx$$

(where $a, b \geq 0$)

Work Done by a Variable Force:

$$W = \int_a^b F(x)\ dx$$

Force Exerted by a Fluid:

$$F = \int_a^b w\,d(y)\,\ell(y)\ dy$$

Taylor Series Expansion for $f(x)$:

$$p_n(x) = f(c) + f'(c)(x-c) + \frac{f''(c)}{2!}(x-c)^2 + \frac{f'''(c)}{3!}(x-c)^3 + \ldots + \frac{f^{(n)}(c)}{n!}(x-c)^n$$

Maclaurin Series Expansion for $f(x)$, where $c = 0$:

$$p_n(x) = f(0) + f'(0)x + \frac{f''(0)}{2!}x^2 + \frac{f'''(0)}{3!}x^3 + \ldots + \frac{f^{(n)}(0)}{n!}x^n$$

Summary of Tests for Series:

Test	Series	Condition(s) of Convergence	Condition(s) of Divergence	Comment
nth-Term	$\displaystyle\sum_{n=1}^{\infty} a_n$		$\displaystyle\lim_{n\to\infty} a_n \neq 0$	This test cannot be used to show convergence.
Geometric Series	$\displaystyle\sum_{n=0}^{\infty} r^n$	$|r| < 1$	$|r| \geq 1$	$\text{Sum} = \dfrac{1}{1-r}$
Telescoping Series	$\displaystyle\sum_{n=1}^{\infty} (b_n - b_{n+a})$	$\displaystyle\lim_{n\to\infty} b_n = L$		$\text{Sum} = \left(\displaystyle\sum_{n=1}^{a} b_n\right) - L$
p-Series	$\displaystyle\sum_{n=1}^{\infty} \dfrac{1}{(an+b)^p}$	$p > 1$	$p \leq 1$	
Integral Test	$\displaystyle\sum_{n=0}^{\infty} a_n$	$\displaystyle\int_{1}^{\infty} a(n)\, dn$ is convergent	$\displaystyle\int_{1}^{\infty} a(n)\, dn$ is divergent	$a_n = a(n)$ must be continuous
Direct Comparison	$\displaystyle\sum_{n=0}^{\infty} a_n$	$\displaystyle\sum_{n=0}^{\infty} b_n$ converges and $0 \leq a_n \leq b_n$	$\displaystyle\sum_{n=0}^{\infty} b_n$ diverges and $0 \leq b_n \leq a_n$	
Limit Comparison	$\displaystyle\sum_{n=0}^{\infty} a_n$	$\displaystyle\sum_{n=0}^{\infty} b_n$ converges and $\displaystyle\lim_{n\to\infty} a_n/b_n \geq 0$	$\displaystyle\sum_{n=0}^{\infty} b_n$ diverges and $\displaystyle\lim_{n\to\infty} a_n/b_n > 0$	Also diverges if $\displaystyle\lim_{n\to\infty} a_n/b_n = \infty$
Ratio Test	$\displaystyle\sum_{n=0}^{\infty} a_n$	$\displaystyle\lim_{n\to\infty} \dfrac{a_{n+1}}{a_n} < 1$	$\displaystyle\lim_{n\to\infty} \dfrac{a_{n+1}}{a_n} > 1$	$\{a_n\}$ must be positive. Also diverges if $\displaystyle\lim_{n\to\infty} a_{n+1}/a_n = \infty$
Root Test	$\displaystyle\sum_{n=0}^{\infty} a_n$	$\displaystyle\lim_{n\to\infty} (a_n)^{1/n} < 1$	$\displaystyle\lim_{n\to\infty} (a_n)^{1/n} > 1$	$\{a_n\}$ must be positive. Also diverges if $\displaystyle\lim_{n\to\infty} (a_n)^{1/n} = \infty$